Polymers in Solar Energy Utilization

Polymers in Solar Energy Utilization

Charles G. Gebelein, EDITOR
Youngstown State University

David J. Williams, EDITOR
Xerox Corporation

Rudolph D. Deanin, EDITOR
University of Lowell

Based on a symposium cosponsored
by the ACS Divisions of
Organic Coatings and Plastics Chemistry
and Polymer Chemistry
at the 183rd Meeting
of the American Chemical Society,
Las Vegas, Nevada,
March 28–April 2, 1982

ACS SYMPOSIUM SERIES **220**

AMERICAN CHEMICAL SOCIETY
WASHINGTON, D.C.   1983

Library of Congress Cataloging in Publication Data

Polymers in solar energy utilization.
(ACS symposium series, ISSN 0097–6156; 220)

Papers derived from a symposium sponsored by the
American Chemical Society's Divisions of Organic
Coatings and Plastics Chemistry and of Polymer
Chemistry.
Includes bibliographies and index.

1. Solar energy—Congresses.   2.   Polymers and
polymerization—Congresses.
I. Gebelein, Charles G., 1929–   . II. Williams,
David J., 1943–   . III. Deanin, Rudolph D.
IV. American Chemical Society. Division of Organic
Coatings and Plastics Chemistry. V. American Chem-
ical Society. Division of Polymer Chemistry.
VI. Series.

TJ810.P78   1983   621.47'028   83–6367
ISBN 0–8412–0776–3

ACS Symposium Series

M. Joan Comstock, *Series Editor*

FOREWORD

The ACS SYMPOSIUM SERIES was founded in 1974 to provide a medium for publishing symposia quickly in book form. The format of the Series parallels that of the continuing ADVANCES IN CHEMISTRY SERIES except that in order to save time the papers are not typeset but are reproduced as they are submitted by the authors in camera-ready form. Papers are reviewed under the supervision of the Editors with the assistance of the Series Advisory Board and are selected to maintain the integrity of the symposia; however, verbatim reproductions of previously published papers are not accepted. Both reviews and reports of research are acceptable since symposia may embrace both types of presentation.

CONTENTS

PREFACE

THE SUN ALWAYS SHINES, even on cloudy days, and pours forth an enormous amount of energy onto the planet Earth. Obviously this solar energy is the cause of the basic meteorological phenomena that we observe daily, and the effect of solar storms and flares on electromagnetic communications systems is also well established. In the final analysis, solar energy is the ultimate source of most of the energy we use. For example, fossil fuels are residues of past animal and plant populations whose basic energy and growth were derived from the sun. Much of the current emphasis on renewable energy resources focuses on sources, such as wood, that store solar energy in some form. In addition, many of the proposed alternate energy sources (e.g., wind power) derive their basic energy from the action of solar energy on our planet. With the exception of nuclear energy, essentially all of our present energy sources are derived from the sun in one way or another. Even nuclear fusion is patterned after the basic mode of energy production in the sun.

Energy is consumed in almost every facet of our lives ranging from food production to recreation, and the energy demand will increase as the results of technology reach out to other portions of the world. Where will the necessary energy come from? Better utilization of solar energy is considered by many to be the best potential source of this needed energy. It has been estimated that all the energy needs of the United States in the year 2000 could be met if we could only harness and utilize the total solar energy that falls on about 15,000 square miles. This land area is less than 0.5% of the area of the United States and is equivalent to about one-tenth the area of California or about one-half the area of Indiana, Maine, or South Carolina.

Two major difficulties arise here, however: (1) the problem of how to harness or utilize this fairly diffuse and intermittent source of energy; and (2) the problem of storing this energy and/or transmitting it to another location where it is needed. Many approaches have been taken to solve both of these difficulties. In this book we will consider the use of polymeric materials in various methods of solar energy utilization. Although some overlap does occur, the 29 chapters in this book have been placed into three sections: General Solar Applications, Polymer Photodegradation in Solar Applications, and Photovoltaic and Related Applications.

Essentially, the 13 chapters in the first section consider the basic economics of solar energy, its collection by various devices, and the storage of the collected energy for short-term use. Chapters include a general survey of the various ways in which polymeric materials have been utilized in solar energy collection and of the problems as well as the potential of polymers in this application; basic economic considerations in solar heating systems with special emphasis on the role of plastics in cost reduction; descriptions of some solar collectors that utilize polymers with consideration of cost, effectiveness, and degradation; use of various polymeric materials for sealants; polymeric coatings on metal mirrors; and use of polymers in heat storage as liners in solar ponds in which water is the thermal storage medium, as well as the storage of thermal energy in the ground itself.

The basic conclusion of the first section is that polymers can be used advantageously in solar applications but that more research is needed. Probably the most pressing problem cited is that of photodegradation, which is the central topic of the seven chapters in the second section. Chapters in this section include theoretical models to predict the utility of a polymer in solar applications by examining the formation of polar groups on the polymer backbone chains—which would make the polymer more water sensitive and less suitable for solar applications, especially photovoltaic applications; the use of UV and fluorescent microscopy to follow photooxidation in some polymers; the use of a laser photoacoustic technique to compare actual outdoor degradation studies with those obtained in an accelerated test; and photodegradation studies of some specific polymers. For example, poly(n-butyl acrylate) degradation and its implications for solar applications are reported and a fairly successful attempt to reduce photodegradation in poly(methyl methacrylate) by the copolymerization of a monomer that contains a UV-screening agent is described. The final chapter of this section considers the effect of uniaxial or biaxial deformation in polyethylene films on the photodegradation. The seven papers in the second section, along with several in the first section, clearly show that photodegradation is a potential problem in any solar utilization of polymers. In addition, they provide some insights into methods of predicting this effect and possible ways to alleviate this difficulty.

Even though the efficiency of the processes may be lower than desired, the early chapters in this book, and other sources, clearly show that solar energy can be captured and utilized to some extent with our current technology. Enthusiasm in this area, however, must be tempered by the facts that storage of this captured energy is limited and that transmission of the solar-derived energy is even more problematic. Trapping the solar energy in some type of thermal storage system may suffice for short-term storage (e.g., for use at night or on rainy days) but cannot be used effectively for truly long-term applications. In addition, thermal energy has a very limited transmission or transportation range. More effective methods of energy storage and/or transmission would involve the conversion of the solar energy into some form of chemical energy (e.g., hydrogen) or into electricity. This problem is considered in the nine chapters of the third section.

Most of the chapters in the third section are concerned with photovoltaic (PV) applications (conversion of light into electrical energy). Because of the diffuse nature of solar energy, the photovoltaic collection devices must be very large or else the light that strikes them must be concentrated. The first chapter in this section gives an overview of luminescent solar concentrators that can be used with the PV collectors. Most PV collectors or modules are multilayered systems containing a photovoltaic cell element. The next four chapters consider the use of various plastics as encapsulant or pottant materials in the PV modules.

The majority of photovoltaic modules use silicon as the photovoltaic cell element, but other materials are, in principle, possible. The last four chapters consider the use of organic polymers (sometimes doped) as the cell element or in some related conducting property: acrylonitrile, some polymeric phthalocyanines; and polymers of 2-vinylnaphthalene that is doped with pyrene and 1,2,4,5-tetracyanobenzene. The study on the last group of polymers was initiated by the idea that they could be used to transfer solar energy to a reaction center and produce some type of chemical reaction. The final chapter carries this approach further in the consideration of polymeric electrodes that could be used to split water into oxygen and hydrogen. The latter could then be utilized as a source of storable, readily transportable chemical energy.

In summary, the 29 chapters in this book point out the vast potential of solar energy utilization and note the advantages and problems of using polymers in this application. Polymers clearly offer advantages in cost, weight, and the variety of materials available but do suffer from varying degrees of photodegradation. This book points out several areas of needed research, but it also shows the promise of an energy-rich future in which solar energy can be used directly as thermal energy or indirectly as solar-derived electricity or chemically stored energy (e.g., hydrogen derived via solar water splitting). This optimistic view is in marked contrast with many of the gloomy prognostications that abound in today's world, but such advances are well within the realm of possibility in the near future.

We wish to express gratitude to the Divisions of Organic Coatings and Plastics Chemistry and of Polymer Chemistry who cosponsored the symposium from which this book is derived. Naturally, we thank all authors for the chapters that they contributed. In addition, we thank the many others who were involved in manuscript preparation and the review process. Finally, we wish to thank our families for their support while this book was being prepared.

CHARLES G. GEBELEIN
Youngstown State University
Department of Chemistry
Youngstown, OH 44555

DAVID J. WILLIAMS
Xerox Corporation
Webster, NY 14580

RUDOLPH D. DEANIN
University of Lowell
Plastics Department
Lowell, MA 01854

January 24, 1983

GENERAL SOLAR APPLICATIONS

Polymers in Solar Energy: Applications and Opportunities

W. F. CARROLL
California Institute of Technology, Jet Propulsion Laboratory,
Pasadena, CA 91109

PAUL SCHISSEL
Solar Energy Research Institute, Golden, CO 80401

Polymers have many potential applications in solar technologies that can help achieve total system cost-effectiveness. For this potential to be realized, three major parameters must be optimized: cost, performance, and durability. Optimization must be achieved despite operational stresses, some of which are unique to solar technologies. This paper identifies performance of optical elements as critical to solar system performance and summarizes the status of several optical elements: flat-plate collector glazings, mirror glazings, dome enclosures, photovoltaic encapsulation, luminescent solar concentrators, and Fresnel lenses. Research and development efforts are needed to realize the full potential of polymers to reduce life-cycle solar energy conversion costs. Problem areas which are identified are the interactions of a material with or its response to the total environment; photodegradation; permeability/adhesion; surfaces and interfaces; thermomechanical behavior; dust adhesion; and abrasion resistance. Polymeric materials can play a key role in the future development of solar energy systems [1]. Polymers offer potentially lower costs, easier processing, lighter weight, and greater design flexibility than materials in current use.

Polymeric materials are used in all solar technologies. In addition to such conventional applications as adhesives, coatings, moisture barriers, electrical and thermal insulation, and structural members, polymers are used as optical components in solar systems. Mirrors on parabolic troughs are made up of metallized fluoropolymers and acrylics. Commercial flat-plate collectors are glazed with fluoropolymers and ultraviolet-stabilized polyester/glass fiber composites. Photovoltaic (PV) cell arrays are encap-

0097–6156/83/0220–0003$06.00/0

sulated with silicones and acrylics for protection from the
weather. In laminated (safety glass) mirrors for central receiver
heliostat systems, polyvinyl butyral is used as a laminating and
encapsulating agent. Cast and molded acrylic Fresnel lenses are
used to concentrate sunlight onto photovoltaic cells and thermal
receivers. The widespread use of polymers is evident from the
information displayed in Table I, where applications of polymers
are listed for each major solar technology.

It is possible to estimate the amount of polymer which might
be used in the optical elements of solar energy systems [1]. The
estimate assumes a market penetration for solar systems equivalent
to 0.4 Quads/year (U.S. 1980 energy use was about 85 Quads, 1
Quad = 10^{15} Btu) starting in the time period 1985-1995. A solar
insolation of 1 kW/m^2 for about 6 hours per day will require an
area penetration of 2 x 10^8 m^2/yr (82 miles^2/yr) or about 5500
metric tons per year per mil of thickness. An average thickness
of 10 mils would require, for example, an appreciable fraction of
the present market for either acrylics or polycarbonate. Presumably, considerably larger amounts of polymer could be used in non-
optical structures but this is system specific and difficult to
estimate. Calculations [1] also suggest that the present low
level of funding for R&D on polymers implies an underestimation of
the potential which polymers have for applications to solar
systems.

The use of polymers in solar equipment will require major
changes from past large-scale applications, especially in
achieving satisfactory performance under all combinations of
stress. Cost, performance, and durability must be optimized. If
early cost-effective commercialization of solar energy is to be
realized, critical delays must be shortened. Service experience
has traditionally guided the evolution of systems toward an
optimum design. The process can be hastened by applying all
available understanding of the basic behavior of materials in the
initial designs of equipment. The stalemate imposed by the lack
of market and supply of solar systems can be broken by government-
supported development of technology that demonstrates the economic
viability of new or modified materials. Such development would
enable materials suppliers and manufacturers of solar equipment to
make knowledgeable business decisions and reduce development cost
and time.

Applications

The optical elements of solar systems are important applications for polymers. The use of polymers for optical elements
will, however, impose several unusual material requirements. Five
examples of the current development of polymeric optical elements
are considered below. Problems such as dirt accumulation and
photodegradation, which are common to most optical elements, are
considered in a later section. More conventional applications are
then noted very briefly.

Table I.

Polymer Application

Solar Energy System	Mirror Glazings	Flat-Plate Glazings	Encapsulation	Seals/Adhesives	Structural Members	Heat Transfer/Energy Storage	Paints/Coatings	Piping	Thermal & Electrical Insulation	Moisture Barriers	Fresnel Lenses	Membranes
Solar thermal conversion	x	x	x	x	x		x	x	x	x	x	
Photovoltaics		x	x	x	x	x	x	x	x	x	x	
Solar heating and cooling of buildings	x	x	x	x	x		x	x	x	x	x	
Wind				x	x	x	x		x			
Ocean thermal				x	x		x	x	x	x		
Biological/chemical		x		x	x	x	x	x	x	x		x

Flat-Plate Collector Glazings. The cover glazings protect the inner elements of the collector from the environment and increase operating efficiency by reducing reradiation and convection. Collectors with single glazings are limited in their operating temperature; however, some recent work suggests that singly glazed collectors can work in the temperature ranges required by desiccant and absorption cooling [2]. Higher operating temperatures are obtained by using two glazings. The outer glazing must withstand the environment while the inner glazing must be temperature resistant, typically up to the stagnation temperature of the device. Results on collector glazings have been reported [3] and environmental degradation studies of materials for glazings are in progress [4]. None of the materials are completely satisfactory either as an outer or inner glazing. The temperature requirement for the inner glazing eliminates most materials other than fluorocarbon polymers and glass. Glass is the most common outer glazing but it suffers from weight, cost, and impact resistance limitations, while lack of environmental durability limits the applications of polymers.

The transparent honeycomb concept is an alternative to the use of a second glazing [5]. The honeycomb is attached to or is an integral part of the outer glazing facing the absorber plate. The honeycomb improves collector performance by suppressing convection and radiation heat losses while only slightly reducing the incoming solar energy. The effectiveness of the honeycomb for improving collector performance is approximately equivalent to that of an inner glazing. Integral honeycombs formed from polycarbonate have good mechanical properties. Other materials tested include polyester, fluorocarbon polymers, and polyimide [5].

Novel approaches to collector fabrication use integral extrusions [6-8] or laminated thin films [2]. Unlike sheet-and-tube designs, some extruded integral units include the transparent glazing, heat-transfer fluid pathways, and backing, all in a configuration which could be rolled out onto a rooftop. Black fluids can act as absorbers and be drained from the collector to prevent excessive stagnation temperatures. The designs vary in detail but a common problem has been the identification of a polymer with acceptable environmental durability and low cost. An extruded polycarbonate collector with an integral optical concentrator has been developed [8]. Other materials that have been used in designs include acrylic and polyethersulfone.

Imaginative applications of polymers to fenestration can also be used for flat-plate collectors. A transparent, coated polymeric glazing which transmits the solar spectrum but returns the infrared radiation effectively increases the insulation provided by the glazing, because the infrared radiation generated inside the structure is retained. A polymeric film which changes from transparent to opaque when heated above a transition temperature acts as an automatic window shade which could help control stagnation temperatures [9].

Polymeric Glazings – Mirrors. The installed price of heliostats is estimated to account for about half of the total capital
cost of a central-receiver solar thermal electric plant and a
larger fraction of the cost of systems for process heat production [10]. Metallized thin polymeric films are one means to make
lightweight mirrors that are less expensive than current design.
Flexible, lightweight mirrors also allow less expensive designs of
auxiliary equipment. Thin, flexible films can be attached with
adhesives to substrates with single or compound curvature.
Earlier studies of aluminized or silvered polymers have included
acrylics, fluorinated polymers, polycarbonate, silicones, and
polyester [11]. Tests at Phoenix, Arizona, showed negligible
degradation of aluminum and silver mirrors protected by acrylic,
Teflon, and glass during exposures exceeding two years, while
similar tests at other sites resulted in severe degradation in
about one year [12]. It was decided that the reliability of
polymer-coated mirrors was insufficient for their use as heliostats at the Barstow demonstration facility [13]. More systematic
environmental degradation studies of some of these materials are
in progress [4] and several mirror configurations, including
aluminized acrylics, are being tested currently at selected locations around the U. S. In recent tests conducted in dry, relatively benign climates, aluminized acrylics have performed well
for up to five years, polymeric glazings that protect silver
surfaces for comparable time periods have not been identified.
    Local irregularities (slope-error) in the shape of reflectors
present a problem with polymer-glazed mirrors. A slope-error
tolerance as low as one milliradian is needed for some point-focus
concentrators [14]. This tolerance has been met with glass
mirrors; however, metallized polymeric films have a poorer tolerance.

Dome Enclosures. An enclosed-heliostat (dome) design envi-
sions large (30-ft diameter) bubbles made of thin, air-supported,
transparent polymeric films as protective covers for metallized
polymeric mirrors. Studies [10] indicate that use of dome-
enclosed solar concentrators may result in significant cost reductions. The air-supported dome configuration is capable of withstanding wind loads and can protect light gauge plastic membrane
heliostats and drive mechanisms which lower costs. The original
concept from the Boeing Engineering and Construction Company used
integral domes while later designs by the General Electric Company
used segments assembled with adhesives.
    A number of transparent polymers have been examined and
tested for this purpose; prototype domes have been fabricated from
polyvinylfluoride which was later determined to be too expensive
and not sufficiently stable. The dominant requirement for this
application is good specular transmission. Several polyesters

were tested and had excellent initial optical transmittance.
However, their environmental durability was too limited. Energy
losses due to absorption or scattering decrease the system efficiency, reducing the cost advantages. In the case of enclosures
for heliostats in central receiver systems, losses are multiplied
because the solar beam must pass through the dome twice. The
material must be compatible with cost-effective dome fabrication
methods and must have suitable mechanical properties and durability to maintain operation under the combined effects of wind,
hail, temperature, sunlight, etc., for several years. Other
materials tested include polyvinylidene fluoride, polycarbonate,
and polypropylene. Biaxially oriented polyvinylidene fluoride is
practical to manufacture commercially and is said to have good
optical properties and excellent weatherability [10].

Flat-Plate Photovoltaic (PV) Encapsulation.* Polymers can
serve several functions in PV encapsulation systems [15]. The
single polymer application common to all configurations and the
core of the encapsulation package is the pottant, which embeds the
solar cells and related electrical conductors. The key requirements for a pottant are high transparency in the range of solar
cell response, mechanical cushioning of the fragile solar cells
from thermal and mechanical stresses, electrical insulation to
isolate module voltage, and cost-effective material and module
fabrication processes.
     Other encapsulation applications of polymers for specific
designs include soil, ultraviolet, and abrasion-resistant front
covers. The cover can serve as a transparent structural superstrate. Substrate support designs require a hard, durable front
cover film to protect the relatively soft pottant from mechanical
damages and excess soil accumulation. A polymeric front cover
must be low in cost, highly transparent, and weather resistant to
compete with glass. For applications out of the optical path
between the sun and the solar cells (adhesives, insulation, edge
seals, gaskets) requirements for polymeric use in encapsulation
are the same as for other applications.

Luminescent Solar Concentrators (LSCs). The LSC uses the
principle of light pipe trapping, transmission, and coupling into
a photovoltaic cell (PV) to concentrate solar radiation [16,17].
This use of a low-cost concentrator can reduce the area requirements of the more expensive PV cells. The LSC has several important advantages. It can be made from inexpensive materials, can be
nontracking and it can concentrate the light input from either
direct or diffuse insolation. The LSC can act as a wavelength

---

*Encapsulation of photovoltaics for concentrator systems depends
on concentration ratio and other system specific parameter tests,
issues that are not discussed in this paper.

matcher between the solar radiation and the spectral response of the PV cell. Also, the solar infrared radiation and the resultant heat load are prevented from reaching the solar cell.

One planar configuration, shown in Figure 1, uses a polymeric host (polymethylmethacrylate) into which dye molecules are dispersed randomly. Photons with wavelengths in the adsorption band of the dye enter the host, are absorbed by the dye, and are reradiated isotropically. Reradiated photons within a certain solid angle are trapped by total internal reflection and guided to the edge of the plate where solar cells are attached. The concentration factor is determined by the ratio of the areas of the face to an edge, by the fraction of reradiated photons which is trapped (75%), and other factors relating to the dye. The trapping efficiency is partly determined by the refractive index of the polymer.

A second planar configuration uses thin (25 µm) polymeric film host (cellulose acetate butyrate) coated onto a support (PMMA, glass). When the support is positioned as a superstrate it acts as a glazing to protect the thin film. Some designs use several thin layers, each containing a dye matched to different solar wavelengths. The physical separation of the dyes can improve their durability. The thin film approach means that more expensive materials may be acceptable. Fluorinated or deuterated dyes can improve optical efficiency and highly concentrated dyes can alter the mechanism and efficiency of energy transport. System lifetime is an important unknown factor principally influenced by dye lifetime. Questions relating to dye-host interactions and the influence of the host on system lifetimes are unanswered.

Fresnel Lenses. Fresnel lens concentrators have been studied for both thermal and photovoltaic systems. The economic viability of their use depends on a large number of system-related factors, including the performance, cost, and durability of the lenses. Performance requirements include minimum absorption, scattering, and surface reflection. Total cost depends on costs of materials and fabrication, or minimum thickness defined by mechanical requirements, and on additional material required for optical design. Like other solar applications of optical polymers, durability for extended periods is required.

Other Applications. Polymers can also be used as edge seals in glass mirrors, films for mirror backings, adhesives, structural members, solar pond liners, and energy storage systems. Glass mirrors are more stable than mirrors with polymeric glazings, but they are expensive, heavy, and probably not stable enough. The state of the art is exemplified by the developments of mirrors for heliostat applications [18]. The structure consists of silvered glass backed with a polyisobutylene film and mounted on an aluminum sheet-paper honeycomb structure. The mirror edges are sealed

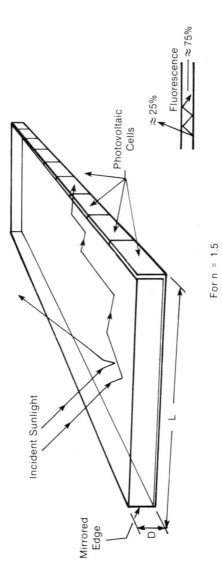

For n = 1.5

Figure 1. Schematic of a Luminescent Solar Concentrator

with polyisobutylene and silicone, and the mirror is held to the supporting structure using a neoprene phenolic adhesive.   Other low-cost heliostat mirror modules have been designed and developed which use plastic to reduce weight and to accommodate high-volume production of complex forms by molding.   Molded rib, extruded panel, or sandwiched honeycomb structures are combined with sprayed silver metallization, sprayed polymeric overcoats, or laminated films [19].   Molded reinforced plastics are also used in parabolic trough module designs [20].   A survey of thermal energy storage projects is available [21].

Research Opportunities

Optical Elements.   Problems which are common to many solar-related optical elements include dirt retention, cleaning, surface abrasion, and photodegradation.   A common feature of some of these problems is that the deleterious effects occur at an interface. Ultraviolet radiation, atmospheric components, mechanical stress, etc., can have a profound effect on performance by changing surface characteristics.   The lifetimes of UV stabilizers can be limited by exudation; permeability can cause harmful reactions at interfaces; and mechanical properties can be influenced by surface crazing.   In other applications mechanical behavior of the bulk polymer is critical and virtually all applications require that the polymer system withstand multiple environmental stresses simultaneously.

Surface/Interface Properties of Polymers

Surface phenomena play a significant role in the major problem areas associated with polymers and, therefore, are basic to most of the studies.   For example, the lifetimes of UV stabilizers can be limited by exudation and accumulation at the surface, permeability can cause harmful reactions at interfaces, adhesion is an interface phenomenon, and mechanical properties can be influenced by surface crazing.   Examples of surface problems affecting mirrors include abrasion, dust adhesion, and cleaning procedures.   Surface interactions also occur during the production of polymers; the subsequent behavior of a polymer can be critically dependent upon the material against which it is formed. Surface measurement techniques will form a general experimental basis for work on specific applications.   Experimental and analytical studies are needed to improve understanding of the chemistry, physics, and morphology of surfaces.   Study of interfaces between polymers and other materials is also needed both for model interfaces and for candidate engineering material interfaces.   Such studies should characterize the inferfaces as originally fabricated and after changes caused by typical environmental exposures.

The accumulation of airborne particulates and aerosols on the

optical surfaces of solar conversion equipment causes unwanted
absorption and scattering which has lowered operating efficiencies
more than 30%. Accumulation is most serious for soft polymers
(e.g., silicone rubber) and least serious for glass. Data indicate that hard polymers like polymethylmethacrylate are nearly as
resistant as glass. An understanding of adhesion mechanisms is
required to develop cost-effective cleaning methods and soil-
resistant polymeric materials [22-24]. Modifications of polymeric
materials, either in bulk or by surface treatment or coating, may
result in materials that are "self-cleaning" or that do not tend
to hold soil, allowing it to be removed easily by natural forces
such as wind and rain.

Regular cleaning probably will be required to maintain high
optical performance of solar systems, and cleaning will be a major
operational cost factor. Automatic cleaning can be made cost-
effective if it is based on an understanding of soil adhesion
mechanisms. Possible problems include mechanical damage to the
surface by cleaning and potential contributions of residual
cleaning agents to material aging. The optical function of polymeric components can be seriously degraded by abrasion due to the
cleaning process or by natural causes. Considerations such as
cost, mechanical compatibility with supporting material, or UV
resistance may preclude the use of inherently abrasion-resistant
materials. Since only a shallow layer of resistant material is
required, adding a coating or using surface processes that produce
a resistant "skin" can yield the necessary resistance. A coating
might have several uses, providing abrasion resistance, improving
antireflective performance, screening UV radiation, or combining
several functions.

Adhesive failure is a problem in solar systems. In the past,
polymers have been used to protect the mechanical integrity of
wood and metal structures in severe outdoor environments and to
protect sensitive electronic components in relatively benign
enclosed environments. Polymers used in solar equipment will have
to protect the optical properties of reflectors, thin-film electrical conductors, and thin-film photovoltaics from the effects of
moisture and atmospheric pollutants in severe outdoor environments
while simultaneously maintaining optical, mechanical, and chemical
integrity. In some systems, the prevention of mechanical failure
is important; frequently, adhesive failure at the metal/polymer
interface is of particular concern because the ensuing corrosion
causes optical failure.

Loss of adhesion may be caused by permeation problems.
However, internal formation of volatile species (outgassing) and
primary bond failure can also contribute to loss of adhesion. All
polymers are inherently permeable, but to widely varying
degrees. Oxygen, moisture, air pollutants, etc., can penetrate
polymer films and attack underlying reflector metalization, conductors, or other functional elements. Furthermore, these gases

can modify the mechanical and optical properties of the polymer and alter mechanical interfaces between layers.

The permeability of a polymer can be sensitive to how the material is processed and used and is sometimes enhanced as the material degrades. Although a fundamental understanding of polymer permeability exists and experimental data are available, current information is inadequate to model or control the effects of permeation of various species in solar equipment. Experimental and analytical studies are needed to consider modern theoretical approaches (e.g., nonequilibrium thermodynamics, non-Fickian diffusion) with the goal of developing models for the transport of H_2O, O_2, SO_x, and other molecules in polymers, and to develop and compile quantitative engineering data on transport through bulk material and across and along interfaces specifically for solar applications [25, 26].

Alternatively, delamination may not be related directly to permeation, but may be due instead to thermal and/or UV effects that are followed by the corrosive failure. Some studies and models indicate that the polymer/metal interface morphology, and the changes in the morphology with exposure to the environment, play a key role in corrosion rates. These characteristics may be even more important in corrosion control than either the diffusion of vapors through the polymer or the inherent corrosion resistance of the metal.

Photochemistry of Polymers

Virtually all polymers deteriorate under exposure to outdoor weathering and solar radiation, but at greatly varying rates. Polymers in solar equipment must maintain optical, mechanical, and chemical integrity despite prolonged exposure to solar ultraviolet radiation. For most outdoor applications of polymers, solar radiation exposure is incidental, but for many solar applications, exposure to solar radiation is deliberately maximized in the equipment design. Transparency is essential for many of the potentially most cost-effective applications, and conventional approaches to ultraviolet protection such as opaque coatings and fillers are unacceptable.

Photodegradation in polymers begins with the primary excited states produced by absorption of ultraviolet photon energy by the polymer. These excited states undergo fast-reaction sequences to form chain radicals which, in turn, decay through chemical reactions within the polymer or with O_2, H_2O, etc. These reactions can produce changes in chemistry or molecular size. The reaction products can absorb additional photons, resulting in further degradation by analogous processes. The cumulative changes may result in yellowing (loss of transparency), change in refractive index, and deterioration of surface properties (e.g., crazing). Changes in the mechanical properties cause increased creep or cracking, while changes in the chemical properties result in increased permeability to H_2O, SO_x, etc., and subsequent cor-

rosive interaction with metallic components in contact with the
polymer.

Polymers that are sensitive to ultraviolet radiation but that
otherwise have desirable properties can be stablized by ultraviolet screens, absorbers, quenchers, radical scavengers, and
antioxidants. Ultraviolet stablizers can be incorporated by
simple addition or by chemical combination with the polymer
molecule. Long lifetime can best be attained by immobilizing the
additive as part of the molecular structure. Using existing
photochemistry, analytical and experimental studies are needed to
develop models of photochemical processes in solar-related
polymers. Recent screening studies using commercial additives
have been directed toward solar applications. They can provide
some input to the system selections [28-31].

Thermomechanical Behavior. Requirements for optical performance impose unprecedented requirements for dimensional stability
of polymers used in high-concentration reflectors. Requirements
for mechanical compatibility are also strict for photovoltaic
systems subjected to moisture and thermal stresses. Moisture,
temperature, and UV, separately and in combination, can change the
volume and thus the stress state of polymers. For example,
temperature and humidity cycles alone do not cause surface microcracks in polycarbonate. However, in the presence of UV radiation, such cycles cause microcracks, while UV alone does
not [32]. An understanding of these relationships is essential to
permit reliable design of equipment that uses polymers.

These factors, coupled with need for reliable design and low
cost, necessitate both a fundamental understanding of mechanical
behavior and reliable mechanical design data. The relationships
between process and environmental effects to mechanical behavior
have been developed for elastomeric polymers to the degree that
these materials can be selected, and their long-term performance
reliably predicted, by a knowledge of some fundamental parameters
determined from a few straightforward experimental measurements.
If the current level of understanding of visco-elasticity of
elastomers can be extended into the range of glassy polymers, then
it will be possible to make comparable predictions of mechanical
stress/time/temperature/strain response and failure relationships. Ultimately, equivalent understanding of glassy polymers
will greatly reduce the need for costly empirical testing each
time a new application is contemplated.

Combined Environmental Effect. Any list of significant
effects of the environment on polymers in solar applications will
include UV degradation, weathering, permeability, high-temperature
performance, delamination/fatigue, dimensional stability, and
soiling/cleaning. These effects are not necessarily
independent: polymers are expected to suffer more serious
degradation during exposure to combined environmental stresses

Effective methods for predicting and verifying the performance of polymers under interactive effects are required, particularly for those applications unique to solar energy systems.

The simultaneous and sequential combinations of environmental stresses that alter the properties and affect the performance of polymers should be identified experimentally. Assessment of the stability of polymers from fundamental rate data or from experimental engineering data requires an understanding of the interactive effects of environmental stress [11, 32]. Testing of all combinations of the stresses would require an unacceptable number of experiments. The number of combinations studied can be limited by recognizing the unique position of optical elements which may use transparent polymers and of structural members; by first ranking the importance of individual stresses; and by using screening tests to identify promising materials [28-30].

Improved analytical and test methods are needed. Accelerated tests stress some parameters to decrease the time before failure occurs, abbreviated tests use analytical techniques to estimate failure rates from incipient degradation during short-term exposures at usual stress levels. Methods are needed to demonstrate correlations between these test results and real-time behavior. Earlier studies [33, 34] provide some basis for this work.

Performance prediction modeling (PPM) is one method for evaluating materials performance that has been defined and is being applied at the Jet Propulsion Laboratory. In its simplest form, PPM can verify the satisfactory performance of a particular material used in a specific design and subject to a defined set of stresses (e.g., temperature, thermal cycling, ultraviolet radiation, mechanical loads). Conversely, it can define limits of stresses for the materials in an available piece of hardware or design. This approach has been used successfully as part of the demonstration of the feasibility of using an ultra-thin polymer film on a solar sail for space propulsion [35], for analytical assessment of an experimental facility to study space radiation effects [36], and for analytical assessment and identification of critical technologies for ceramic receivers [37]. The method is currently being applied to photovoltaic encapsulation [38].

A second procedure, using the methods of thermodynamics applied to irreversible processes, offers another new approach for understanding the failure of materials. For example, the equilibrium thermodynamics of closed systems predicts that a system will evolve in a manner that minimizes its energy (or maximizes its entropy). The thermodynamics of irreversible processes in open systems predicts that the system will evolve in a manner that minimizes the dissipation of energy under the constraint that a balance of power is maintained between the system and its environment. Application of these principles of nonlinear irreversible thermodynamics has made possible a formal relationship between thermodynamics, molecular and morphological structural parameters, than from a simple sum of effects of individual stresses. and their rate of change.

Experimentally, these principles emphasize dynamic measurements that make possible the separation of the dissipative and the conservative components of energy incident upon the system. Dynamic mechanical analysis has been an important area of research for over 40 years. Computer-controlled experimentation now makes it possible to apply analogous techniques to the measurement of many other thermodynamic stresses. One example currently under investigation, dynamic photothermal spectroscopy, is expected to provide a new approach to predicting the long-term effects of ultraviolet radiation on materials [39].

Acknowledgments

The authors wish to express their gratitude for the support of the Materials Research Branch and, in particular, Barry Butler, A.W. Czanderna, and R.F. Reinisch for their contributions. Thanks also are due to members of the Energy and Materials Research Section of the Jet Propulsion Laboratory for their help in preparing this document. This document was prepared for the U.S. Department of Energy under Contract No. EG-77-C-01-4024. The Jet Propulsion Laboratory is a National Aeronautics and Space Administration facility, and the Solar Energy Research Institute is a Department of Energy facility.

Literature Cited

1. Carroll, W. F.; Schissel, P. "Polymers in Solar Technologies: An R&D Strategy"; SERI/TR-334-601; Solar Energy Research Institute: Golden, CO 1980.
2. Wilhelm, W. G. "Low Cost Solar Energy Collection for Cooling Applications"; BNL51408; Brookhaven National Laboratory: Upton, NY, 1981.
3. Clark, Elizabeth; Roberts, W. E.; Grimes, J. W.; Embree, E. J. "Solar Energy Systems—Standards for Cover Plates for Flat-Plate Solar Collectors"; NBS Technical Note 1132; National Bureau of Standards, Center for Building Technology: Washington, D.C., 1980.
4. DSET Laboratories, Inc. "Properties and Durability Data for Solar Materials"; Contract XH-9-8215-1; Solar Energy Research Institute: Golden, CO, 1981.
5. Lockheed Missiles and Space Co., Inc. Optimization of Thin-Film Transparent Plastic Honeycomb Covered Flat-Plate Solar Collectors. SAN/1256-78/1. Lockheed: Palo, Alto, CA, 1978.
6. Acurex Corp., Mountain View, CA. DOE Contract No. DE-ACO4-79L12032. 1980.
7. Fafco Inc., Menlo Park, CA. DOE Contract No DE-ACO3-78CS32241. 1980
8. Ramada Energy Systems, Inc. Technical Bulletin RES TB 180.
9. Suntex Research Associates. "An Energy Efficient Window

System: Phase 1 Technical Report;" Suntex: Corte Madera, CA, 1976.

10. Mavis, C.L.   "Status and Recommended Future of Plastic-Enclosed Heliostat Development; SAND 80-8032.   Sandia National Laboratories: Albuquerque, NM, 1980.

11. Schissel, Paul; Czanderna, A.W.   "Reactions at the Silver/Polymer Interface: A Review."   Solar Energy Materials. 1980, 3; 225-245.

12. University of Minnesota; Honeywell, Inc.   "Research Applied to Solar Thermal Power Systems";   NSF/RANN/SE/GI-34871/PR/74/4; 1975.

13. McDonnell-Douglas Astronautics Co.   "Central Receiver Solar Thermal Power System, Phase 1 Vol. III (Book 1, Collector Subsystem), Preliminary Draft, MDCG 6776 (May 1977).

14. Jet Propulsion Laboratory.   "Annual Technical Report: Fiscal Year 1980";   JPL 81-39; Pasadena, CA, 1981.

15. Jet Propulsion Laboratory.   "Photovoltaic Module Encapsulation Design and Materials Selection"; LSA Encapsulation Task, Project Report 5101-177.

16. Batchelder, J. S.; Zewail, A. H.; Cole, T.   "Luminescent Solar Concentrators--1:   Theory of Operation and Techniques for Performance Evaluation; Applied Optics.   1979, 18, 3090-3110.

17. Hermann, A. M.   1981.   "Luminescent Solar Concentrators--A Review"; to be published in Solar Energy.

18. Martin Marietta Corp.   "Second-Generation Heliostat Development"; SAND 79-8192; Denver, CO., 1980.

19. Hobbs, Robert B., Jr.   "Solar Central Receiver Heliostat Mirror Module Development"; SAND 79-8189; General Electric Co.: Philadelphia, PA, 1981.

20. General Electric Co.   "Design and Development of a Reinforced Plastic Trough Module"; SAND 80-7150; Sandia National Laboratories: Albuquerque, NM, 1981.

21. Baylin, F.; Merino, M.   "A Survey of Sensible and Latent Heat Thermal Energy Storage Projects"; SERI/RR-355-456; Solar Energy Research Institute; Golden, CO, 1981.

22. Berg, R. S.   "A Survey of Mirror-Dust Interactions" presented at the ERDA Concentrating Solar Collector Conference; Georgia Institute of Technology; 26-28 Sept. 1977.

23. Berg, R. S.   1978 (March).   SAND 78-0510.   Albuquerque, NM: Sandia Laboratories.

24. Hampton, H.L.; Lind, M. A.   "The Effects of Noncontact Cleaners on Transparent Solar Materials"; Battelle Pacific Northwest Laboratory: Richland, WA, 1979.

25. Carmichael, D. C.   "Studies of Encapsulation Materials for Terrestrial Photovoltaic Arrays"; 5th Quarterly Progress Report"; ERDA/JPL-954328-76/6; Battelle Columbus Laboratories: Columbus, OH, 1976.

26. Carmichael, D. C. et al.   "Evaluation of Available Encapsulation Materials for Low-Cost Long-Life Silicon Photovoltaic

Arrays"; DOE/JPL-954328-78/2; Batelle Columbus Laboratories: Columbus, OH, 1978.

27. Roberts, F. R.; Schonhorn, H. "Polymer Preprints"; Amer. Chem. Soc., Div. Polym. Chem. 1975, 16, 146.

28. Baum, B.; Binnette, Mark. "Solar Collectors--Technical Status Report No. 14, 5 Oct.-5 Nov. 1979"; Contract EM-78-C-04-5359; Springborn Laboratories, Inc.: Enfield, CT, 1979.

29. Baum, B.; Cambron, R.; White, R. "Technical Report No. 2, 19 Oct.-19 Nov. 1979"; Contract DE-AC01-79ET21106; Springborn Laboratories, Inc.: Enfield, CT, 1979.

30. Willis, P. B.; Baum, B. "Investigation of Test Methods, Material Properties, and Processes for Solar Cell Encapsulants"; DOE/JPL-954527-79-10; Springborn Laboratories, Inc.: Enfield, CT, 1979.

31. Willis, P. B. et al. "Investigation of Test Methods, Material Properties, and Processes for Solar Cell Encapsulants: Annual Report"; ERDA/JPL-954527-77/2; Springborn Laboratories, Inc.: Enfield, CT.

32. Blaga, A; Yamasaki, R. S. Journal of Material Science. 1976, 11, 1513.

33. Kolyer, J.M.; Mann, N.R. "Accelerated/Abbreviated Test Methods: Interim Report, Apr. 1-Oct. 24, 1977; ERDA-JPL-954458-77/2; Rockwell International: Anaheim, CA, 1977.

34. Thomas, R.E.: Carmichael, D.C. "Terrestrial Service Environments for Selected Geographic Locations." ERDA/JPL-954328-76/5; Battelle Columbus Laboratories: Columbus, OH, 1976.

35. Jet Propulsion Laboratory. "Sail Film Materials and Supporting Structures for a Solar Sail"; JPL Report 720-9; Pasadena, CA:, 1979.

36. Jet Propulsion Laboratory. "Effects of Space Environment on Composites: An Analytical Study of Critical Experimental Parameters"; JPL Report 79-47; Pasadena, CA, 1979.

37. Jet Propulsion Laboratory. "Performance Prediction Evaluation of Ceramic Materials in Point Focusing Solar Receivers"; DOE/JPL-1060-23; Pasadena, CA. 1079.

38. Jet Propulsion Laboratory. "Physical/Chemical Modeling for Photovoltaic Module Life Prediction"; presented at the International Photovoltaics Conference, Berlin. Jet Propulsion Laboratory: Pasadena, CA, 1979.

39. Lindenmeyer, P.H. "Principles of Nonlinear Irreversible Thermodynamics Applied to the Testing of Materials"; D180-25583-1; Boeing Co.: Seattle, WA.

RECEIVED February 9, 1983

Economics of Solar Heating Systems

JOHN W. ANDREWS

Brookhaven National Laboratory, Solar and Renewables Division,
Upton, NY 11973

The development of solar space heating systems has required assessment of the allowable cost
of such systems. Such assessments have most often
utilized life-cycle costing with a 20-year period
of analysis. The use of this method, especially
in conjunction with high energy cost escalation
rates, makes it possible to justify theoretically
almost any system cost. Thus, a Gresham's Law has
come in play: in the competition for research dollars, costly, material-intensive, technically safe
systems have crowded out more innovative but technically more risky approaches keyed to more realistic cost goals.

Methods of Economic Analysis

Three methods of analysis commonly used in evaluating residential solar systems are 1) simple payback; 2) positive cash
flow; 3) life cycle costing. Simple Payback is an answer to the
question, "If I spend more now for a solar system than for a
conventional alternative, how long will it take for my cumulative fuel savings to equal my extra initial outlay?" The HVAC
industry generally considers a three- to five-year payback to be
necessary in order to justify new, more efficient products. The
solar industry is accustomed to much longer periods, often 20
years or longer. It has been suggested that seven or eight
years is a reasonable compromise which allows for an increased
energy consciousness on the part of the public, but does not di-
verge totally from current practice. If no allowance were made
for the increasing cost of energy, this would mean that the incremental solar system cost, or difference between the first
cost of the solar system and that of the conventional alternative, should be no more than eight times the first year's fuel

0097-6156/83/0220-0019$06.00/0

savings. When both fuel cost escalation and time value of money
are taken into account, this limiting ratio might be raised to
about 10. The required payback time is a matter of judgement,
and upon this judgement can depend the entire direction of a
development program. Let us therefore look at some additional
ways of addressing this problem.

Positive Cash Flow is based upon the idea that people will
buy a solar system if their total payments for mortgage, maintenance, and energy are less for the solar system than for the conventional one. Depending on the rate of interest, this criterion
could be compatible with relatively long payback periods. In
order to assess this criterion, the capital recovery factors
(CRF) for nine mortgages of terms 10, 20, and 30 years and interest rates 6, 10, and 14% have been calculated. The CRF is the
ratio of the annual mortgage payment (interest plus principal) to
the face amount of the loan. The loan amount is taken as the
incremental solar system cost, that is, the difference between
the first cost of the solar system and that of the conventional
one. This is then modified by subtracting the tax savings due to
the deductibility of the interest and by adding the incremental
maintenance and miscellaneous costs, where these are assumed to
equal 2% of the incremental system cost. The net capital recovery factor represents the amount of fuel savings required, per
dollar of incremental system cost, to achieve zero initial cash
flow relative to the competing conventional system. The inverse
of the CRF is the ratio of incremental solar system cost to first
year's fuels savings needed to achieve zero relative cash flow in
the first year. Values of this Cost/Savings Ratio are displayed
in Table I. A Cost/Savings Ratio of 10 is consistent with a 20-
year 10% loan and a marginal tax rate of 37%, or else a 20-year
14% loan with a 50% marginal tax rate.

Table I
Ratios of Incremental Solar System Cost to First Year Fuel
Savings Consistent with Zero Relative Cash Flow for the First
Year. Values Given for Marginal Tax Brackets of 30% and 50%.

Term of Loan (years)	Interest Rate (%)	Cost/Savings Ratio	
		30% Marginal Tax Rate	50% Marginal Tax Rate
10	14	5.9	7.0
	10	6.5	7.5
	6	7.2	7.9
20	14	7.8	9.9
	10	9.3	11.5
	6	11.2	13.0
30	14	8.3	10.8
	10	10.4	13.2
	6	13.3	15.9

Life-Cycle Costing is a method of taking into account all of the various costs and benefits which occur in the course of achieving an objective over time. A cost which is incurred in the future is not as great a liability as the same cost incurred now. This is true even in the absence of inflation. Qualitatively it is human nature to defer pain as long as possible. Quantitatively, by deferring a cost one can in the meantime earn interest on the money set aside to pay the cost, and the longer one defers the payment, the more interest will be earned. It may also happen, however, that the further in the future a cost may be deferred, the greater it will be. In an era of rising energy prices, the cost of a given amount of fuel will increase over time. A way is needed to put on an equal footing costs incurred at different times. This is accomplished by means of the present value function (PVF) which is defined (1) as the amount of money which must be set aside now, at an annual rate of return d, to cover over the next N years an annually occurring expense which now costs one dollar but is expected to escalate at an annual rate e:

$$PVF\ (d,\ e,\ N)\ =\ \frac{1}{d-e}\left[1-\left(\frac{1+e}{1+d}\right)^{N}\right] \text{if } d{\neq}e$$

$$=\ \frac{N}{1+d} \qquad\qquad \text{if } d{=}e \tag{1}$$

In evaluating the relative merits of a solar and a conventional HVAC system the present values of the various costs (or benefits) are added (or subtracted) to obtain a total present value (TPV) of the life-cycle costs of the solar system and of the conventional system. The system having the lower TPV is the more cost effective under the given assumptions. For residential systems the following costs and benefits must be considered: 1) system first cost; 2) maintenance and miscellaneous costs; 3) energy costs; and 4) tax benefits due to deductibility of interest payments. The solar tax credit, because it is expected to expire soon, is not considered. (It could be included in the analysis by subtracting its value from the system first cost.)

To illustrate what can be done with life-cycle costing the PVF was calculated for periods of analysis of 5 to 25 years and fuel cost escalation rates from 5% to 30% with a 10% discount rate. The values so calculated are shown in Table II.

The PVF is approximately equal to the allowed cost-to-savings ratio under the simplifying assumptions of equal discount, interest, and inflation rates and of canceling effects of maintenance costs and of the benefit due to tax deductibility of

Table II
Values of the Present Value Function for Various
Fuel Escalation Rates and Periods of Analysis

Fuel Escalation Minus Discount Rate	Fuel Cost Escalation	Periods of Analysis (Years) Present Value Function (Calculated for d = 0.1)				
		5	10	15	20	25
−0.05	0.05	4.2	7.4	10.0	12.1	13.8
0	0.10	4.5	9.1	13.6	18.2	22.7
0.05	0.15	5.0	11.2	19.0	28.7	40.1
0.10	0.20	5.5	13.9	26.9	47.0	78.0
0.20	0.30	6.5	21.6	56.3	136.2	320.6

mortgage interest.* The point that is illustrated by Table II is
that it is possible theoretically to justify very high cost-to-
savings ratios and hence very high solar system costs by using
high fuel cost escalation rates and long periods of analysis.

The use of long system lifetimes such as 25 years and large
fuel escalation rates such as 30% results in very high PVF's and
appears to suggest that one should rationally pay 320 times the
first year's energy savings for a solar system. In addition to
noting that these assumptions imply a 700-fold increase in the
price of fossil fuels, one may observe as well that this apparent
price freedom is limited by certain additional constraints:

1. The period of analysis should not exceed the system
life. Indeed, it should probably be enough shorter than the expected system life that purchases will be induced with the expectation of a profit. Since experience with solar systems is
limited, the use of system lifetimes as long as 20 years is probably not warranted. Even 15 years may be too long.

2. The expectation of high fuel cost escalation rates will
probably not be borne out as much for electricity as for fossil
fuels. Fuel costs are only a portion of the total cost of electric power generation, and cheaper solid fuels are in any case
expected to displace oil in electric power generation.

3. Insofar as fuel costs do escalate at a rapid rate, this
will likely lead to relatively rapid innovation in solar heating
and other forms of energy conservation. Thus a long-life high-

*The latter two effects are of the same order of magnitude and
cancel exactly, for example, if first-year maintenance costs are
1.7% of system first cost, the consumer is in the 50% tax brack-
et, and a 15-year period of analysis is used.

cost solar system may be rendered obsolete well before the end of its service life.

Recently there has been evidence of a change of heart in the solar community concerning cost criteria. At a meeting of the U.S. Department of Energy (DOE) solar contractors held in September, 1981, the consensus moved away from 20-year life cycle costing. For residential systems, payback was seen as the most appropriate criterion with a median recommended payback time of 6 years. For commercial systems, where the purchaser is more sophisticated, life-cycle costing was seen as appropriate, but with a 10-year time horizon.(2) Additionally, a recent DOE marketing study (3) indicated that a definite relationship exists between payback and market penetration, with penetration dropping below 20% for payback periods greater than 8 years.

In line with the above discussion, a Cost/Savings Ratio of 10 was selected as an upper limit, consistent with a fuel cost escalation rate of 5% above inflation and a breakeven time of under 10 years.

Consequences for Solar System Engineering

The amount of energy obtained annually from a solar collector in a well-designed system depends on the collector performance characteristics, the type of load to which the collector is matched, the size of the collector array relative to the load, and the solar availability and ambient temperatures experienced by the system. For a residential space-and-water heating system using flat-plate collectors, however, 100,000 Btu/ft^2-yr (1.14 GJ/m^2-yr) represents a reasonable average over much of the United States.(4) Current gas and oil prices are typically ~$0.50/therm and ~$1.20/gallon respectively. If an average conversion efficiency of 70% is assumed, then the energy savings per unit area will equal $0.71/ft^2 ($7.69/m^2) against gas or $1.22/ft^2 ($13.18/m^2) against oil. Allowing for the likelihood that the price of gas will rise to near parity with oil, we chose $1.20/ft^2 ($13/m^2) as the benchmark value of the energy savings, based on unit collector area.

The factor-of-ten rule then requires a system cost of $12/ft^2 ($130/m^2) of collector. Is there any hope of meeting such a cost goal using current technology based on extruded metal and glass collectors? The answer is no. The materials costs alone for such collectors are $5-$6/ft^2 ($55-$65/m^2).(5) By the time manufacturing, distribution, and installation costs are incurred, the price rises to $20-$25/ft^2 ($220-$270/m^2). The balance of system (storage, piping, pumps, heat exchangers, valves, and controls) contributes another $20/ft^2 ($220/m^2). Thus the overall system cost is over budget by nearly a factor of four.

It is clear that both the collector and the balance of system must experience drastic cost reductions before active solar space heating can be said to be cost-effective. The approach taken at Brookhaven National Laboratory (BNL) was to allocate a budget of $5/ft^2 ($55/m^2) for the installed cost of the collector and the remainder of $7/ft^2 ($75/m^2) for the balance of system. The design objectives were then set to: 1) reduce the materials costs of the collector to ~$1/ft^2 ($11/m^2), and 2) design the collector to be compatible with balance-of-system cost savings.

Materials cost reduction has been achieved through the use of thin-film polymeric materials in both the absorber and glazing portions of the collector. The films, attached to a lightweight bent-metal frame, form a set of stressed membranes that contribute to the overall strength of the panel.

In this design water is used as the heat-transfer medium. This water flows through the absorber at atmospheric pressure from the top to the bottom of the panel. The use of ordinary water without antifreeze makes possible the elimination of at least one heat exchanger and two pumps and can lead to an improvement in system efficiency. The nonpressurized operating mode relaxes design requirements in the piping and storage as well as in the collector.

A companion paper (6) describes the Brookhaven thin-film collector in greater detail.

Acknowledgment

Work performed under the auspices of the U.S. Department of Energy under Contract No. DE-AC02-76CH00016.

Literature Cited

1. Perino, A. M., A Methodology for Determining the Economic Feasibility of Residential or Commercial Energy Systems, SAND 78-0931, 1979, p.11.
2. Active Solar Heating and Cooling Contractors' Review Meeting, U.S. Dept. of Energy, Roundtable Discussion on Solar Systems Evaluation, Washington, D.C., September, 1981, Proceedings in preparation.
3. Lilian, G. L. and Johnston, P.E., A Market Assessment for Active Solar Heating and Cooling Products, Category B: A Survey of Decision Makers in the HVAC Market Place, OR/MS Dialogue, Inc., Final Report DO/CS/30209-T2, September 1980.
4. For example, using the Balcolm-Hedstrom load-collector ratio method (Solar Engineering, January 1977, p.18) the median collectable solar energy was 99,000 Btu/ft^2-yr over a sample of 84 cities, for systems providing 50% of the building heating load from solar energy. All but 6 cities fell in the range 70,000 to 160,000 Btu/ft^2-yr. See also the editorial by Bruce Anderson in Solar Age, January 1978: "... there is no evidence, and little possibility, that any solar space-heating system will ever deliver more, annually, than

100,000 Btus per square foot of aperture area in a climate where sunshine is possible 50 percent of the time."

5. "Cost Reduction Opportunities: Residential Solar Systems," Booz, Allen & Hamilton, Bethesda, MD, 1981 (Draft subject to revision).

6. Wilhelm, W. G., "The Use of Polymer Films and Laminate Technology for Low Cost Solar Energy Collectors," submitted to the Am. Chem. Soc. Polymers in Solar Energy Symposium, Las Vegas, NV, March 28– April 2, 1982.

RECEIVED November 22, 1982

Polymer Film and Laminate Technology for Low-Cost Solar Energy Collectors

WILLIAM G. WILHELM

Brookhaven National Laboratory, Solar and Renewables Division, Upton, NY 11973

Solar energy collector panels using polymer film and laminate technology have been developed which demonstrate low cost and high thermal performance for residential and commercial applications. This device uses common water in the absorber/heat exchanger portion of the device which is constructed with polymer film adhesively laminated to the aluminum foil as the outer surfaces. Stressed polymer films are also used for the outer window and back surface of the panel forming a high strength structural composite. Rigid polymer foam complements the design by contributing insulation and structural definition. This design has resulted in very low weight ($3.5 Kg/m^2$), potentially very low manufacturing cost ($\sim\$11/m^2$), and high thermal performance. The development of polymer materials for this technology will be a key to early commercial success.

The potential for large reductions in capital cost for residential and commercial solar energy collectors can be realized with a design strategy that utilizes polymer films and high speed production equipment. Development work performed at Brookhaven National Laboratory (BNL) under U.S. Department of Energy (DOE) (1) contract has demonstrated that the concept can be applied to solar flat plate collector designs (Figure 1) that exhibit high thermal performance (Figure 2) for summer cooling (2) and winter space heating and hot water applications. Several collectors have been built and tested both at BNL and at the Florida Solar Energy Center with very encouraging results. The major contribution of this design towards high performance and low cost (3)

Figure 1.   BNL lightweight polymer film solar collector panel.

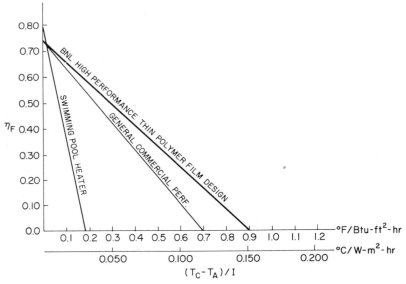

Figure 2.   Performance of BNL solar collector design compared to that of commercial solar collectors.

evolves from the use of high performance polymer films in the
window and absorber/heat exchanger portions of the solar panel.
Present efforts in the development of this technology have identified polymer films which can meet the requirements of performance and cost.   Of even greater significance is the potential
within the industry for the creation of polymer films engineered
specifically for the design.

Solar Absorber/Heat Exchanger

Presently the absorber/heat exchanger is the most unique
contribution in the development because it permits the use of
non-freeze-protected water to be used in a vapor enclosed package
(Figure 3).   In addition, it allows the water to pass through the
package at atmospheric pressure while permitting effective heat
exchange to the liquid.   This has been accomplished by the development of a channeled envelope consisting of a symmetrically
bonded laminate with an outer foil layer.   The foil is important
here because it transfers dimensional stability to the polymer
while providing good lateral heat transfer for the various operating conditions.   The polymer film layer insures good corrosion
protection while improving tear resistance and overall package
integrity.   The outer foil surfaces further contribute by providing a back surface with low optical emissivity for low heat loss
and a suitable front surface for deposition of a solar selective
coating.   This package demonstrates economy by requiring a minimum of material (<0.16mm) for a cost many times lower than provided by conventional technology even when high-performance high-
cost polymers are used (Table I).   The poylmer presently being
considered for use in the absorber is fluorinated ethylene propylene ("Teflon" FFP).   Many other polymers may be candidates
particularly   if   low   cost   for   given   performance   can   be
demonstrated.

Currently, studies are underway (Trian Company, San Fran-
cisco) (4) to explore the synthesis of new polymer film formulations which can offer the required thermal performance at substantially reduced cost (Table II).   The greatest challenge and
reward remains to be exploited with additional formulations which
can meet the requirements of high service temperature (up to 96°C
in normal operation, but as high as 230°C under thermal stagnation), high liquid barrier properties, general chemical inert-
ness, and relatively low cost.   The incentive to the chemical industry may be derived from a high volume, high-value-added material product.

Printable Optical Selective Absorber Coating

The black surface that lies below the solar collector window
functions by converting the radiant energy from the sun into
thermal   energy   for   conduction   through   the   heat   exchanger

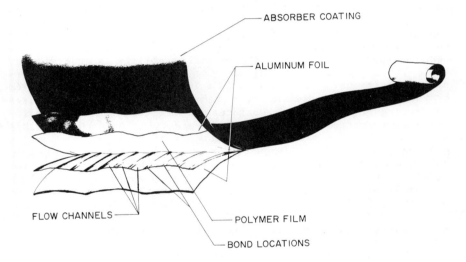

Figure 3.   BNL absorber/heat exchanger incorporating a polymer film and foil laminate.

Table I
AUTOMATED PRODUCTION* COST OF POLYMER FILM SOLAR COLLECTOR BASED
ON CONVENTIONAL POLYMER FILMS
Collector Area = $3m^2$

Materials	Cost/ Collector	Cost/ m^2
Aluminum sheet metal frame	$ 7.00	
Polymer film glazing (PVF)	13.00	
Polymer film back surface (oriented PET)	5.00	
Laminate Absorber/heat exchanger (FEP)	28.50	
Optical selective coating (treated nickel foil)	12.80	
Insulation (structural foam)	4.80	
Adhesives	0.30	
Mounting hardware	0.75	
TOTAL MATERIAL COST	$ 72.15	$ 24.09
DIRECT LABOR COST, 4 men at $10/hr, 1 collector/min	0.67	
FACTORY OVERHEAD COST (450%)	3.02	
TOTAL PRODUCTION COST	$ 75.82	$ 25.27
"SHRINKAGE" (2%)	1.52	
TOTAL MANUFACTURING COST	$ 76.34	
GENERAL, ADMINISTRATIVE AND PROFIT (43%)	32.83	
FACTORY SELLING PRICE	$109.17	
DISTRIBUTOR-INSTALLER REPRESENTATIVE (15%)	16.38	
DISTRIBUTOR-INSTALLER FIRST COST	$125.55	
WARRANTY COST (10%)	12.56	
DISTRIBUTOR-INSTALLER SECOND COST	$138.11	
INSTALLATION COST PER COLLECTOR	50.00	
INSTALLED COST TO THE CUSTOMER	$188.11	$ 62.70

*Based on 210,000 solar collectors per year.

Table II
AUTOMATED PRODUCTION* COST OF POLYMER FILM SOLAR COLLECTOR BASED
ON DEVELOPMENT OF ADVANCED POLYMER FILMS
Collector Area = $3m^2$

Materials	Cost/ Collector	Cost/ m^2
Aluminum sheet metal frame	$ 7.00	
Polymer film glazing (oriented PVDF)	7.50	
Polymer film back surface (oriented PET)	5.00	
Laminate absorber/heat exchanger (oriented H.T. PAN)	6.80	
Optical selective coating (printable)	1.60	
Insulation (structural foam)	4.80	
Adhesives	0.30	
Mounting hardware	0.75	
TOTAL MATERIAL COST	$ 33.75	$ 11.25
DIRECT LABOR COST, 4 men at $10/hr, 1 collector/min	0.67	
FACTORY OVERHEAD COST (450%)	3.02	
TOTAL PRODUCTION COST	$ 37.44	$ 12.48
"SHRINKAGE" (2%)	0.75	
TOTAL MANUFACTURING COST	$ 38.19	
GENERAL, ADMINISTRATIVE AND PROFIT (43%)	16.42	
FACTORY SELLING PRICE	$ 54.61	
DISTRIBUTOR-INSTALLER REPRESENTATIVE (15%)	8.19	
DISTRIBUTOR-INSTALLER FIRST COST	$ 62.80	
WARRANTY COST (10%)	6.28	
DISTRIBUTOR-INSTALLER SECOND COST	$ 69.08	
INSTALLATION COST PER COLLECTOR	50.00	
INSTALLED COST TO THE CUSTOMER	$119.08	$ 39.69

*Based on 210,000 solar collectors per year.

surfaces. Such surfaces, if not optically selective, are good emitters of radiant energy. Because polymer film is used as a window material and because the optical transmission bandwidth of these materials tends to be high a fair amount of reradiation can be lost out through the polymer window. This can be corrected by preparing the absorber surface so that it has a low emissivity (<0.3) while maintaining good optical absorption (>0.94). Such optically selective surfaces are currently available to the solar industry but at a cost of about $10/m^2.

The optically selective surface is important to the BNL solar collector design because high thermal performance is possible while maintaining structural simplicity. Because low cost goals are part of the BNL development effort, new technologies are being explored which suggest that good optical selective surfaces can be deposited at low cost. Current studies (Honeywell Corporation) (4) suggest that such surfaces can be made economically ($0.10 to $0.20/m^2). The design mode proposed by BNL suggests incorporating of the selective coating by gravure printing as a serial part of the laminating of the channeled absorber envelope (Figure 4). It is expected that such a method will insure good economy while maintaining the required performance. Such work is presently being explored.

Polymer Film Window

A polymer film is also used in the window portion of the solar collector. The window admits solar radiation to the absorber while serving as a thermal barrier to inhibit upward heat loss. In the BNL design it also contributes structurally to the panel's rigidity because of its stressed state and high tensile modulus. This film (~0.01mm) requires properties different from those of the absorber/heat exchanger. This polymer material must be very transparent (>90%) to the incident solar radiation but immune to the destructive effect of its ultraviolet component. The polymer film must maintain its optical and physical properties for at least 8 years at temperatures approaching 100°C. Presently the BNL collector design uses polyvinyl fluoride (DuPont's Tedlar) (5) which generally meets the requirements stated above. Other materials also appear attractive such as oriented polyvinylidene fluoride. The best options will come from polymer films specifically designed for the collector requirements and at costs which demonstrate additional economic advantage.

Insulation and Compression Substrate

The insulation in the BNL solar collector design uses rigid insulation as a structural member. This member is presently made from glass reinforced polyisocyanurate with a aluminum foil facer (Thermax, Celotex). The sheet metal frame is parametrically coupled to the rigid insulation and is under compression when the

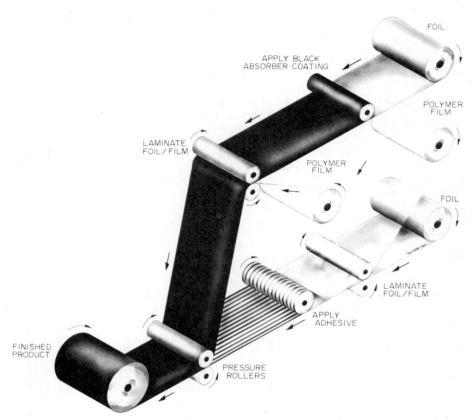

Figure 4. Mass production techniques applied to thin film absorber/heat exchanger.

front and the back film are stress relieved. An additional layer of fiberglass insulation is placed over the rigid insulation to minimize high temperature excursions on the rigid insulation during thermal stagnation (the condition when a dry collector is permitted to go up to its highest equilibrium temperature due to no interval fluid mass flow, high solar intensity and high ambient temperature). While these materials function well under most tests there is still considerable room for improvement. As with the other polymer materials mentioned the collector design could benefit from material compositions specifically designed for the solar collector with particular emphasis on shrink properties, higher environmental tolerance and low cost. We look forward to exploring material options in this area.

Thermal Performance

When plastics are associated with solar energy collecting it is generally perceived that they are low temperature devices such as might be used for swimming pool heating. The current solar collector designs being developed at BNL have on the contrary very high thermal performance comparable to the very best commercial flat plate collectors. These high thermal efficiencies permit summer (7) and winter water collection temperatures as high as 96°C with good efficiency depending on the ambient conditions. This high efficiency results from the use of high temperature materials like fluorinated ethylene propylene which has a service temperature as high as 230°C. Such polymer film permits tolerance of very high stagnation temperatures. Since the stagnation temperature is increased when the heat losses from the solar collector are reduced, there is a correlation between stagnation tolerance and overall thermal performance. The consequence of this is that if you can find suitable high temperature material you can design for proportionally lower heat losses and consequently high thermal performance. This has been attempted in the BNL design. Materials which have even higher service temperature (dry temperature under thermal stagnation conditions) will permit designs with even higher thermal performance.

The graph in Figure 2 indicates the relative performance of the current BNL design compared with the average commercial grade solar flat plate collector. It has thermal performance characteristics of solar flat plate collectors normally applied to applications requiring relatively high operating temperatures (85°C) as in solar cooling.

Application

The two major limiting factors to solar applications are cost and performance. Demonstrations of solar applications with traditional technology have been disappointing primarily due to

very high cost and modest performance. With the potential of new
material development, thin polymer film solar collector technology could change the application picture dramatically. New
low cost material will permit this technology to complement applications which require high temperature and to utilize locations which have previously been considered inappropriate.
Northern climates (i.e. Canada) (8) can benefit from the high
thermal efficiency and low cost for applications requiring space
heating and hot water. These locations have previously not been
considered good candidates for solar energy application because
of the high relative cost of efficient solar energy collection.
Commercial applications like restaurants, laundries, and motels
can also benefit from the favorable economics. Light industrial
applications may also benefit but are usually limited by high
energy volume requirements and limited solar collection space.
Another area which may be of very significant importance is the
effect this low cost technology will have on the growth of third
world nations which have previously been energy starved due to
poor economic equity.

High Speed Mass Production

    The major key to low cost is a solar collector design which
has low material intensity and is capable of being produced at
very high speed to reduce labor intensity. The low material intensity results from the use of polymer films and film laminates.
The development of an efficient non-pressurized thin film absorb-
er/heat exchanger also contributes to minimum material requirements. The use of a simple roll-formed sheet metal frame with a
cost one third that of a conventional extruded frame further reduces the cost and can be produced as one piece to be bent around
the insulating substrate. The weather seals are integrated into
the film/adhesive attachment for simplicity, reliability and very
low additional material requirements.

    With the material reduced to a minimum the remaining key
factor to affect production cost is the number of units that can
be produced in a given period of time. The laminating process
shown in Figure 4 can be accomplished at over one hundred linear
meters per minute, reducing the unit labor cost of the absorber/
heat exchanger manufacture to a small fraction of that required
by conventional technology. In addition, since the BNL collector
is constructed from four basic pieces (insulating substrate,
absorber/heat exchanger, one piece sheet metal frame and the
polymer film for the front and back surfaces), final assembly can
be automated for a high assembly rate (one unit per minute). The
projected result is a production cost as low as $11/m^2$, (see
Tables I and II).

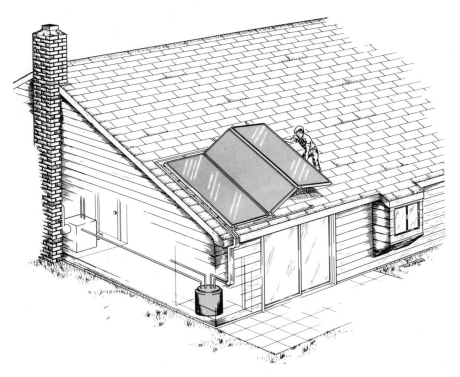

Figure 5.   Example of low-cost installation.

Conclusion

The material presented here suggests that cost competitive solar heating systems can be developed which could have a large effect on marketability. The large solar applications market that could develop as a result of addressing the correct market economics would have a beneficial effect on society by providing a competitive energy alternative that is clean and preserves natural fossil resources. The petroleum product which would normally be used in a combustion process for heat can now be used more effectively as a material in the solar energy collector and the balance-of-system components. Such materials application will add high value to the petroleum end product and provide a service to the community by stabilizing a social need for energy. With the identification of this new technology the plastics industry has the opportunity to complement the new designs with the development of the desired low cost polymer film materials. The net result can be very large market penetration with both social and commercial benefits.

Acknowledgment

Work performed under the auspices of the U.S. Department of Energy under Contract No. DE-AC02-76CH00016.

Literature Cited

1.   Andrews, J. W.; Wilhelm, W. G. "Thin Film Flat-Plate Solar Collectors for Low-Cost Manufacture and Installation," BNL 51124, March 1980.
2.   Wilhelm, W. G. "Low Cost Solar Energy Collection for Cooling Applications," BNL 51408, June 1981.
3.   Wilhelm, W. G. "Flat Plate Solar Collectors Utilizing Polymeric Film for High Performance and Low Cost," BNL 30148, August 1981.
4.   Fouser, J. P. (Trian Company), Personal communications on advanced polymer film development.
5.   McKevey, W. D.; Zimmer, P. B.; Lin, R. J. H. "Solar Selective Coating Development," Honeywell Inc., Final Report, DOE Contract No. DE-AC04-78CS14287, December 1979.
6.   deBussy, R. P. "Tedlar 400 SE PVF Film for Glazing Solar Collectors - A Cost Effective Break-Through," E. I. duPont deNemours μ Co., Inc., Wilmington, DE.
7.   Merrick, R. (Arkla Industries Inc.), Personal communication, May 1981.
8.   Wilhelm, W. G. "Low Cost, High Performance Solar Flat Plate Collectors for Applications in Northern Latitudes, BNL 30633, September 1981.

RECEIVED November 22, 1982

Stability of Polymeric Materials in the Solar Collector Environment

M. A. MENDELSOHN, F. W. NAVISH, JR., R. M. LUCK, and F. A. YEOMAN

Westinghouse Research and Development Center, Pittsburgh, PA 15235

Several types of materials consisting of fluoro-carbon, silicone, acrylic, ethylene-acrylic, ethylene-propylene (EPDM), and butyl elastomers, which could be employed as gaskets, and silicone, acrylic, butyl, and chlorosulfonated polyethylene caulking compositions were subjected to studies of their durability to the environment inside a thermal solar collector. Degradation characteristics of these materials on exposure to severe hydrolytic and thermal-oxidative conditions are described. Although none of the gasket type elastomers tested were found to be entirely satisfactory, the fluorocarbon (Viton) displayed the best durability and thermal stability overall. The silicones were second best. Unfortunately, fluoro-carbons tend to exhibit excessive low temperature compression set, a characteristic which could be a serious problem in geographic zones having relatively cold winters. The silicones show very poor resistance to compression set on thermal aging and, while the fluorocarbon is considerably superior in this respect, it nevertheless displays undesirably high values. The polyacrylate and ethylene-acrylic copolymers and one of the ethylene-propylene terpolymers (Nordel) were the best of the intermediate temperature elastomers. Except for resistance to compression set, these materials were inferior to the silicones in thermal stability as measured by their retention of tensile properties. The other EPDM compounds and butyl rubber were considerably inferior to the above-mentioned elastomers. It is not expected that the service life of the tested materials will be limited solely by their ability to resist hydrolytic degradation. The only caulking compositions which retained moderate physical integrity on thermal aging were the silicones.

0097–6156/83/0220–0039$11.25/0

Elastomeric materials are used in thermal solar collectors as gasket and caulking compounds. In addition to these sealant applications, polymeric materials are also widely employed as thermal insulation and occasionally as glazing and frame components. This paper provides a supplement and a continuation of a previously reported study(1) of the endurances of several commercially available sealants to the harsh environment of the collector.

In order to function adequately over many years of use, a sealant must exhibit good long-term resistance to air at high temperatures. Inside the cell temperatures may exceed 200°C during the relatively brief periods, possibly as long as several weeks, while the collector is under no-flow or stagnation conditions. During normal operation, the units attain temperatures between 125 and 200°C. Depending on the collector design, the sealant can experience prolonged exposure to relatively high temperatures. In addition, the sealants are exposed to such environmental stresses as high humidity, ozone, ultraviolet radiation, etc.

Until very recently almost no new materials were developed for the solar collector industry, which is essentially in its infancy. This investigation was performed to help identify deficiencies of the available sealant materials with the expectation, which we believe is beginning to be realized, that this would lead to development of superior materials for solar collector applications.

Thus we investigated the effects of thermal-oxidative and hydrolytic degradation on various elastomers which could be employed as collector sealants. Although such sealants are generally not required to possess high strength, we nevertheless monitored the effects of severe environmental stresses on tensile properties since these characteristics provide an indication of degradation of the polymer. However, the effects of aging on compression set were studied since this important property measures the ability of a gasket to retain its sealing capability. Outgassing of products from degrading polymers is another very important manifestation of the degradation processes since it affects the solar light transmittance of the glazing. This area was studied in considerable detail and is discussed in the following paper(2). Other effects which we plan to describe in future publications include: resistance to weathering, formation of corrosive products, ozone resistance, and fungal susceptibility.

Degradation Processes

Thermal Aging in Air

Early in the thermal aging process, the elastomers generally experience an increase in degree of cure, which results in an overall improvement of their tensile strength (Figure 1). Simultaneously, low molecular weight materials, such as plasticizers, processing aids, and stabilizers diffuse out of the polymer matrix. This has the overall effect of increasing the tensile strength while reducing the elongation. In addition, loss of stabilizers such as antioxidants, antiozonants and ultraviolet absorbers causes the polymer to become more susceptible to the ensuing degradation. As the degradation process continues, the stabilizers are also lost through chemical processes.

In many polymers, additional cross-linking continues throughout much or all of the aging period with the effect of increasing the modulus and eventually causing embrittlement. However, some polymers undergo essentially the opposite effect--that of chain scission which reduces the strength of the material. Further complicating the picture is that both processes occur simultaneously in many elastomers and that some of the reactions are essentially pyrolytic and others oxidative in nature. The degradation process becomes even more complex when the elastomer consists of copolymers or terpolymers since the different types of chain units can degrade by different mechanisms, can interact, and also influence the reactivity of their neighbors.

Simultaneous cross-link formation and chain scission, which occur during the thermal-oxidative degradation of many elastomers, transform the polymer into a harder but weaker material having a consistency that can best be described as cheesy.

As linkages in the primary chains are being broken and low molecular weight chain fragments and oxidized degradation products are diffusing out of the polymer the material experiences a loss in weight. In an actual collector, the very low molecular weight products will escape from its confines; however, the less volatile degradation products can coat the absorber plate and the glazing, thus decreasing the efficiency of the collector(2).

When elastomers are subjected for extended periods of time at elevated temperatures to a compressive load while being permitted to distort, they will not recover to their original dimensions when the load is removed. Measuring the compression set of compressed elastomers during thermal aging is a very useful and realistic means of evaluating materials for gasket applications.

Permanent set or distortion of a polymer can be an essentially physical or combination of chemical and physical phenomena. Under the application of heat and pressure, the polymer chains rearrange to accommodate the applied forces. When the pressure is released, the newly formed alignments undergo some shifting, but nevertheless generally do not permit a complete recovery of the material

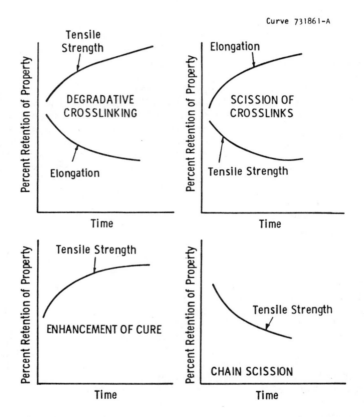

Figure 1. Typical effects of various types of degradation processes on physical properties.

to its original configuration. Polymers, such as polyurethanes and polyamides, generally are quite susceptible to exhibiting high compression set values because of dislocations of hydrogen bonds.

Chemical changes frequently accompany the shifting of the polymer chains. Cross-links are broken under the combined influence of thermal stress and the induced shear forces and are often reformed in a manner so as to accommodate the newly formed alignment of the polymer chains. When cross-links break and reform to an appreciable extent, polymers will exhibit very poor recovery.

Scission of primary linkages in the polymer chain also leads to high compression set. For example in the case of silicones, an intermolecular interchange of siloxane linkages, which does not affect the number average molecular weight of the polymer, permits the network to rearrange to a new configuration. In addition, intramolecular siloxane interchange reactions, which not only account for evolved products, also contribute to the lack of recovery of the stressed polymer.

Intermolecular Siloxane Interchange

Intramolecular Siloxane Interchange

As a result of such reactions silicone elastomers are noted for their ease of undergoing permanent deformations when subjected to loads at elevated temperatures.

In another type of example polyurethanes, which are generally quite susceptible to taking permanent sets, not only undergo changes in their configurations by undergoing realignments of their hydrogen bonds, but also are susceptible to the reversible dissociation reactions between the urethane linkage and isocyanate and hydroxyl groups (analogous reactions also occur for allophante, urea, and biuret linkages present in many polyurethanes). The dissociated linkages continually reform; however, if this occurs between a hydroxyl and isocyanate group which were not combined previously the polymer will suffer loss of recovery. As a result of this phenomenon, polyurethanes are not recommended for certain gasketing applications.

Thermal Aging in Water

The flat-plate solar service environment is likely to involve not only intermittent exposure to high moisture levels (from high atmospheric humidities, from dew or from rain), but also concurrent elevated temperatures as the collector unit warms up immediately after a period of exposure to high moisture levels. Therefore, we are concerned with the hydrolytic stability of the sealant.

Elastomeric systems immersed in hot water can, depending upon their chemical and physical characteristics, undergo several simultaneous processes, all of which can alter the mechanical properties of the system. These processes, involving a variety of chemical or physical mechanisms, may, in some cases, reinforce each other in causing a particular effect and in other cases compete with each other because they tend to produce opposing effects. During the early stages of exposure to water at elevated temperatures, many polymers experience the previously described increase in tensile strength because of further polymerization (Figure 2). As water permeates into the polymer matrix, it behaves as a plasticizer causing a dimnution in tensile strength and often an increase in ultimate elongation. Concurrently, plasticizers, processing aids, and other additives are leached out causing a decrease in elongation and an increase in tensile strength. This effect is compounded by the penetration of water into the interface between the polymer and reinforcing filler. Which of the preceding effects dominates is dependent upon the nature of the polymer and the other ingredients in the compound.

The most harmful hydrolytic effect arises when the polymer contains an appreciable concentration of such hydrolyzable linkages as ester, urethane, or siloxane. Cleavage of hydrolyzable linkages manifests itself as a drop in tensile strength and an increase in ultimate elongation eventually leading to a loss of thermosetting characteristics.

Experimental

Specimens of Class PS (gasket type materials, preformed seals) elastomers were obtained as vulcanized, post-cured sheets, approximately 0.18 cm in thickness, from their manufacturers with the exception of silicone rubbers I and J which were supplied as calendered rolls requiring a 4 h bake at 200°C before using (Table I).

Sheets of Class SC (caulking compounds) elastomers, 0.16 cm thick, were prepared by casting the caulks in open molds, 15 cm X 2.5 cm. The bottom of the mold was covered with a thin sheet of Mylar, which had been coated with a silicone and then a fluoro-

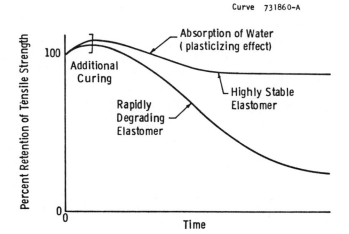

Figure 2.   Typical effects on combination of high humidity and high temperature on elastomers.

TABLE I

Elastomers Evaluated

Code	Supplier's Designation	Class	Type Elastomer	Supplier
A	Silicone rubber sealant (also known as DC 732 white)	SC	Silicone	Dow Corning
B	790 building sealant (also known as DC 790)	SC	Silicone	Dow Corning
C	RTV 103	SC	Silicone	General Electric
D	Mono	SC	Acrylic terpolymer	Tremco
E	Eternaflex Hypalon sealant	SC	Hypalon	Gibson-Homans
F	Tremco butyl sealant	SC	Butyl	Tremco
G	SE-7750	PS	Silicone	General Electric
H	Silastic 747	PS	Silicone	Dow Corning
I	HS-70	PS	Silicone	Dow Corning polymer compounded by North American Reiss
J	NPC 80/40	PS	Silicone	"
K	Hycar 4054, 210-103-35-1	PS	Acrylic	Goodrich
L	Vamac, 3300-12A	PS	Ethylene-acrylic	DuPont
M	Viton, PLV 1008	PS	Fluorocarbon	DuPont, compounded by Pelmor
N	Viton, 31-323-0731A	PS	Fluorocarbon	DuPont

O	Nordel, 3300-11	PS	Ethylene-propylene terpolymer (EPDM)	DuPont
P	Vistalon 78E-09-28-2	PS	EPDM	Exxon
Q	E-633	PS	EPDM	Pawling
R	8EX-122	PS	Bromobutyl	Polysar
S	8EX-123 (Butyl 100)	PS	Butyl	Polysar
T	SR 35020	PS	Butyl	Stalwart
U	Hyaplon 3300-10	PS	Chlorosulfonated polyethylene	DuPont
V	Hydrin 100 (HM 13-27-1)	PS	Hydrin	Goodrich
W	Hydrin 200 (HM 13-SEC-3-2)	PS	Hydrin	Goodrich
X	Hydrin 400 (HM 14-10-1)	PS	Hydrin	Goodrich

carbon mold release agent. The caulking compound was inserted
into the mold cavity and spread evenly with a single-edge razor
blade until it conformed to the dimensions of the cavity and had
a smooth appearing surface. Each casting was permitted to cure
under room conditions for 4 to 6 weeks before die cutting of
tensile specimens. Prior to the initiation of the aging exposures,
the specimens were post-cured for 24 h at the temperature at which
they were to be aged.

The procedures for monitoring the thermal aging of the sealants in air and water have been described previously(1,3).
Briefly they involve aging dumbbell tensile specimens of the Class
PS and SC elastomers at a series of elevated temperatures while
suspended in air and immersed in water and monitoring their tensile properties (The terms elongation and tensile strength are
used in this paper for ultimate elongation and tensile strength
at break, respectively). Compression set measurements were
performed in evaluating the recovery characteristics of materials
subjected to continual compressive deflection while held at elevated temperatures. Aging times and temperatures were not those
suggested in the ASTM D-395-69 Method B procedure, but were selected to provide useful Arrhenius data.

The description of the screening tests is presented in the
text.

Materials Evaluated

Three principal criteria were initially employed to select
materials for evaluation. The first consisted of expectations,
based upon an extensive computerized literature survey and
manufacturers' recommendations, that the materials would offer
superior resistance to the environment of the collector. The next
factor, that the sealant should not exhibit an extremely high cost,
eliminated the very expensive high temperature specialty compounds.
Thirdly, certain materials were examined because of their current
extensive use in solar collectors although we did not expect them
to display high thermal stability or good endurance characteristics.

The above criteria were employed to select several commercially supplied Class PS elastomers for laboratory screening by
employing selected tests taken from National Bureau of Standards
NBSIR 77-1437(4) and ANSI/ASTM D-3667-78 specifications for "Rubber Seals Used in Flat-Plate Solar Collectors". Four silicone,
three EPDM, two fluorocarbon, three epichlorohydrin, one ethylene-
acrylic, one polyacrylic, one chlorosulfonated polyethylene, one
bromobutyl and two butyl rubbers were studied in these screening
tests. These materials are identified in Table I and those
compositions which were revealed by their manufacturers are shown
in Table II. Undoubtedly some materials which should have been
included were omitted; however, we hope that this sampling will
provide an indication of the applicability of a wide range of
materials for use as sealants in thermal solar collectors.

Table II. Chemical Formulations of Compounded Elastomers*

Compound and Cure Conditions	Code	Ingredient	Parts by Weight	Supplier	Chemical Description	Function**
Viton (31-323-0731A) Cure Conditions 24 hr/168°C	L	Viton E60	94.4	DuPont	Fluoroelastomer	Base polymer
		MT Black (N908)	25	Cabot	Medium thermal carbon black	Reinforcement
		Mag D	3	Whittaker, Clark & Daniels	Magnesium oxide	Vulcanizing agent, stabilizer
		Ca(OH)$_2$	6	(Multiple source of supply)	Calcium hydroxide	Drying agent, acid acceptor
		VPA #1[2]	1.5	DuPont	33% organophosphonium salt (proprietary)	Viton processing aid
		VC #20	1.7	DuPont	67% fluoroelastomer	Viton curative
		VC #30	4	DuPont	50% dihydrooxyaromatic compound + 50% fluoro-elastomer	Viton curative
Viton (31-323-0731) Cure Conditions 10 min/168°C + 24 hr/171°C	HH	Viton E60C	100	DuPont		
		MT (N908)	25	Cabot	Medium thermal carbon black	Reinforcement
		Maglite D	3	Whittaker, Clark & Daniels	Magnesium oxide	Vulcanizing agent, stabilizer
		Ca(OH)$_2$	6	(Multiple source of supply)	Calcium hydroxide	Drying agent, acid acceptor
		VPA #1	1.5	DuPont		Viton processing aid
Hypalon Cure Conditions 30 min/152°C	BB	Hypalon 40	100	DuPont	Chlorosulfonated polyethylene	Base polymer
		TLD-90	32	Wylough & Loser	90% dispersion of litharge	Polymeric carrier
		Polyethylene 617A	2	Allied Chemical	Low molecular wt. polyethylene	Processing aid
		SRF-LM (N762)	35	Witco Chemical Co., Inc.	Semireinforcing furnace black	Reinforcement
		MDTS	0.5	DuPont	Benzothiazole disulfide	Activator
		Tetrone A	2	DuPont	Dipentamethylene thiuram tetrasulfide	Very active accelerator
		NBC	1	DuPont	Active ingredient, nickel dibutyldithiocarbamate	O$_3$, weathering, cracking inhibitor
Nordel (3300-11) Cure Conditions 30 min/160°C	N	Nordel 1320	100	DuPont	Ethylene propylene terpolymer (EPDM)	Base polymer
		Zinc oxide	5	American Zinc Sales Co.	Zinc oxide	Activator + reinforcement
		Agerite Resin D	2	R.T. Vanderbilt	Polymerized trimethyl dihydroquinoline	Antioxidant
		FEF (N550)	60	Cabot	Fast extruding furnace black	Reinforcement
		DiCap 40C	8	Hercules Powder Co., Inc.	40% dicumylperoxide on calcium carbonate	Nonsulfur vulcanizing cross-linking agent
		HVA-2	2	DuPont	N,N-m-phenylene dimaleimide	Curative
Vamac (3300-12A) Cure Conditions 3 hr/177°C	J	Vamac (N124)	124	DuPont	Ethylene-acrylic copolymer	Base polymer
		SRF (N774)	30	Cabot	Semireinforcing black	Reinforcement
		MDA	1.25	DuPont	Methylene dianiline	Curing agent
		DPG	4	American Cyanamid Co.	Diphenyl guanidine	Accelerator

Continued on next page

Table II. Chemical Formulations of Compounded Elastomers—
Continued

Compound and Cure Conditions	Code	Parts by Weight	Ingredient	Supplier	Chemical Description	Function
Vistalon (28E-09 28-2)	CC	100	Vistalon 5600	Exxon	Ethylene-propylene terpolymer	Base polymer
		100	Carbon black N550	Cabot	Fast extruding furnace black	Reinforcement
Cure Conditions		95	Cirosol 4240 oil	Sun Oil Co.	Naphthenic oil	Processing aid
20 min/160°C		1.0	Stearic acid	City Chemical Co.	Stearic acid	Primary activator, plasticizer
		5.0	Zinc oxide	American Zinc Sales Co.	Zinc oxide	Activator, reinforcement
		1.5	TMTDS	Monsanto	Tetramethylthiuram disulfide	Accelerator
		0.5	MBT	DuPont	2-mercaptobenzothiazole	Activator
		1.5	Sulfur	Akron Chemical Co.	Sulfur	Vulcanizing agent
Bromobutyl (8EX-122)	DD	100	Polysar bromobutyl X-2	Polysar	Brominated butyl rubber	Base polymer
		1.0	Stearic acid	City Chemical	Stearic acid	Primary activator, plasticizer
Cure Conditions		1.0	Maglite D	Whittaker, Clark & Daniels	Magnesium oxide	Vulcanizing agent, stabilizer
15 min/180°C		10.0	Heliozone wax	DuPont	Selected blend of petroleum waxes	Sun-checker, tracking resistor
		55.0	N550 carbon black	Cabot	Fast extruding furnace black	Reinforcement
		10.0	Sunpar 2280	Sun Oil Co.	Paraffinic oil	Softener
		3.0	Zinc oxide	American Zinc Sales	Zinc oxide	Activator, reinforcement
		0.3	TMTD	Akron Chemical	Tetramethylthiuram disulfide	Accelerator
		2.0	Permalax	DuPont	Active ingredient, di-ortho-tolylguanidine salt of dicatechol borate	Accelerator
Butyl (8EX-123)	P	100	Polysar butyl 100	Polysar	Butyl rubber	Base polymer
		1.5	Stearic acid	City Chemical	Stearic acid	Primary activator, plasticizer
Cure Conditions		4.0	SP 1045	Schenectady Chemical Inc.	Phenol formaldehyde resin	Curative
45 min/180°C		10.0	Heliozene wax	DuPont	Blend of petroleum waxes	Sun-checker, cracking resistor
		55.0	N550 carbon black	Cabot	Fast extruding furnace black	Reinforcement
		7.0	Sunpar 2280	Sun Oil Co.	Paraffinic oil	Softener
		2.67	Stannous chloride presperison	Ware Chemical Corp.	Stannous chloride dihydrate (75% in 25% inert oil)	Catalyst
Acrylic (210-108-35-1)	K	100	Hycar 4054	B. F. Goodrich	Polyethylacrylate elastomer	Base polymer
		2.0	Acrawax C	Charles L. Huskling Co., Inc. (Glycol Chem. Div.)	Synthetic wax	Finishing dusting agent
Cure Conditions		2.0	TE-80	Technical Processing, Inc.	(Proprietary)	Lubricant processing aid
4 min/170°C +		60.0	Phil black N550	Phillips Petroleum Co.	Fast extruding furnace black	Reinforcement
8 hr/175°C		3.0	Witco sodium stearate	Witco Chemical	Sodium stearate	Curative, dispersing agent
		2.0	Adogen 345D	Ashland Chemical Co.	Hydrogenated tallowamine	Curative
		1.0	NBS stearic acid	City Chemical Co.	Stearic acid	Activator, processing aid

Table II. Chemical Formulations of Compounded Elastomers—

Continued

Compound and Cure Conditions	Code	Parts by Weight	Ingredient	Supplier	Chemical Description	Function
Hydrin 100 (HM 13-27-1)	EE	100	Hydrin 100	B. F. Goodrich	Epichlorohydrin based elastomer	Activator, vulcanizer
		5.0	Red lead	Eagle Picher Co.	> 90% red lead oxide	
		1.0	NBC	DuPont	Active ingredient, nickel dibutyl dithiocarbamate	O_3, weathering, cracking inhibitor
Cure Conditions 30 min/175°C		40.0	N550, FEF black	Cabot	Fast extruding furnace black	Reinforcement
		1.0	TE-70	Technical Processing Co.		Lubricant, processing oil
		1.5	2-mercapto-imidazoline (NA-22)	DuPont	2-mercapto-imidazoline	Fast general purpose accelerator
Hydrin 200 (HM-13-SEC-3-2)	FF	100	Hydrin 200	B. F. Goodrich	Epichlorohydrin based elastomer	Activator, dusting agent
		1.0	Zinc stearate	Witco Chemical Co.	Zinc stearate	Activator, vulcanizer
		5.0	Red lead	Eagle-Picher Co.	> 90% red lead oxide	
		1.0	NBC	DuPont	Active ingredient, nickel dibutyl dithiocarbamate	O_3, weathering, cracking inhibitor
Cure Conditions 30 min/175°C		40.0	N550, FEF black	Cabot	Fast extruding furnace black	Reinforcement
		1.0	ZO-9	Yerzley Co.	Blend of waxes	Mold lubricant, plasticizer
		1.5	2-mercapto-imidazoline	DuPont	2-mercapto-imidazoline	Fast general purpose accelerator
Hydrin 400 (HM 14-10-1)	GG	100	Hydrin 400	B. F. Goodrich	Epichlorohydrin based elastomer	Primary activator, plasticizer, softener
		0.8	Stearic acid	City Chemical Co.	Stearic acid	
Cure Conditions		2.0	Dyphos	National Lead	Dibasic lead phosphite	Heat, light, weathering, chemical stabilizer
		0.9	NBC	DuPont	Active ingredient, nickel dibutyl dithiocarbamate	O_3, weathering, cracking inhibitor
20 min/160°C		0.5	Methyl niclate	R. T. Vanderbilt	Methyl niclate	Accelerator
		20.0	N326 black	Cabot	High abrasion furnace black, low structure	Reinforcement
		30.0	N550 black	Cabot	Fast extruding furnace black	Reinforcement
		10.0	DOP	Monsanto	Dioctyl phthalate	Plasticizer, softener
		0.3	ZO-9	Yerzley Co.	Blend of waxes	Mold lubricant, plasticizer
		2.5	SR-350	Sartomer Corp.	Polyfunctional methacrylate	Auxiliary cure material
		3.5	Di-Cup 40C	Hercules Powder Co., Inc.	~ 40% dicumylperoxide on calcium carbonate	Nonsulfur, vulcanizing, cross-linking agent

* Formulations that were supplied by manufacturers are presented in this table. Suppliers of the other compounds viewed their formulations as proprietary and therefore did not reveal them.

** Most of the terminology used in describing the functions of the ingredients is that employed in the product descriptions by the suppliers.

Properties measured in the screening tests included compression set after 70 h at 150°C, compression set after 166 h at -10°C, ultimate elongation, tensile strength and hardness and changes in these three properties after aging 166 h at 150°C and volatiles lost during the latter aging period (Table III).

Among the silicone elastomers screened, compound H was eliminated from consideration for the more stringent testing because its weight loss and drop in tensile strength during aging were excessive. The remaining silicones performed very well in all categories. Nordel 3300-11 was the only EPDM rubber of the three formulations tested which exhibited good retention of physical properties on aging at 150°C. The two fluorocarbon elastomers screened appear to be essentially identical in all respects. Their performance was excellent in all categories except for their proclivity to develop crystallinity at low temperatures as evidenced by their high compression set values at -10°C. We selected composition N for the extended testing, due primarily to its immediate availability. None of the three epichlorohydrin elastomers performed well enough to merit further testing, and, in consequence, the entire category was eliminated from further consideration. The ethylene-acrylic copolymer did not perform well on the screening tests, however, the supplier, DuPont, indicated that the initial batch provided had been under cured and that a second batch, which they later provided, would be far superior. This batch was placed in the list of compositions for more exhaustive testing. The one chlorosulfonated polyethylene screened was eliminated because it showed poor compression set at both low and high temperatures; however, it performed very well in other areas.

The one polyacrylic elastomer screened performed excellently in all respects save for ultimate elongation where it failed by a small margin. It was selected for further testing. The bromobutyl elastomer performed poorly and was eliminated from further testing. However, the better butyl rubber S, even though it failed by a substantial margin in low temperature compression set and in total volatiles, was continued into the final testing because butyl rubbers have found extensive application in the solar collector industry. Thus three silicones, one fluorocarbon, one ethylene propylene terpolymer, one ethylene acrylic, one polyacrylic and one butyl rubber were selected for more extensive testing. Their tensile properties are shown in Table IV.

Since our survey disclosed only a very limited number of viable candidates in the Class SC (caulks) category, comparable screening tests were not performed in this area. Six candidates were made available for the more extensive testing. Three of these were silicones, one an acrylic, one a butyl and one a chlorosulfonated polyethylene (Table I, Code A-F).

Table III. Class PS Elastomer Screening Tests

Class PS Material	Hardness Grade [1]	Ultimate Elongation	Compr. Set 70 h, 150°C [2]	Compr. Set 166 h, 10°C	Hardness Change [2]	Ultimate Elongation Change [2]	Tensile Strength Change [2]	Volatile Lost [2]
Silicone								
G	5	P+	P+	P+	P+	P+	P+	P+
H	7	P	P+	P+	P+	P	P	F
I	7	P+	P	P+	P+	P+	P+	P+
J	7	P+	P+	P+	P+	P	P	P-
EPDM								
Q	7	P+	F-	F-	P	F-	F	F-
P	6	P+	F-	F-	F-	F-	P+	F
O	8	P+	P+	P+	P+	P+	P+	F
Fluorocarbon								
M	8	P+	P+	F-	P+	P+	P+	P+
N	8	P+	P+	F-	P+	P+	P+	P+

[1](Hardness grade x 10) ± 5 = Shore A durometer hardness
[2]Materials exposed to 150°C/70 h

P = pass by relatively small margin
P+ = pass by substantial margin
F = fail by relatively small margin
F- = fail by substantial margin

Continued on next page

Table III. Class PS Elastomer Screening Tests -- Continued

Class PS Material	Hardness[1] Grade	Ultimate Elongation	Compr. Set 70 h, 150°C[2]	Compr. Set 166 h, 10°C[2]	Hardness Change[2]	Ultimate Elongation Change[2]	Tensile Strength Change[2]	Volatile Lost[2]
Epichlorohydrin								
V	6	P+	F⁻	P+	P	F	P	F⁻
W	7	P+	F	P+	P	P	P+	F⁻
X	8	P+	F	P+	F	P	P	F⁻
Ethylene Acrylic								
L	7	P+	F	F⁻	P+	F	P+	F⁻
Chlorosulfonated Polyethylene								
U	7-8	P+	F⁻	F⁻	P+	P	P+	P+
Polyacrylic								
K	6-7	F	P+	P	P+	P+	P	P+
Bromobutyl								
R	6	P	F	F	P+	F	P	F⁻
Butyl								
S	5-6	P+	P+	F⁻	P+	P+	P+	F⁻
T	5	P+	F⁻	P+	P+	P	F⁻	F⁻

[1](Hardness grade x 10) ± 5 = Shore A durometer hardness
[2]Materials exposed to 150°C/70 h

P = pass by relatively small margin
P+ = pass by substantial margin
F = fail by relatively small margin
F⁻ = fail by substantial margin

Table IV

Initial Physical Properties of Aged Sealants[a,b,c]

Code	Material Category (Class)	Tensile Strength (lb/in^2)	Ultimate Elongation (%)	Tensile Stress at 100% Elong.
A	Silicone (SC)	220	360	90
B	Silicone (SC)[d]	>120	>800	30
C	Silicone (SC)	400	390	120
D	Acrylic (SC)	100	280	110
E	Hypalon (SC)	110	80	---
F	Butyl (SC)	450	~0	---
G	Silicone (PS)[d]	>880	>750	120
I	Silicone (PS)	1280	700	220
J	Silicone (PS)	1060	280	460
K	Acrylic (PS)	1570	170	660
L	Ethylene-acrylic (PS)	2080	420	430
N	Fluorocarbon (PS)	1690	220	650
O	EPDM (PS)	2010	180	800
S	Butyl (PS)	1470	500	150

[a] Elastomers selected for further aging studies.

[b] Values for caulking (SC) compounds were determined after 4–6 weeks aging under room conditions and then baking for 24 h at the following temperatures: Silicones–225°C, acrylic, Hypalon, butyl–150°C. Since caulks were postcured for 24 h at the aging test temperatures, their original values will differ for each test. For example, on postcuring at 125°C caulks A, B, and C display tensile strengths of 160, >60, and 340 and elongations of 210, 800, and 290 respectively.

[c] The PS compounds had been postcured by their suppliers and were tested as received.

[d] Ultimate elongation exceeded capabilities of test apparatus, specimens did not break.

Hydrolytic Aging

Resistance of elastomers to degradation in hot and humid
environments was studied by employing accelerated aging tests in
which tensile properties of specimens immersed continuously in
water at elevated temperatures were monitored (Table V; Examples
of hydrolytic aging data for several materials are shown in Figures
3-6). It must be emphasized that the test conditions were severe
since it is quite unlikely that 100% relative humidity will be
achieved for any significant length of time inside a collector
which is at an elevated temperature. Furthermore, the high
service temperature will have a drying effect on the polymer.
Thus the results of these accelerated tests should be considered
to be essentially of comparative value for the specific conditions
under which the tests were performed. Since fillers and other
additives will affect the pH of the water penetrating the polymer
matrix and may provide a catalytic effect on the hydrolysis
reactions, the results of this study may differ from those
obtained if the polymers were in an essentially pure state.

Generally during the very early stages of aging the sealants exhibited further curing as evidenced by concomitant increases
in tensile strength and elongation above their original values.
Physical effects of penetration by water and degradative effects
of hydrolysis became apparent after aging had progressed
especially at the elevated temperatures.

The degradative processes described for the various sealants
are based solely on observations of the changes of tensile properties of the aging specimens. The risk in this approach is that
one could be mislead by a simultaneous occurrence of different
degradative phenomena that present opposing physical effects.
There is concern that in some cases this may lead to erroneous
conclusions that a material is significantly more stable than
it actually is. However, the likelihood of a major interpretive
error of this nature is low because of the relatively long period
of aging in which a given mode of degradation will probably begin
to dominate and thus tensile strength and elongation are not
likely to be affected to the same extent. In addition the relative effects on the tensile properties will differ at the various
test temperatures. Further work involving compositional analysis,
and measurements of molecular weight and cross-link density of the
exposed materials would help substantiate the suggested natures
of the degradation processes described below. The results of
these processes are summarized in Table V.

Silicone Class PS, G, I, and J. The moderate decline of
both tensile strength and elongation of composition J on immersion
at 125°C indicates that hydrolysis of the polymer is taking place
(Figure 3). A slight rise in tensile strength for the 67 and 83°C
aging temperatures, which peaked about midway during the aging
period, indicated that cross-linking or chain formation reactions

Table V

Retention of Tensile Properties After Immersion in Water

Code	Material[a] Category (Class)	Water Temp. °C	Days Aged	Percent Retention of Original Property	
				Tensile Strength	Ultimate Elong.
A	Silicone (SC)	125	16	10	70
		100	64	90	140
		83	128	90	110
		67	128	160	110
B	Silicone (SC)[b]	125	4	Too tacky to measure	
		100	64	< 50	< 80
		83	64	< 70	< 100
		67	128	< 120	< 90
C	Silicone (SC)	125	32	40	150
		100	64	100	120
		83	128	130	120
		67	128	110	100
G	Silicone (PS)[b]	125	32	< 80	< 70
		100	128	< 80	< 50
		83	128	< 130	< 80
		67	128	< 140	< 80
I	Silicone (PS)	125	32	50	20
		100	128	40	10
		83	128	70	50
		67	128	90	60
J	Silicone (PS)	125	32	70	60
		100	128	90	60
		83	128	90	60
		67	128	110	70
K	Acrylic (PS)	125	2	Almost total loss of physical integrity	
		100	64	110	80
		83	128	80	70
		67	128	90	90

[a] Caulks D, E and F lacked adequate physical integrity for testing after one day immersion.

[b] Ultimate elongation of unaged specimens exceeded measuring capability of test apparatus (c.f. Table IV).

Continued on next page

Table V (Cont'd)

Retention of Tensile Properties After Immersion in Water

Code	Material[a] Category (Class)	Water Temp. °C	Days Aged	Percent Retention of Original Property	
				Tensile Strength	Ultimate Elong.
L	Ethylene-Acrylic (PS)	125	32	60	0
		100	64	110	70
		83	128	100	160
		67	128	90	80
N	Fluorocarbon (PS)	125	32	80	100
		100	128	70	90
		83	128	80	100
		67	128	90	110
O	EPDM (PS)	125	32	100	100
		100	128	100	110
		83	128	110	110
		67	128	100	100
S	Butyl (PS)	125	32	80	80
		100	128	70	80
		83	128	80	80
		67	128	90	90

[a]Caulks, D, E and F lacked adequate physical integrity for testing after one day immersion.

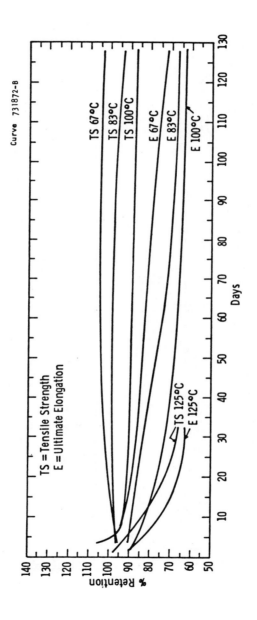

Figure 3. Hydrolytic aging of a silicone rubber.

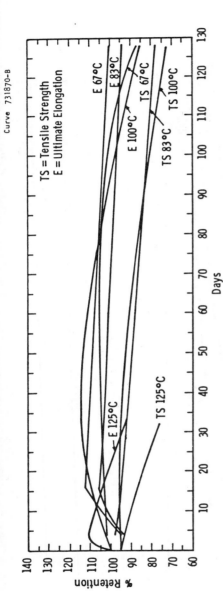

Figure 4. Hydrolytic aging of a fluorocarbon elastomer.

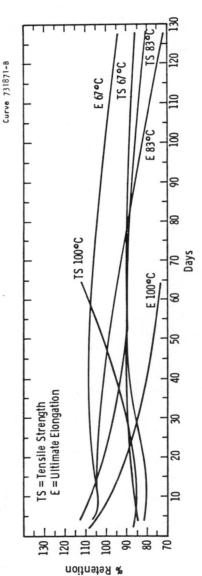

Figure 5. Hydrolytic aging of an acrylic elastomer.

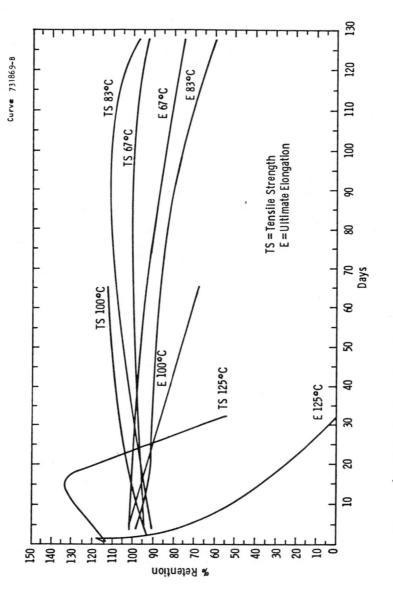

Figure 6. Hydrolytic aging of an ethylene-acrylic copolymer.

were dominant over hydrolytic effects. Even at 100°C, the tensile
strength did not decline very much. Compound I, which was much
more susceptible to hydrolytic degradation than J, displayed considerable decreases in its tensile strength and elongation during
the hydrolytic aging period. It is difficult to make a direct
comparison of the effect of aging on retention of original properties of G, since its elongation was so great that it exceeded the
capability of the test apparatus and thus did not break until
the aging had progressed to a considerable extent. By the 32nd
day of immersion at 125°C the specimen broke at a tensile stress
of 700 lb/in^2 and elongation of 510%,and on the 64th day at 100°C
it displayed a tensile strength of 940 lb/in^2 and an elongation
of 530%. Further testing would be required to permit a valid
comparison between G and J.

Silicone Caulks, A, B and C. After an initially sharp rise
in tensile strength and elongation during the 125°C immersion,
which indicated that additional curing of material A was taking
place, both tensile properties dropped precipitously as a consequence of severe hydrolysis. The stability at the lower aging
temperature was fairly good. During the aging at 125°C, compound
C also displayed after a brief initial increase a sharp drop in
tensile strength; however, its elongation continued to rise. At
100°C this material also displayed good stability. Compound B was
more difficult to assess because its initial elongation exceeded
the testing capability of the test apparatus. However, during
the course of aging at 83°C and 100°C its properties appeared to
drop faster than those of the other two compounds.

Fluorocarbon, N. After an early enhancement of cure its
tensile strength and elongation appeared to fall slowly, perhaps
due to hydrolysis; however, the relatively greater drop of tensile
strength suggests that placticization by water is occurring Figure
4).

Acrylic, K. After one day's immersion at 125°C the specimen
did not possess adequate physical integrity for testing. However,
it was much less affected at the three lower test temperatures.
On immersion in 100°C water the increase of tensile strength and
concomitant decrease of elongation indicated that cross-linking
reactions associated with continuing thermal exposure were quite
important Figure 5.

Ethylene-acrylic, L. On aging at 125°C an initial increase
of short duration in elongation and of much longer duration in
tensile strength, followed by a rapid drop in both properties,
indicated that the polymer initially underwent additional crosslinking which was followed by hydrolysis. On immersion at 100°C
the increase in tensile strength and decrease in elongation indicate that cross-linking is the dominant phenomenon for the first

64-day period. At the lower temperatures it appears that additional cross-linking predominates initially and that hydrolysis later becomes the controlling factor (Figure 6).

Butyl, S. The elongation and tensile strength dropped slowly at 125°C and 100°C after an initial increase, indicating an initial enhancement of cure followed by absorption of water. The behavior at 67°C in which a peak tensile strength was attained after about 64 days immersion is also consistent with the above described phenomena. Overall this material appeared to display good resistance to hydrolytic aging.

EPDM, O. This material appeared to display excellent resistance to hot water.

Other Materials. The acrylic D, Hypalon E, and Butyl F caulking compounds were too weak and tacky for the tensile testing technqiues employed and thus were not evaluated in these tests. (Peel adhesion tests, a topic of a future paper, were employed in monitoring the hydrolytic stability of these soft caulking compounds).

Thermal Aging in Air

Class PS Elastomers

Measurements of tensile properties of specimens suspended in forced air circulating ovens provided a convenient means of determining the stability of sealants to a hot air environment (Table VI; Examples of thermal aging data for several materials are shown in Figures 7-9).

Unlike the previously described hydrolytic degradation processes, which involved not only chemical reactions, but also the physical phenomena resulting from the entrance of water into the polymer, the effects found here are attributed primarily to chemical degradative processes. A brief summary of the aging characteristics of several Class PS sealants is presented below.

Fluorocarbon N. This elastomer displayed a slight enhancement of cure followed by excellent resistance to degradation at the test temperatures of 200°C and 225°C (Figure 7).

Silicones G, I, and J. These silicones display similar behavior patterns on aging in air. After a brief enhancement of cure, both the tensile strength and elongation continue to decrease during the remainder of the aging at all exposure temperatures. Comparing the results obtained from the least severe aging temperature of 175°C, J (Figure 8) appears slightly superior to I. Since the initial elongation of G exceeded test capability it is difficult to compare precisely its thermal endurance

Table VI. Retention of Tensile Properties After Aging in Air

Code	Material Category (Class)	Aging Temp. °C	Days Aged	Percent Retention of Original Property	
				Tensile Strength	Ultimate Elong.
A	Silicone (SC)	225	100	60	40
		200	100	70	120
		175	100	100	140
B	Silicone (SC)[a]	225	100	<140	<3
		200	100	<50	<3
		175	100	<150	<50
C	Silicone (SC)	225	100	70	50
		200	100	70	80
		175	100	110	100
D	Acrylic (SC)	150	100	340	<0
		125	100	630	50
		100	100	440	180
E	Hypalon (SC)	150	4	350	<0
		125	64	400	7
		100	100	1570	90
F	Butyl (SC)	150	4	50	<0
		125	100	17	<0
		100	100	60	<0
G	Silicone (PS)[a]	225	64	<40	<10
		200	100	<40	<20
		175	100	<60	<30
I	Silicone (PS)	225	16	50	20
		200	100	60	20
		125	100	70	20
J	Silicone (PS)	225	100	120	<0
		200	100	10	<0
		175	---	80	40
K	Acrylic (PS)	175	64	120	<0
		150	100	340	4
		125	100	630	50
L	Ethylene-acrylic (PS)	175	32	50	5
		150	100	100	50
		125	100	120	90

Table VI (Cont'd). Retention of Tensile Properties After Aging in Air

Code	Material Category (Class)	Aging Temp. °C	Days Aged	Percent Retention of Original Property	
				Tensile Strength	Ultimate Elong.
N	Fluorocarbon (PS)	225	100	100	100
		220	100	100	90
O	EPDM (PS)	175	16	40	6
		150	100	40	6
		125	100	90	90
S	Butyl (PS)	175	16	3	<0
		150	16	25	110
		125	100	30	120

[a]Ultimate elongation of unaged specimens exceeded measuring capability of test apparatus (c.f. Table IV).

Table VII. Estimated Life of Sealants in Air

Code	Material	Days to Reach 60% Retention of Original Tensile Property			Property[b]
		100°C	125°C	150°C	
S	Butyl	520 (650)	49 (59)[a]	7 (8)[a]	TS
O	EPDM	---	180	30	TS
			220	35	E
K	Acrylic	---	590	91	TS
			480	72	E
L	Ethylene-	---	790 (970)[a]	82 (96)[a]	E
J	Silicone	---	4300 (6800)[a]	430 (630)[a]	E

[a]( ) indicates time to reach 50% retention of original property value.

[b]Property studied is that which was considerably more severely affected by the thermal aging. Both properties are listed in cases where the extent of their decline was similar.

TS - Tensile Strength

E - Ultimate Elongation

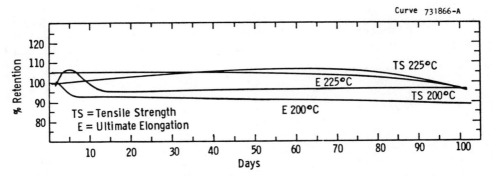

Figure 7.   Thermal aging in air of a fluorocarbon elastomer.

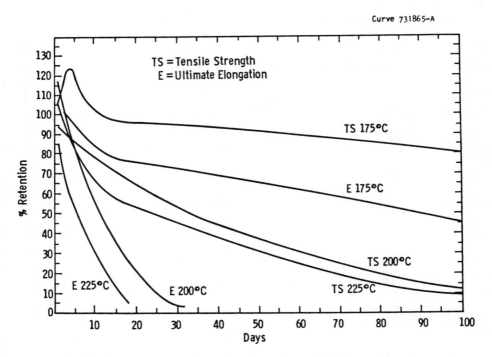

Figure 8.   Thermal aging in air of a silicone rubber.

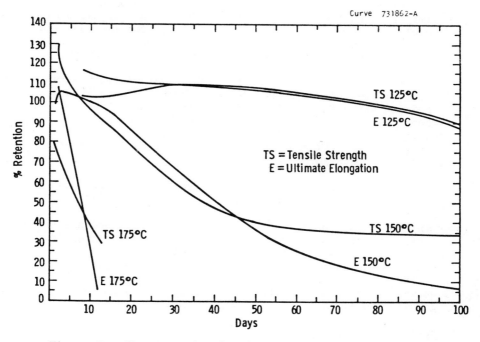

Figure 9.   Thermal aging in air of an ethylene–propylene
terpolymer (EPDM).

with that of other materials. However, overall its performance appears to be somewhat comparable to that of I.

Ethylene-acrylic, L. Before the onset of severe degradation this elastomer displays a considerable increase in tensile strength, especially on aging at 125°C and 150°C. At 125°C the elongation increases slightly until about the 70th day and then decreases slowly. However, on aging at the higher temperatures the elongation drops rapidly. On exposure to 175°C both tensile properties decline rapidly.

Acrylic, K. This material exhibits a behavior pattern that is similar to that of compound L with the exception that during the early stages of aging at 125°C its elongation increases to a greater extent than its tensile strength. Overall, compositions K and L display a very similar resistance to thermal aging.

EPDM, O. During the early aging this material exhibits a slight improvement in properties attributed to increased cure. Its tensile strength and elongation appear to have been affected to about the same extent over much of the range of the test. At exposure temperatures of 175°C and 150°C rapid degradation is evident, however, this composition displays good stability at 125°C (Figure 9).

Butyl, S. The butyl rubber displays the least resistance to thermal aging of all the materials tested. Aging at 150°C was discontinued after 16 days as a result of a drop of tensile strength to about 25% of its original value. However, the retention of elongation of over 100% of its initial value indicated that the material was undergoing chain scission or "reversion". This effect also appeared at the 125°C test temperature in which after 100 days the elongation was at about 115% of its original value while the tensile strength had dropped to about 30% of its original value.

Arrhenius Treatment of Data

Whenever feasible, the retention of either the tensile strength or elongation was subjected to an Arrhenius treatment in order to obtain predictions of the lives of the materials at temperatures other than those employed in these tests. The logarithm of the time required for the percent retention of the tensile strength or elongation of the sealant to drop to 60% of its original value is plotted against the reciprocal of the absolute temperature of the test. Examples of two such plots are shown in Figures 10 and 11. We feel that a 40% decline of either tensile property indicates that the sealant has undergone substantial degradation. Nevertheless, in some cases, Arrhenius plots were also made for 50% retention. When the tensile strength and elon-

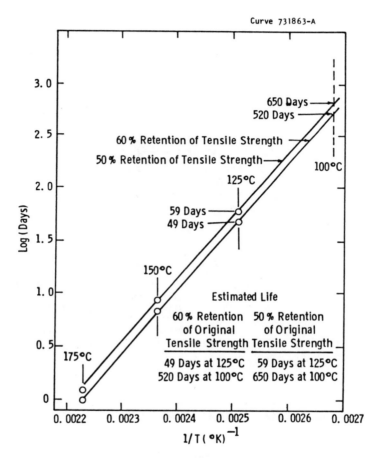

Figure 10.   Arrhenius plot of thermal aging in air of a butyl rubber.

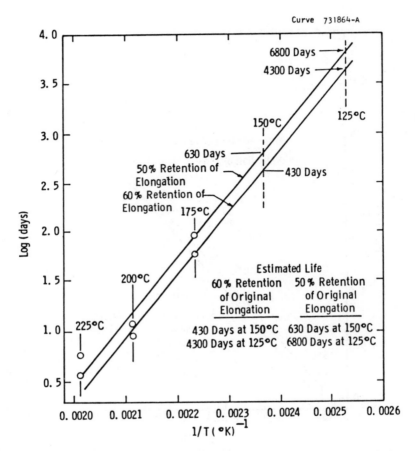

Figure 11.   Arrhenius plot of thermal aging in air of a silicone elastomer.

gation underwent comparable declines both properties were subjected to an Arrhenius treatment; otherwise the property which
underwent the most severe decline was employed to evaluate the
thermal endurance of the sealant.

Generally, data points at at least two test temperatures are
required to draw the straight line for the Arrhenius plot. Some
investigators will employ a single point and assume a slope based
on results from a chemically similar material undergoing the same
process. Feeling that such an approach is too speculative, we
did not employ it in this investigation.

The Arrhenius treatment provides the most reliable predictions
when the length of extrapolation is minimized, data are collected
at several temperatures, and the points fall on a straight line.
When the points do not fall on a straight line, those at the
lowest temperatures are weighted the most heavily in drawing the
plots. For example, in cases where the graph contains three points
essentially equidistant in temperature, we arbitrarily weight the
highest temperature point 50% as high as the middle temperature
point and draw the line through the lower temperature point and
the weighted average of the two upper temperature points.

The validity of the Arrhenius treatment is based upon the
assumption that the activation energy for the rate limiting step
in the degradation process is essentially independent of temperature. Since the mechanism of the degradation process can change
with temperature, the validity of this assumption decreases as the
temperature range of the study and the extrapolation become larger.
Thus, points on an Arrhenius plot occasionally deviate too far
from a straight line to permit any reasonable linear extrapolation.
For example, it was not feasible to apply the Arrhenius treatment
to silicone I since the mechanism for its degradation process
appears to differ appreciably between 225°C and 175°C, thus giving
three log time versus reciprocal absolute temperature points which
deviate considerably from a straight line.

In addition to the silicone I several other materials were
excluded from the Arrhenius treatment for the following reasons:
(1) initial properties on the material were too poor for performing
meaningful tensile tests; (2) monitored properties of the material
did not exhibit any clearly defined base points (e.g., some caulking compounds continued to give off solvents and other volatile
compounds during much of the aging period); (3) ultimate elongation
exceeded the capability of the testing apparatus; (4) material did
not exhibit failure at at least two test temperatures (e.g., the
fluorocarbon).

The results of the Arrhenius treatments are shown in Table VII.
As expected, the life of the butyl rubber held continuously at
125°C is extremely short, 49 days. The EPDM is considerably better
showing a life of 180 days. Next are the acrylic and ethylene-
acrylic having estimated lives of 480 and 790 days, respectively.
These are followed by the silicone which displays an extrapolated
life of 4300 days. The fluorocarbon, which was not subjected to

this treatment because it did not exhibit failure on testing, is
undoubtedly superior. At a temperature of 150°C (302°F), which
is readily attainable in many collectors, the order remains the
same; however, the thermal endurances of the sealants are quite
poor. Butyl gives 7 days; EPDM, 30 days; acrylic, 72 days;
ethylene-acrylic, 82 days; and the silicone, 430 days.

Caulking Compounds

Performance of caulking compositions during thermal aging was
found to be very difficult to interpret, due in major degree
to the rather large quantities of volatile compounds normally
employed in these materials to provide suitable working properties.
These volatiles remain present in substantial quantity even after
long periods of curing at room temperature. Thus, considerable
amounts of volatile components continue to evolve during the early
portion of any thermal aging program. Loss of these volatiles,
primarily plasticizers and solvents, has the effect of increasing
tensile strength and modulus quite sharply while reducing ultimate
elongation. Meanwhile, degradative reactions involving chain
scission which result in reduction of tensile strength and rigidity, produce an opposing effect. In consequence, it is not easy
to determine to what extent an observed change in properties is
due to relatively small progress in one process or to substantial
progress in both with one predominating. The picture is further
complicated by the presence of other processes such as increased
cure and degradative cross-linking which may proceed concurrently
with the two just discussed. As a result of the numerous processes
occurring simultaneously, it was not feasible to employ an Arrhenius treatment to the aging data for caulks. A summary of the test
results is presented in Table VI.

The silicone caulks were aged at 175°C, 200°C and 225°C,
whereas the acrylic, Hypalon, and butyl caulks were aged at 100°C,
125°C and 150°C. The lower aging temperatures of the latter
materials reflected their known lower thermal stabilities. All of
these materials were post-cured for 24 h at their aging temperatures in order to minimize the very large changes that occur at
the outset of elevated temperature exposure as volatiles are
driven off. In general, aging intervals began at one day and
doubled with each consecutive exposure up to 64 days. The final
aging interval was 100 days.

Among the silicone caulks A and C display a superior retention of tensile properties on aging over compound B. They show better retention of tensile properties after 100 days aging at 175°C
than does the acrylic D, the best of the organic class SC elastomers, after 100 days aging at 150°C. The acrylic caulk still retains respectable tensile strength after 100 days at 150°C (well
above the initial tensile strength, although it appears to be de-

clining), but its elongation has declined almost to zero, indicating loss of elastomeric properties. The butyl caulk F showed essentially zero elongation at the start of the test at all aging
temperatures indicating that even the 24-hour post-cure at 100°C
had been sufficient to cause it to lose its elastomeric properties.
Hypalon E performed better, but was remarkably inferior to the
acrylic. Its ultimate elongation had declined to zero after 100
days at 125°C. This compares to 80% elongation, which represents
a significant property retention, for the acrylic after 100 days
at 125°C. After 100 days at 100°C, the acrylic exhibited 210%
elongation while the Hypalon had declined to 60%

Thermal Aging Under Compression

For class PS or gasketing type sealants, compression set is
one of the most pertinent properties for solar collector applications. It is a measure of the ability of an elastomer to maintain a tight seal as a gasketing material by retention of its
original configuration and resilience. A gasket material which
undergoes a severe set will not maintain pressure against the
collector frame to provide an effective seal in many collector
unit designs. Thus low test values are desirable.

All of the materials evaluated for gasketing applications
exhibited similarly shaped families of compression set versus
time curves and differed essentially in the magnitude of the set
for the same time and temperature conditions. For purposes of
illustration, we have shown compression set curves for fluorocarbon N and silicone J (Figures 12 and 13).

The compression set data, which we had obtained, reflect far
poorer performances than we had anticipated on the basis of con-
versations with manufacturers of solar collector units and suppliers of sealants. A summary of thse data are presented in
Table VIII. All three silicones, G, I, and J, tested exhibited
compression sets of the order of 100% after only one day of
aging at either 250°C or 225°C. (Compression set values greater
than 100% are attributed to essentially complete loss of resilience by the specimen combined with an additional reduction in
sample thickness due to shrinkage associated with thermal degradation). For example, the best of the three silicones, J, showed
compression set values of 94% after one day at 225°C and 89%
after 28 days at 175°C. We feel that a sealant should not exceed
50% in compression set if it is to retain good sealing ability.

Table IX displays aging times required at 175°C, 150°C and
125°C for the eight elastomeric compositions tested to reach compression set values of 50% and 75%. These data were obtained by
interpolation of the compression set aging curves and were used to
plot the Arrhenius curves displayed in Figure 14. The Arrhenius
plots again show that the fluorocarbon elastomer far outperforms
all other materials tested. It is interesting to note that the plots
on Figure 14 for the EPDM, the polyacrylic and the best silicone

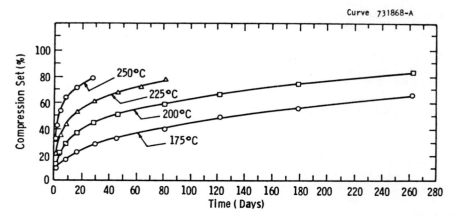

Figure 12.   Aging of a fluorocarbon elastomer under compression.

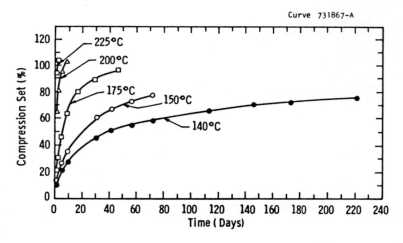

Figure 13.   Aging of a silicone rubber under compression.

Table VIII

Effect of Thermal Aging on Compression Set of Class PS Materials

Code	Material Category	Temp. (°C)	Time(days) to Reach Compression Set of 50%	75%
G	Silicone	225	<1	<1
		200	<1	<1
		175	1-2	4-6
		150	4-8	25-35
		140	17-19	60-70
I	Silicone	225	<1	<1
		200	<1	<1
		175	1-2	5-7
		150	4-8	20-30
		140	8-13	40-50
J	Silicone	225	<1	<1
		200	<1	1-2
		175	4-6	12-15
		150	14-22	57-67
		140	35-45	195-215
K	Acrylic	175	9-13	37-47
		150	25-35	>85
		125	95-115	>225
L	Ethylene-Acrylic	175	3-5	20-30
		150	7-9	45-55
		125	28-38	185-205

Table IX

Summary of Compression Set Thermal Aging Data

| Material | Code | Time (days) to Reach 50% Compression Set | | |
		175°C	150°C	125°C
Fluorocarbon	N	135	510*	7400[a]
Silicone	J	5	20	100
Silicone	G	2	7	50
Silicone	I	2	7	33
EPDM	O	28	50	103
Acrylic	K	11	28	105
Ethylene-acrylic	L	4	8	35
Butyl	S	4	8	23

[a]Extrapolation of Arrhenius Plot.

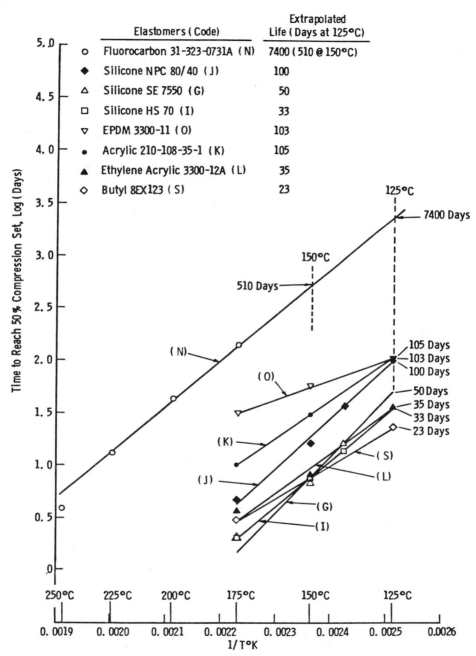

Figure 14.   Arrhenius plot of compression set aging data.

cross at about 125°C. The indication is that both the EPDM and
the polyacrylic will outperform the silicone at temperatures
higher than 125°C with the EPDM being considerably superior to
the polyacrylic, and the silicone will perform better than these
two organic materials at temperatures below 125°C. The superior
performance of the organics in the higher temperature region is
also supported by actual data in the compression set aging curves.
The remaining two organic elastomers, ethylene-acrylic and butyl,
perform more poorly than EPDM and polyacrylic. The Arrhenius plot
for butyl crosses that for silicone G at about 140°C, indicating
that it will have greater (but very short) life at temperatures
above that level. Below about 140°C, the butyl is the poorest
composition tested. The ethylene-acrylic plot crosses that for
the silicone I, at 125°C. Hence, its performance is better than
that of the poorer silicones at temperatures above that level, but
poorer below. The ethylene-acrylic must be considered the
second poorest composition tested from the standpoint of compression set retention.

It may be of interest to point out that extrapolation of the
Arrhenius plot for the EPDM rubber shows it crossing the plot for
the fluorocarbon elastomer at about 225°C, a circumstance which
would indicate that the EPDM should outperform the fluorocarbon
at temperatures higher than 225°C. We know that this is not true
and the circumstance is pointed out merely to illustrate fallacies
which may be encountered by reliance on Arrhenius extrapolation
over too great a temperature range. Conclusions drawn from
Arrhenius data treatment are valid only so long as the chemical
reactions involved in the polymer degradation under consideration
remain the same. In the case of EPDM degradation, new degradation
mechanisms would become significant before the 225°C temperature
level was reached, thus invalidating the extrapolation.

Conclusions

Among the materials evaluated the one displaying the best
overall properties for use as a sealant in a thermal solar collector is the fluorocarbon, Viton. However, it has the disadvanatages
of high cost and high compression set at temperatures of -10°C and
lower which could present problems in northern climate.

Where the criterion of life at a select aging temperature is
taken as exceeding a compression set value of 50%, it was found
that none of the elastomers tested were completely adequate. Viton, which displayed the greatest resistance to elevated temperature compression set of all materials tested, provided data which
extrapolated to a 510 day life for continuous exposure at 150°C.
This is roughly equivalent to four years service life under conditions of maximum severity and possibly three or four times that
long under more normal service conditions. The best silicone by
the compression set criterion failed after 100 days at 125°C.
Also the other materials tested were not good. It is evident that

satisfactory service life can be obtained in the flat-plate solar collector application with presently available rubbers only if the collector unit is so designed that its performance does not depend upon low compression set values in order to maintain a seal.

Where retention of tensile properties during thermal aging in air is the criterion of service life, the fluorocarbon elastomer is again the best performer. Failures did not occur with this material at the aging times and temperatures of our tests, but we judge that it would perform satisfactorily at temperatures as high as 175°C if compression set is not a consideration. The class PS silicones were the next best performers by the tensile criterion and we believe that they are acceptable for service at 150°C or slightly lower, if low compression set is not required. The relatively economical intermediate temperature elastomers, acrylic (Hycar), ethylene-acrylic (Vamac) and EPDM (Nordel), are regarded as serviceable at 125°C or slightly lower. The class PS butyl is the least stable and is not considered useful at temperatures above 100°C. The class SC silicone caulks are considered to be about as good as the class PS silicones, but the class SC organic caulks, acrylic (Mono), chlorosulfonated polyethylene (Hypalon) and butyl are much less stable and lose their elastomeric properties quite rapidly at temperatures in the 100°C to 125°C range.

Although most of the materials subjected to the hydrolytic stability tests displayed some propensity towards degrading under the severe test conditions, we feel that the service life of all of the treated materials will not be limited by hydrolytic degradation processes. However, in some cases, the combination of hydrolysis and thermal-oxidative degradation processes could diminish the life of the sealant to an extent beyond that predicted by studying each factor separately.

Acknowledgment

This work has been supported by the Solar Heating and Cooling Research and Development Branch, Office of Conservation and Solar Applications, U. S. Department of Energy.

Literature Cited

1. Mendelsohn, M. A., Luck, R. M., Yeoman, F. A., and Navish, F. W., Jr., "Sealants for Solar Collectors", I&EC Product Research & Development, 1981, 20, 508.
2. Luck, R. M., and Mendelsohn, M. A., "The Reduction of Solar Light Transmittance in Thermal Solar Collectors as a Function of Polymer Outgassing". (following paper in this symposium).
3. Mendelsohn, M. A., Luck, R. M., Yeoman, F. A., and Navish, F. W., Jr., "Collector Sealants and Breathing", Contract No. DE-AC04-780515362, NTIS, PO A09/MF A01, 1980.
4. Stiehler, R. D., Hockman, A., Embree, E. J., and Masters, L. W., "Solar Energy Systems-Standards for Rubber Seals", NBSIR 77-1437, National Bureau of Standards Report, 1978.

RECEIVED November 22, 1982

The Reduction of Solar Light Transmittance in Thermal Solar Collectors as a Function of Polymer Outgassing

R. M. LUCK and M. A. MENDELSOHN

Westinghouse Research and Development Center, Pittsburgh, PA 15235

The outgassing of polymeric materials used in the construction of thermal solar collectors is of great importance to solar collector manufacturers. The chemical products evolved through thermal degradation and outgassing usually condense on the glass or plastic glazing surfaces where they significantly reduce the transmittance of solar light and thereby reduce the efficiency of the solar collector This work was performed in an endeavor to determine the amount and type of condensable compounds given off by polymeric materials during thermal aging, and to relate these with the reduction in solar light transmittance and reduced solar collector efficiency.

Sources of Outgassing Products

Inside each solar collector two types of dew (condensation) points exist. One is a moisture dew point and the other a series of dew points resulting from organic compounds. The amount of a given vapor present in the air at any specific temperature/pressure condition is essentially a function of its relative volatility and its thermodynamic activity.

Air will hold more vapor, whether it be moisture or an organic compound, at a high temperature than it will at a low temperature. When the temperature of the collector glazing or frame drops below the moisture or chemical vapor dew point, condensation occurs. As the temperature of the collector rises above the dew point, the condensation on the glazing can evaporate. The evaporation of moisture is clean and complete, and no residues or deposits are left on the glazing. However, over long periods of time, moisture condensates slowly leach out sodium and other metal salts from the glass and a white deposit slowly forms. These salt deposits greatly reduce the transmittance of solar light through the glass(1).

0097–6156/83/0220–0081$06.00/0

Some condensates do not evaporate from the glazing in a clean manner and often leave a colored viscous oily residue, a white powdery deposit or a continuous solid film. These chemical deposits are produced by three different processes occurring simultaneously within the solar collector. First, condensable volatiles are being produced by diffusive loss of low molecular weight compounds originally present in the plastic or elastomeric materials. Second, the polymer is being degraded and depolymerized as a result of exposure to elevated temperatures, water, ultraviolet light, oxygen and ozone. Third, the liquid, thin film chemical condensate that is being formed on the glazing undergoes thermal, oxidative, and ultraviolet induced reactions to form a solid polymeric deposit (2,3,4). During the many condensation and evaporation cycles that a solar collector experiences, these solid deposits slowly build up reducing solar light transmittance.

The sources for outgassing include preformed seals, seal caulks, room temperature vulcanizing polymers, thermal insulations, polymeric coatings, and polymeric materials used in solar collector structural applications.

Experimental

A test was developed to study the outgassing and degradation of polymeric materials. A sample of the polymeric material to be evaluated is placed in the bottom of a glass tube, and an infrared sodium chloride crystal is mounted in the open end of the tube. This assembly (Figure 1) is then positioned vertically into a closely fitting hole in the top of the oven so that only the lower two-thirds of the tube is inside the oven. Essentially, all of the condensable outgassing products condense on the bottom surface of the sodium chloride crystal. Condensable volatiles are observed as weight increase of the crystal and noncondensables by difference between this and the total weight loss of the test sample. The sodium chloride crystal is placed directly into the infrared spectrophotometer for determination of the chemical nature of the condensable product. The heat vulcanized preformed seals, glazing plastics, absorber plate polymers and insulation materials were evaluated as received. The seal caulks and the room temperature vulcanizing (RTV) materials were cast in approximately 1.5 mm thick sheets on fluorocarbon film and room temperature vulcanized for four to five weeks prior to testing.

In the standard outgassing test, the oven was operated at 150°C for a period of 225 hours. This is not an unrealistic test temperature since thermal solar collectors can reach 125-200°C. The sodium chloride crystal, during the test period, reached an equilibrium temperature of 65 ± 2°C. The compositions of the condensable products were identified by infrared analysis using a Perkin-Elmer Infrared Spectrophotometer. Relative light transmittance values were obtained on several condensable products

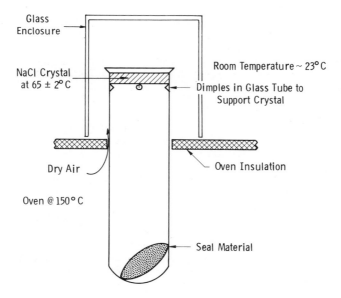

Figure 1.   Outgassing measurement apparatus.

collected on glass disks. These, as well as several glass glazing
samples taken from operating solar collectors, were measured
for their relative light transmittance over the range of 400 to
900 nm using a Coleman Spectrophotometer with a tungsten power
supply.

The composition and source of the polymeric materials
evaluated in this study are shown in Table I.

Results and Discussion

Condensables are defined in this work as compounds which
will condense on a surface maintained at $65 \pm 2°C$. The noncondensables are those which will not condense on that heated surface.

The noncondensable chemicals generally do not adversely
affect solar light transmittance in a solar collector. They
remain in a gaseous state while inside the working collector and
usually escape to the surrounding atmosphere. A large quantity of
noncondensable and condensable products indicates that changes
within the polymeric material, which can adversely affect its
physical and mechanical properties, have taken place.

  Outgassing of Silicone Polymers. The percent condensable
versus time curves at 150°C for the heat vulcanized (HV) preformed
silicone seals, and for the room temperature vulcanized (RTV)
silicone sealants are shown in Figure 2. Samples of each class
of silicone were obtained from several sources for evaluation in
this study. For ease of discussion, however, only the high and
low extremes for each class of silicone are plotted in Figure 2.
All of the other silicones evaluated fell between these extremes.

The RTV silicones, in general, exhibit considerably greater
outgassing of both condensable and noncondensable materials than
the HV silicones. Their condensable-time curves were usually
linear and only one condensable product was being evolved. The
infrared spectra of the condensable products from all of the
silicones evaluated indicated these materials to be predominately
low molecular weight alkyl linear and/or cyclic polysiloxanes
(Table II). Silicone A gave a nonlinear condensable-time curve
and evolved two types of condensable products; an alkyl linear
and/or cyclic polysiloxane, and an aromatic ester.

The HV silicones evolved less condensable and noncondensable
compounds than their RTV counterparts. Their condensable-time
curves were linear and essentially only one type of condensable
compound was evolved. The condensable products from the HV silicones were also identified as alkyl linear and/or cyclic polysiloxanes.

None of the silicone polymers studied produced a maximum
amount of condensables during the 225 hour test. The outgassing
products were being evolved and condensed at the same rate at the
end of the test period as were observed during the early stages
of the test. Linear curves usually indicate a slow continual

TABLE I

Source and Composition of Polymers
for Outgassing Studies

Sample	Composition	Source/Identification
A	Silicone (RTV)	Dow Corning 93-067-2
B	Silicone (RTV)	Dow Corning 732
C	Silicone (HV)	N.A. Reiss 747
D	Silicone (HV)	Pawling 96-B-24
E	Butyl (HV)	Polysar 100
F	Fluorocarbon (HV)	DuPont Viton
G	Acrylic (HV)	Goodrich Hycar 4054
H	Ethylene-acrylic (HV)	DuPont Vamac
I	Acrylic terpolymer (RTV)	Tremco Mono
J	Butyl (RTV)	Tremco Butyl
K	Chlorosulfonated polyethylene (RTV)	Gibson-Homans Hypalon
L	Ethylene-propylene terpolymer (HV)	DuPont Nordel (EPDM)
M	Ethylene-propylene terpolymer	Bio Energy System Sola Roll (EPDM)
N	Polypropylene	Comco (PP)
O	Polyvinylchloride	Comco (PVC)
P	Polymethylmethacrylate	Rohm & Haas Plexiglas (PMMA)
Q	Polycarbonate	General Electric Lexan (PC)
R	Cellulose acetate butyrate	Eastman Chemical (CAB)
S	Glass reinforced polyester	Fillon (GRP-558)
T	Fluorocarbon	DuPont (FEP)
U	Polyurethane	Mobay Baytherm 851
V	Fiberglass	Corning Glass

RTV = Room temperature vulcanized

HV = Heat vulcanized

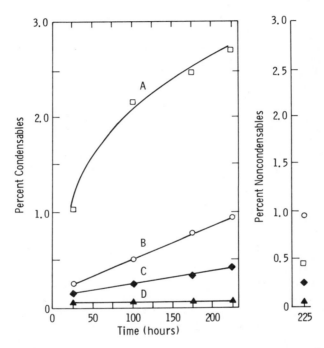

Figure 2. Percent condensables and noncondensables of silicones as a function of time at 150 °C: (□,○) RTV silicones; and (♦,▲) preformed silicones.

degradation (5,6) of the polymer chain rather than a simple outgassing of low molecular weight materials, such as plasticizers
and processing oils, which are incorporated into the polymer but
are not chemically bonded.

TABLE II.   Infrared Spectrophotometric Analysis of The
            Condensable Volatiles From Silicone Polymers

Sample	Major Bands Wave Number cm^{-1}	Interpretation
A	800, 1000-1100, 1260, 1490, 1570, 1590, 1720, 2800-3000	Alkyl linear or cyclic polysiloxane plus an aromatic ester
B	800, 1000-1100, 1260, 2800-3000	Alkyl linear or cyclic polysiloxane
C	800, 1000-1100, 1260, 2800-3000	Alkyl linear or cyclic polysiloxane
D	800, 1000-1100, 1260, 2800-3000	Alkyl linear or cyclic polysiloxane

The noncondensable volatiles were usually identified as
moisture which had been adsorbed by the silicones. The moisture
content of the silicones was less than 0.5% by weight. Noncondensables values greater than 0.5% by weight indicate the
presence of other low molecular weight, highly volatile materials
such as acetic acid and methanol. These materials are generated
and evolved during the room temperature vulcanization of RTV
silicones. Several RTV silicones evolved as much as 7% by weight
of these noncondensable materials. Apparently, many RTV silicones
do not completely vulcanize even after four to five weeks exposure
to air prior to testing.

Outgassing of Preformed Organic Polymers. The percent condensable-time curves for several heat vulcanized (HV) preformed
organic seals are shown in Figure 3. The curves are generally
linear and only one condensable product was evolved. The fluorocarbon polymers usually produced very little if any condensable
products as indicated by sample F. The acrylic polymers, such as
sample G, produced a very small quantity of unidentified products
and gave linear, low slope condensable-time curves.

The ethylene-acrylic polymer (H) produced 4.2% by weight of
total volatiles. The condensables were identified as acrylic
fragments or oxidized ethylene fragments (Table III). The noncondensables were not identifed, but were considered to be moisture if their concentration was less than 0.5% by weight.

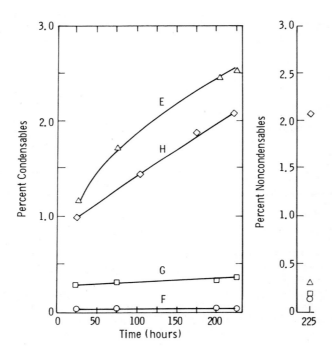

Figure 3.   Percent condensables and noncondensables of preformed seals as a function of time at 150 °C: (o) fluorocarbon; (□) acrylic; (◊) ethylene acrylic; and (△) butyl.

TABLE III. Infrared Spectrophotometric Analysis of The
Condensable Volatiles From Preformed Organic Polymers

Sample	Major Bands Wave Number cm^{-1}	Interpretation
E	1300, 1460, 1700, 2800-3000	Stearic acid
F	Insufficient sample for analysis	
G	Insufficient sample for analysis	
H	1380, 1420, 1710, 2800-3000	Acrylic or oxidized ethylene fragments

The butyl polymer (E) gave a nonlinear curve indicating that
possibly two or more products were being condensed. Only one
product, stearic acid, was identified by infrared analysis, however, it is possible that other oxidized hydrocarbons were present
in lesser concentrations. Stearic acid is used in many rubber
formulations as an accelerator activator and as a lubricant processing aid (7), and usually exists in the free state in the rubber. It is therefore easily outgassed, as are the plasticizers,
and condenses on the cooler surfaces of the collector. The stearic
acid was still being evolved after 225 hours at 150°C.

<u>Outgassing of Polymeric Caulking Compounds</u>. The organic
caulking compounds evolved large quantities of volatile materials
even after four to five weeks exposure to air prior to testing.
The condensable-time curves for the polymeric caulking materials
are shown in Figure 4. The chlorosulfonated polyethylene (K)
liberated 28.4% by weight of volatile materials. The condensable
portion, which accounted for almost half of the volatile products
was identified as an alkyl sulfonic acid ester type material.
The noncondensables were not identified. The butyl compound (J)
evolved about 15% by weight of volatile materials. The condensables were identified by infrared analysis to be either oxidized
paraffinic oil or oxidized butyl fragments or a combination of
both (Table IV).

The acrylic terpolymer (I) produced 0.8% by weight of condensable material which was identified as acrylate fragments and/
or oxidized paraffinic oil.

The noncondensable volatiles from this group of organic
caulking compounds were not identified but are believed to consist
mostly of solvents and monomeric diluents.

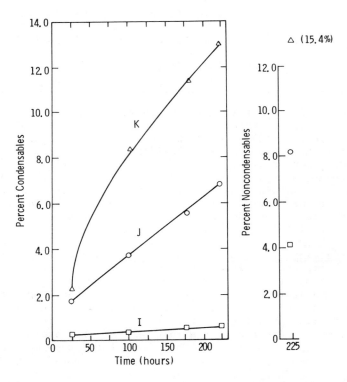

Figure 4. Percent condensables and noncondensables of caulking compounds as a function of time at 150 °C: (□) acrylic
terpolymer; (○) butyl; (△) chlorosulfonated polyethylene.

TABLE IV. Infrared Spectrophotometric Analysis of Condensable
Volatiles From RTV Organic Polymers

Sample	Major Bands Wave Number cm^{-1}	Interpretation
I	1160, 1220, 1380, 1460, 1720, 2800-3000	Acrylate or oxidized oil
J	1160, 1220, 1380, 1460, 1700, 2800-3000	Oxidized butyl fragments or oxidized oil
K	1180, 1240, 1380, 1460, 1720, 2800-3000	Alkyl sulfonic acid ester

Outgassing of Absorber Plate Polymers. Several polymers
have been used as absorber plates for thermal solar collectors.
The condensable-time curves for a few selected polymers are shown
in Figure 5. The curves for these polymers are linear indicating
only one type of condensable volatile product. The high density
polypropylene (N) showed little outgassing, and produced only a
small amount of condensable material. The polyvinylchloride
polymer (O) produced 0.5% by weight of condensable product
identified as oxidized hydrocarbons (Table V). The noncondensables evolved from PVC were acidic and consisted of a mixture
of HCl and low molecular weight hydrocarbons, such as ethylene
and butylene.

TABLE V. Infrared Spectrophotometric Analysis of Condensable
Volatiles From Absorber Plate Polymers

Sample	Major Bands Wave Number cm^{-1}	Interpretation
L	1260, 1380, 1460, 1600, 2800-3000	Naphthenic oil
M	1260, 1380, 1460, 1700, 2800-3000	Oxidized paraffinic processing oil or oxidized EPDM fragments
N	Insufficient sample for analysis	
O	1260, 1460, 1700-1720, 2800-3000	Oxidized hydrocarbons

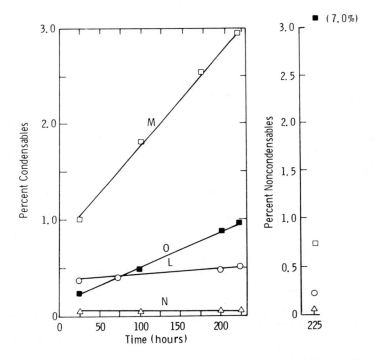

Figure 5. Percent condensables and noncondensables of
absorber plate polymers as a function of time at 150 °C:
(o) EPDM source L; (□) EPDM source M; (△) PP; and (■) PVC.

The EPDM polymers (L and M) produced different condensable materials. Sample L evolved a small quantity of naphthenic oil, whereas sample M yielded a high amount of oxidized hydrocarbons.

Outgassing of Polymeric Glazing. In Figure 6 are shown the condensable-time curves for several polymers that are being used as glazing (cover plates) for thermal solar collectors. These polymers produced linear curves and only one condensable material was evolved from each polymer. The polycarbonate (Q), the polymethylmethacrylate (P), and the fluorinated ethylene-propylene (T) evolved essentially no condensable materials. The glass reinforced polyester (S) gave a linear curve indicating a single type of condensable material identified as an aromatic ester (Table VI). The cellulose acetate butyrate (R) also produced a linear condensable-time curve which consisted of CAB fragments.

TABLE VI. Infrared Spectrophotometric Analysis of Condensable Volatiles From Glazing Polymers

Sample	Major Bands Wave Number cm^{-1}	Interpretation
P	Insufficient sample for analysis	
Q	Insufficient sample for analysis	
R	1080, 1180, 1730, 2800-3000	CAB fragments
S	1070, 1130, 1280, 1570, 1590, 2800-3000	Aromatic ester
T	Insufficient sample for analysis	

Outgassing of Thermal Insulation. Very little condensable products were evolved from either the glass wool or the polyurethane thermal insulations during thermal aging. The polyurethane foam was stoichiometrically balanced and uniformly mixed to insure essentially complete and total reaction of the isocyanate and the polyol components. The polyurethane foam, however, did exhibit moderatly high noncondensables which were identified to be predominately the fluorocarbon blowing agent which slowly diffuses through the foam cell walls (Figure 7).

The glass wool exhibited about 1% by weight of noncondensables which consisted of adsorbed moisture and low molecular weight oil fragments.

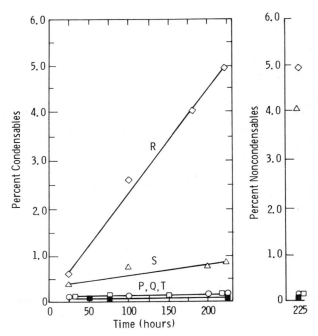

Figure 6. Percent condensables and noncondensables of glaze polymers as a function of time at 150 °C: (o) PMMA; (□) PC; (◊) CAB; (△) GRP; and (■) FEP.

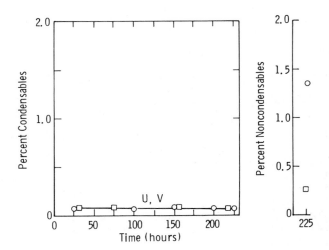

Figure 7. Percent condensables and noncondensables of thermal insulations as a function of time at 150 °C: (o) polyurethane; and (□) fiberglass.

Effects of Condensable Deposits on Relative Light Transmittance

The evolved compounds initially condense on the glazing material as liquids, and produce little adverse effect upon the relative light transmittance. However, as these thin condensate films are exposed to heat and oxygen, over long periods of time, they slowly degrade and oxidize, and are eventually converted into solid materials which greatly reduce light transmittance. The effects of condensate deposits on the relative light transmittance are shown in Table VII. Several of these values were obtained from working solar collectors while others were obtained from laboratory prepared samples.

Salt Deposits. The salt deposits formed on a glass glazing due to water leaching, over a period of two to three years, produce a considerable reduction in light transmittance. A salt deposit having a density of 0.237g/square meter produces a 34% reduction in relative light transmittance. This reduction is caused by the reflection and absorption of the light by the salt crystals (8).

Silicones Polymers. The outgassing products from silicone polymers initially condense on the glazing as liquid alkyl linear and/or cyclic polysiloxanes. These liquid condensates have been observed to form on solar collectors after several days of testing in the Arizona desert. The liquid polysiloxanes do not appreciably affect the relative light transmittance. However, as the thin film is exposed to the harsh environmental conditions encountered during three months of desert testing, it is slowly converted into a white colloidal silica powder through the following type of overall reaction (9).

$$(CH_3)_2Si \underset{O - Si(CH_3)_2}{\overset{O - Si(CH_3)_2}{\Big\langle}} O \quad \xrightarrow[\Delta]{O_2} \quad \left[\begin{matrix} CH_3 \\ | \\ Si-O \\ | \\ O \\ | \end{matrix} \right]_n \; + \; \left[\begin{matrix} | \\ Si-O \\ | \\ O \\ | \end{matrix} \right]_n$$

A colloidal silica deposit having a density of 0.129g/square meter can reduce the relative light transmittance by as much as 18%.

Butyl Polymers. The outgassing products from the butyl polymers also initially condense out as liquids and are slowly converted into tan colored solid varnish-like deposits. These resinous deposits are formed through a thermal oxidative degradation process as follows:

TABLE VII

Effect of Glaze Deposits
on Relative Light Transmittance

Deposit	Quantity (g/m^2)	Relative Light Transmittance % Reduction
Water Leaching of Glass		
Salts	0.151	20
Salts	0.237	34
Silicone Rubber Fragments		
Liquid	0.4	∿1
Powder	0.129	18
Butyl Rubber Fragments		
Liquid	0.2	∿1
Solid Varnish	0.08	4
Acrylic Fragments		
Solid Film	0.02	∿1
Stearic Acid		
Liquid	0.4	∿1
Solid	0.4	5
Processing Oils		
Liquid	0.4	∿1
Solid Varnish	0.2	10
EPDM Fragments		
Liquid	0.1	∿1
Solid Varnish	0.1	4
PVC Fragments		
Liquid	0.05	∿1
Varnish	0.1	3

$$
\begin{array}{rclcll}
R\cdot & + & O_2 & \longrightarrow & RO_2\cdot & & (2,3,10) \\
RO_2\cdot & + & RH & \longrightarrow & ROOH & + & R\cdot \\
ROOH\cdot & & \longrightarrow & RO\cdot & + & \cdot OH \\
RO\cdot & + & RH & \longrightarrow & ROH & + & R\cdot \\
HO\cdot & + & RH & \longrightarrow & H_2O & + & R\cdot \\
R\cdot & + & R\cdot & \longrightarrow & R\text{--}R &
\end{array}
$$

Molecular enlargement, crosslinking, and polymer termination occur during the last step to form a resinous material, which produces a great reduction in light transmittance.

Acrylic Polymers. The acrylic polymers usually produce a clear transparent and continuous film on the glazing. These films produced relatively little effect on light transmittance, and did not appear to change with time.

Stearic Acid. Many rubber formulations contain stearic acid as a processing aid. Stearic acid is easily outgassed, and condenses on the glazing in either the liquid or solid form depending on the temperature of the solar collector and the glazing. The liquid stearic acid produces little effect on relative light transmittance, whereas the crystalline form produces a significant reduction.

Processing Oils. The various oils used in the manufacture of rubber products as processing aids and extenders are not chemically bonded to the basic polymer. During thermal aging they slowly diffuse out of the rubber, and condense on the glazing as liquids. They then undergo thermal oxidative degradation, in the same manner as the butyl condensates, to form resinous, varnish-like deposits which greatly reduce the light transmittance.

EPDM Polymers. The condensable materials from the EPDM polymers, though different in composition, still formed resinous, varnish-like products through a thermal oxidative degradation process. The liquid condensates did not adversely affect light transmittance. The solid products produced a significant reduction similar to the solid product obtained from the butyl polymers and the processing oils.

PVC Polymers. The condensable products from the PVC also formed a solid, varnish-like material. The percent light transmittance reduction was essentially equivalent to that obtained from the butyl polymers, the processing oils, and the EPDM polymers.

Conclusion

Outgassing of polymeric materials is one of the major factors reducing solar light transmittance in solar collectors. Various materials are evolved during thermal aging and most of these adversely affect the efficiency of the solar collector.

In order to produce solar collectors, which will display high efficiency for many years, it is extremely important to utilize those polymeric materials which are thermally and oxidatively stable, and do not produce materials which have an adverse effect on the performance characteristics of the solar collector.

Acknowledgment

This work has been supported by the Solar Heating and Cooling Research and Development Branch, Office of Conservation and Solar Application, U.S. Department of Energy.

Literature Cited

1. Mendelsohn, M. A.; Luck, R. M.; Yeoman, F. A.; and Navish, F. W., Ind. Eng. Chem. Prod. Res. Dev., 1981, 20, 508-514.
2. Gibroy, H. M. "Durability of Macromolecular Materials", Eby, R. K., Ed., American Chem. Soc., Washington, DC, 1979, pp 63-74.
3. Shelton, J. R.; Pecsok, R. L.; and Koenig, J. L. ibid, pp. 75-96.
4. Grassie, N. "The Chemistry of High Polymer Degradation Processes", Interscience Publisher, Inc., New York, 1956, pp. 68-80.
5. Flynn, J. H.; and Dickens, B. "Durability of Macromolecular Materials", Eby, R. K., Ed., American Chem. Soc., Washington, DC, 1979, pp. 108-115.
6. Licari, J. J.; and Wegand, B. L. "Resins for Aerospace", May C. A. Ed. American Chem. Soc., Washington, DC, 1979, pp. 127-137.
7. Winspear, G. G. "Rubber Handbook", R. T. Vanderbilt Co., Inc., New York, 1968, pp. 68 and 237-238.
8. Jaffe, H. H.; and Orchin, M. "Theory and Application of Ultraviolet Spectroscopy", John Wiley and Sons, Inc., New York, 1962, p. 4.
9. Rochow, E. G. "Chemistry of the Silicones", John Wiley and Sons, Inc., New York, 1951, p. 95.
10. Holmstrom, A. "Durability of Macromolecular Materials", Eby, R. K., Ed., American Chem. Soc., Washington, DC, 1979, pp. 45-62.

RECEIVED November 22, 1982

Optical, Mechanical, and Environmental Testing of Solar Collector Plastic Films

M. J. BERRY [1] and H. W. DURSCH

Boeing Engineering and Construction Company, Solar Systems, Seattle, WA 98124

Optical and mechanical testing of several commercially available plastic films indicated that
while desired initial performance was attained,
long term weatherability and stability in specularity (specular reflectance or specular transmittance) and strength were observed only in the fluoropolymers studied. The objective of the described
testing (characterization – environmental exposure –
retest) was to identify or stimulate the development
of reflective ($\geq$ 93%) and transmitting ($\geq$ 92%) films
possessing low cost, high specularity, adequate
strength and weatherability (moisture, UV, hail).
Kynar and Tedlar (PVF_3 and PVF) nearly met the
desired specular transmittance and showed little or
no optical and mechanical degradation after the equivalent of 16 years exposure. Metalized polyesters
and polycarbonates met initial optical and mechanical requirements but showed severe property degradation after relatively short environmental exposure.

The Boeing Engineering and Construction Company (BEC) has been
involved in the study and use of plastic film for solar
collectors since 1975. Several studies resulted in designs that
require the use of highly specular plastic films. Solar
collector designs for which hardware was fabricated and tested
include a thin film heliostat (Figure 1), a parabolic linear
trough concentrator (Figure 2), a pressure stabilized point
focusing concentrator and a plastic film/steel sheet laminate
point focusing concentrator (Figure 3). The designs for the
heliostat, trough concentrator and the pressure stabilized point
focusing concentrator all involve the use of lightweight
reflectors and structures which are protected from the

---

[1] Current address: Resources Conservation Co., P.O. Box 3766, Seattle, WA 98124

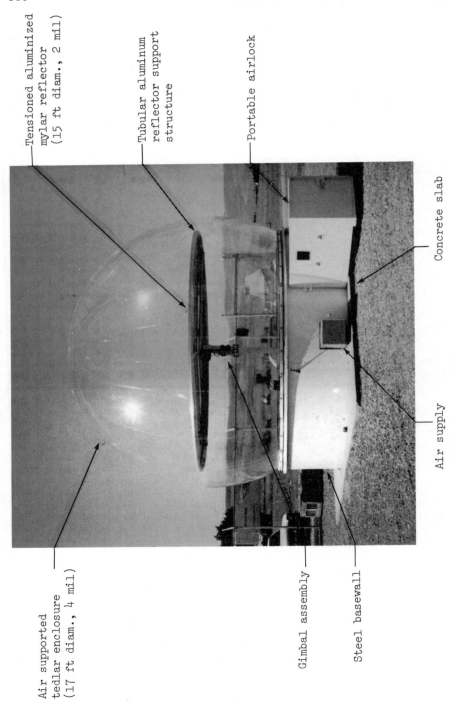

Tensioned aluminized mylar reflector (15 ft diam., 2 mil)

Tubular aluminum reflector support structure

Portable airlock

Air supported tedlar enclosure (17 ft diam., 4 mil)

Gimbal assembly

Steel basewall

Air supply

Concrete slab

Figure 1. Thin Film Heliostat.

Figure 2. Parabolic Linear Trough.

Figure 3. Point Focusing Concentrator
(Plastic Film/Steel Sheet Laminate).

environment by pressurized thin film enclosures (fabricated with
panels and seams, or one piece thermoforming process).

APPROACH

Thirty plastic film suppliers were contacted with 21
companies actively responding. Tables I and II list those
suppliers that have been most active, along with the materials
supplied.

The materials were screen tested in Boeing laboratories.
The screen testing eliminated film candidates with poor optical
and/or mechanical properties and provided control values for
materials that were exposure tested. Microtensile coupons were
tested per ASTM D1708 for determination of yield strength,
ultimate strength and ultimate elongation. Specular reflectance
is measured using a modified bi-directional reflectometer
utilizing a 633 nm wavelength laser source and a variable
aperture system (0.08° to 1.17° cone angles) (Figure 4). Specular
transmittance is measured using a Beckman DK-2A spectrophotometer
and a Gier-Dunkle integrating sphere to provide transmittance
within an acceptance cone angle of 0.5° for wavelengths of 250
through 2500 nm (Figure 5) . The results are integrated over an
air-mass 2 solar spectrum.

Materials showing promise after screen-testing were sent to
Desert Sunshine Exposure Testing Facility (DSET) located near
Phoenix, Arizona. Real time exposure testing is performed on 45°
elevation, south facing racks, providing direct (1 sun) exposure.
Accelerated testing is performed on EMMA (equatorial mount with
mirrors for acceleration). EMMA acceleration factors average out
to approximately 8 times direct (8 suns) over a year's period of
exposure. The bare transparent films were placed in the
environment as received while the reflective material coupons
were placed inside of transparent (fluorocarbon) film envelopes
to simulate BEC's plastic film heliostat design. The same
testing techniques that were used to screen test the samples were
used to test the samples after outdoor exposure. (Optical
testing of exposed samples was performed after cleaning).

Simulated hail testing was performed on a limited number of
materials that had passed initial screening; (fluorocarbon,
polyester and polycarbonate films were included). A "sling shot"
type launcher consisting of an ice ball holder, an elastic
propelling system, and a commercial velocity measurement
instrument was used (see Figure 6). Potential damage was assessed
by measuring the loss in mechanical strength and transmittance at
the point of contact of the ice balls.

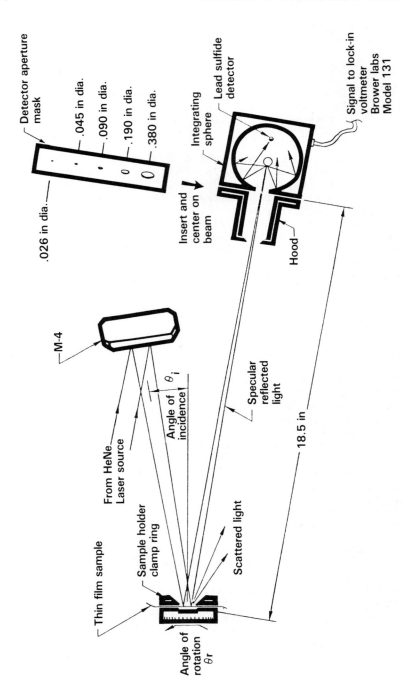

Figure 4. Bi-directional Reflectometer.

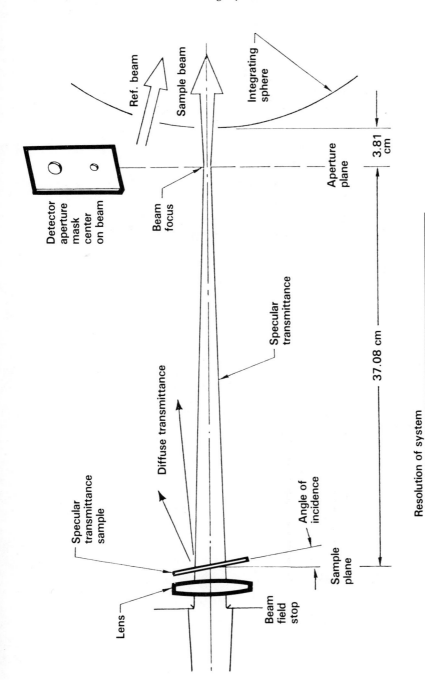

Resolution of system
0.32 cm dia. aperture passes 98 percent of beam (without sample), 0.5° cone

Figure 5, Spectrophotometer (Transmittance Measurement).

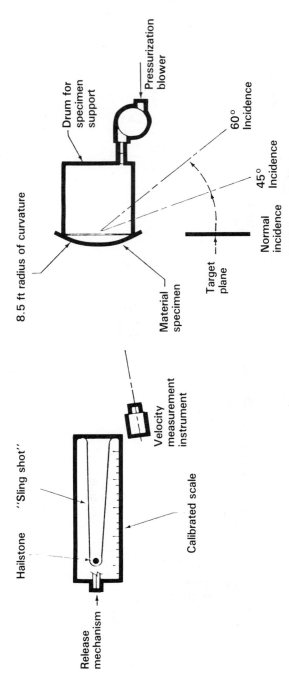

Figure 6. Hail Testing Apparatus.

PROGRAM RESULTS

Test results are based upon single coupon optical measurements and tensile measurements using 4-8 coupons/sample. (Successive measurements on same material, not same sample). This limited number of coupons, coupled with apparent variations in the plastic film process variables such as non-uniformity of orientation, coating thicknesses, results in property data scatter. When a small number of truly promising materials have been isolated, large numbers of coupons should be tested, providing sufficient data for statistical treatment.

Shown in Table I are the enclosure (transparent) film materials currently being exposure tested at DSET. After 24 months of exposure, the two fluoropolymers, Kynar and Tedlar, have proved to be the most promising enclosure films. Figures 7 and 8 give the ultimate elongation and specular transmittance of these and other materials. Experience has shown ultimate elongation to be the most sensitive indicator of degradation. After an equivalent of 16 years of solar exposure, Kynar has shown a 2% decrease in specular transmittance and 33% increase in elongation. During the same time, Tedlar decreased 5% in specular transmittance and 11% in elongation. Data for coupons cut from a Tedlar enclosure, exposed for 2 years in northeastern Oregon, is shown in Figure 7. Also shown in Figures 7 and 8 are the results from exposure testing of a typical polyester and a typical polycarbonate. Both materials lost considerable mechanical strength in EMMA after 6 months (4 years equivalent). In both cases, elongation was reduced to near zero. The losses in transmittance were 29% for 3M polyester and 60% for the Cryovac polycarbonate. It was decided to discontinue accelerated testing of the polyester and polycarbonate after 6 months since the damage had been so severe. Real time testing was continued to determine if the acceleration effects of EMMA are causing damage at a rate that is in excess of 8 times direct. The 3M and Cryovac samples were uv stabilized, but as results indicate, the stabilization was inadequate.

Listed in Table II are the reflective film materials undergoing outdoor exposure testing at DSET. Typical results for coupon testing of these materials are shown in Figures 9 and 10. Also shown are data from coupons cut from an uncoated aluminized polyester reflector supported in a Tedlar enclosure that was exposed in northeastern Oregon. Of the reflector material candidates studied, only the coated, silverized, stabilized polyester (OCLI) maintained its specular reflectivity for 18 months on EMMA (equivalent of 12 years). All reflective films showed severe reduction in mechanical properties.

None of the 4 films investigated for hail resistance failed when impacted with 25mm (1 in) iceballs travelling at terminal

Table I *Enclosure Materials Undergoing Exposure Testing at DSET*

Supplier	Material	Solar specular transmittance (%) at 0.59° cone angle (control value)
3M*	Anti-reflective coated/internally stabilized/polyester	93
Dupont*	Fluorocarbon (Tedlar)	90
Pennwalt*	Fluorocarbon (Kynar, lab material)	89
Allied Chemical*	Polyester (Petra A)	89
Mobay	UV stabilized polycarbonate	89
3M	PMMA (Acrylar)	89
3M	Coated/internally stabilized/polyester	89
XCEL	Acrylic (Korad)	88
Pennwalt	Fluorocarbon (Kynar, production)	86
Allied Chemical	Fluorocarbon (H-2)	86
W.R. Grace (Cryovac)*	UV stabilized polycarbonate	86
National Metalizing*	UV stabilized polyester	86
HOECHST	Fluorocarbon (Hostaflon)	86
3M	Acrylic/polyester lamin. (Flexigard)	86
Martin Processing*	UV stabilized polyester (Llumar)	85
Celenese*	UV stabilized polyester (Celenar)	85
ICI	Internally UV stabilized polyester	85
ICI*	Internally UV stabilized polyester (Melinex OW)	83
Morton Chemical*	AR coated fluorocarbon (Tedlar)	83
Martin Processing*	UV stabilized polyester (Llumar)	82
W.R. Grace (Cryovac)*	UV stabililized polycarbonate	73

* Materials with exposure test results

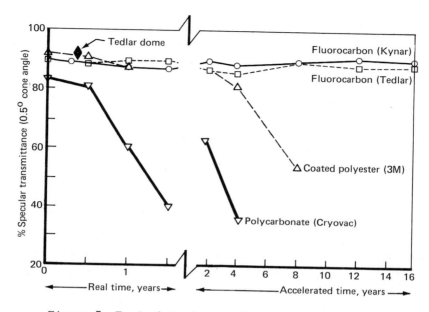

Figure 7. Typical Enclosure Material Optical Data.

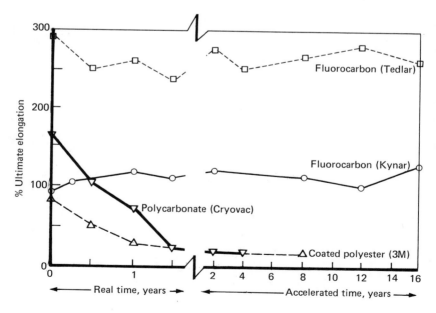

Figure 8. Typical Enclosure Material Mechanical Data.

Table II *Reflector Materials Undergoing Exposure Testing at DSET*

Supplier	Material	Reflectance (%) at 0.14° cone angle (control value)
OCLI*	Silverized/UV stabilized/polyester	94
Dunmore*	Aluminized/internally UV stabilized/ polyester	88
National Metalizing*	Coated/aluminized/polyester	86
Sheldahl*	UV stabilized polycarbonate/ silverized/polyester	85
3M	Acrylic coated/aluminized/polyester	85
Dunmore*	Coated/aluminized/polyester	84
Morton Chemical*	Coated/aluminized/polyester	83
Dunmore*	Aluminized polyester	76
Morton Chemical*	Coated/aluminized/polyester	73
Sheldahl*	UV stabilized polycarbonate/ aluminized/polyester	66
Mobay*	Aluminized polycarbonate	58
Dunmore*	Coated/aluminized/polyester	43
XCEL	Aluminized acrylic	27
* Materials with exposure test results		

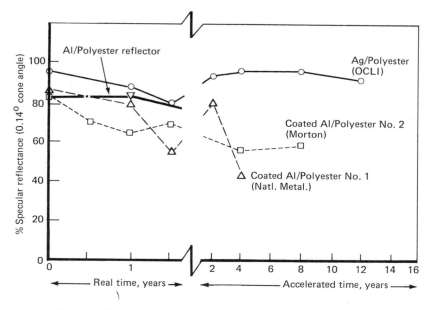

Figure 9. Typical Reflector Material Optical Data.

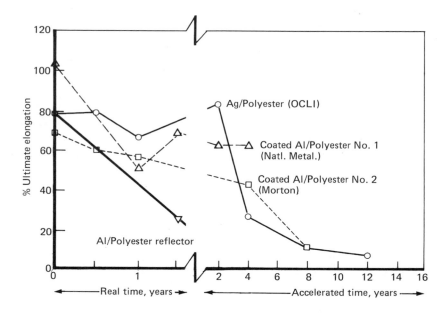

Figure 10. Typical Reflector Material Mechanical Data.

TABLE III TEST RESULTS

MATERIAL	Film Thickness (Mils)	Hailstone Weight Average (Grams)	Angle From Normal Incidence	Velocity (FPS)	DAMAGE				
					Indentation Diameter (Inches)	Specular Transmittance (%)	Loss In Yield Strength (%)	Loss In Ultimate Strength (%)	Velocity At Failure (FPS)
TELAR (Polyvinyl Flouride) • In Laboratory	4	7.9	0°	76.3	.40 [1]	56	0.7	0.9	122
		8.8	45°	74.8	NONE	--	--	--	--
		8.6	60°	74.8		--	--	--	--
• On Heliostat In Field	4	8.0	0°	76.2	.39 [1]	--	--	--	106
		8.1	45°	77.8	NONE	--	--	--	--
		7.9	60°	77.8		--	--	--	--
MELINEX O (Polyester)	2	8.4	0°	74.8	.37	77	2.9	0	113
		7.3	45°	74.8	NONE	--	--	--	--
		9.1	60°	73.4	NONE	--	--	--	--
PETRA A (Polyester)	5	7.5	0°	74.8	.23	71	0	0	104
		9.0	45°	74.8	NONE	--	--	--	--
		--	60°	--	--	--	--	--	--
POLY CARBONATE	3.5	7.8	0°	73.4	.46	24	0	9.5	>117
		7.0	45°	73.4	NONE	--	--	--	--
		--	60°	--	--	--	--	--	--

[1] Fine Scratch Lines, ½ inch long

velocity (see Table III). The maximum mechanical damage was 9.5% loss in ultimate strength for the polycarbonate; others showed less than 3% loss of ultimate or yield strength. The specular transmittance was reduced within the affected areas. An analysis, which considered collector geometry, hailstone incidence angle, and storm severity and frequency for the southwestern U.S. was used to predict the useful area that would be lost through transmittance reduction in a 15 year period. Losses of 1 to 3% were estimated.

CONCLUSIONS

Test results to date indicate that the fluoropolymers, Kynar and Tedlar, exhibit the best weatherability of the enclosure materials studied. Little or no optical or mechanical degradation was observed after the equivalent of 16 years. The metalized polyesters and polycarbonates, although stabilized, were severely degraded in mechanical properties in both real time and accelerated exposures. There was some evidence that reflectivity can be stabilized through the use of protective coatings. Metalizing of fluorocarbons is under investigation. Several films exhibited a good resistance to damage from 25mm hail falling at terminal velocity.

RECEIVED November 22, 1982

Protective Coatings and Sealants for Solar Applications

K. B. WISCHMANN

Sandia National Laboratories, 7472, Albuquerque, NM 87185

An aging study has been completed which evaluated
a number of polymeric materials for potential use
as 1) protective coatings for back surfaces of
mirrors and 2) solar heliostat edge seals. These
investigations were conducted in an artificial
weathering chamber that accelerated thermal cycling.
We observed the primary mirror failure mode to be
silver corrosion resulting from moisture exposure.
To increase mirror longevity in current heliostat
designs, intimate bonding at all the composite
interfaces is essential to minimize moisture pathways to the silvered surface. With good adhesion,
a KRATON rubber was found to exhibit superior back
surface mirror protection, 12 months in environmental chamber with no corrosion. An ultraviolet
stabilized butyl rubber appeared to be the best
edge seal. All heliostats edge sealed with silicone
materials showed silver corrosion which indicated
either poor bonding or moisture permeation.

Many current solar heliostat and flat plate collector designs
which have been field tested show a vulnerability to long-term
outdoor weathering (1). Therefore, the purpose of this investigation was to evaluate potential protective polymeric materials,
i.e., coating, sealants, that could be used in various solar
applications. Depending on the specific design, material
weathering can occur in a variety of ways.

For example, the reflectivity of silvered mirror heliostats can deteriorate with time in the presence of moisture (1,2). The mechanisms of silver deterioration are currently an area of conjecture (3). In flat-plate solar collectors high temperatures, humidity, ozone and ultraviolet (UV) radiation all contribute to aging processes which limit the lifetime of sealants, adhesives and gasket materials (4).

This work does not address the fundamental question of mirror corrosion mechanisms. Rather, our study was limited to visual observations of weathering effects upon commercial products exposed to extreme conditions in an environmental test chamber. The principal parameters investigated were 1) the materials' ability to protect silvered mirrors from moisture and 2) the effect of mechanical stress (coefficients of expansion mismatches) due to temperature-humidity cycling.

Materials

Back Surface – Materials examined for the protection of the back surface of mirrors were: 1) ESTANE 5714 (a polyurethane) from B. F. Goodrich, 2) Pro Seal 890 (a polysulfide) from Essex Chemical Corporation, 3) SARAN (polyvinylidene chloride) from Dow Chemical Company, and 4) KRATON (a styrenebutadience block copolymer) from Shell Chemical Company.

Edge Seals – The edge seals employed in this study were DC-790 and DC-738 from Dow Corning, GE 1200 from General Electric; all three are silicones. One butyl rubber sealant, Adcoseal B-100 was supplied by Adhesive Development and Chemical Operations, Inc.

Test specimens were prepared by bonding a 4" x 4" x 1/8" Gardner mirror to an appropriate substrate with EC-3549 (amine cured polyurethane from 3M Co.) and then sealed around the mirror edge with a bead of the respective sealants. Substrates included in this test were: 1) cellular glass (Solaramics Co.), 2) polystyrene foam (Dow Corning), 3) polystyrene foam with added butyl rubber pad, 4 and 5) paper honeycomb seal with epoxy-fiber glass and melamine respectively, 6 and 7) sine-wave fiberglass sealed with epoxy-fiber-glass and melamine respective (Items 4-7 were supplied by Parabolite, Inc.)

Experimental

Artificial weathering was conducted in a Conrad, Inc., Environmental Test Chamber. The chamber was programmed to cycle from -29° C to 50° C three times during 24 hours, a cycle consists of a two hour hold at each temperature extreme with a two hour ramp in between. The humidity was measured at 82% at 50° C and 50% at 7° C, thus a distinct freeze-thaw cycle occured when a test passed through 0° C (5). This chamber does not incorporate UV radiation as an environmental stress.

Results and Discussion

Back Surface Protection

As mentioned, one of the primary factors contributing to mirror deterioration is water in its various forms (1). In an effort to retard the deterioration of silvered mirrors, several different commercial polymeric materials were evaluated. Our selection was governed by the materials' hydrophobic nature or known use in weather resistant applications. Specifically, these choices were: 1) a polyurethane, 2) a polysulfide, 3) polyvinylidene chloride and 4) a styrene-butadiene copolymer (hydrocarbon). One must realize that any polymeric coating will permeate water in time, consequently, predicting "protected" mirror lifetimes becomes very difficult. The coatings were evaluated on: 1) mirrors which contain only the silver and standard sacrificial copper layer and 2) mirrors having the familiar protective gray paint. The first experiment was designed to eliminate the contributing effect of gray paint.

The coatings will be discussed in order of their increasing effectiveness. The polysulfide (Pro Seal 890) almost immediately attacked the silver. This illustrates one of the first problems with material selection, materials compatibility. The sulfur in this class of compounds was simply incompatible with silver. In fact, the painted mirror showed evidence of attack only a few days after the unpainted sample. The SARAN coating began to peel after 7 days indicating very poor adhesion. This peeling exemplified another critical problem associated with protecting any surface from moisture, the necessity for intimate adhesion between coating and substrate. We believe a key to mirror longevity is the integrity of the various bond interfaces. If there is intimate bonding, only molecular water permeates to the silver and negligible damage occurs (6). Even though SARAN has outstanding moisture resistance, if poor bonding exists then a void or delamination can occur. As a result, water will congregate and mirror degradation follows by whatever corrosion mechanism. The polyurethane (ESTANE 5714) appeared to offer some degree of protection; however, with time it also began to peel, followed by mirror corrosion. The styrenebutadiene copolymer (KRATON 1101) afforded the greatest protection, exceeding all other coated samples in durability. Assuming good bonding, this result was not unreasonable to expect since the coating was a hydrocarbon and would not be expected to have an affinity for water.

From this limited study, hydrocarbon coatings with good bonding characteristics would be the materials of choice when attempting to protect against moisture. These materials are not intended as substitutes for the traditional gray paint protective coating, but simply as added moisture barriers.

Edge Seal Protection

Heliostats for long term field use will probably require edge seal protection. Materials for this application must have outstanding weatherability. A review of existing candidate edge seal materials indicates that silicones possess some of the best overall weathering properties. Inasmuch as there are no edge seals specifically designed for solar hardware, the following commercially available building sealants were selected for evaluation: silicones from Dow Corning (DC-790, 738), General Electric (GE 1200) and a butyl rubber termed Adcoseal B-100. Some understanding of the chemistry of these compounds is required to assure compatibility with the heliostat design. Most silicone building sealants are one component systems that rely on moisture to cure. As a result, they eliminate by-products which can be potentially detrimental to the longevity of aluminum or silver mirrors. The DC-790 liberates amine products, DC-738 emits alcohol, GE 1200 yields acetic acid and butyl rubber outgasses toluene (sealant solvent). By-products such as amines and acetic acid would be considered incompatible with silvered mirrors. Model heliostat reflector segments along with controls were placed in the temperature-humidity cycling chamber for evaluations.

Degradation was found to occur as a variety of pitting or spots, silver delaminations, dendritic growths, color variations and many forms of shadowing or fogging. Earlier beliefs that deterioration would begin at the edges and propagate inward are not necessarily true; degradation appeared to start at virtually any location with no systematic pattern.

Cellular glass was considered a promising heliostat substrate because the coefficients of thermal expansion between the mirror and substrates were matched. Test mirrors were bonded and sealed with the above mentioned silicone sealants and aged. Within three months all the tested samples crumbled. The substrates became filled with water and crumbled during the freeze-thaw cycling due to coefficient of expansion mismatches between the water and glass (5). Obviously to use a cellular glass substrate, it must be sealed. A KRATON coated cellular glass appears to be a promising candidate. This sample showed no evidence of crumbling after 12 months in the environmental chamber (7).

The use of polystyrene foam as a heliostat substrate has been a popular choice because of low cost and ease of fabrication. However, varying forms of degradation or corrosion occurred regardless of what edge seal was employed. The DC-790 and GE 1200 were anticipated to be inferior because of their outgassing products (amines, acetic acide), yet the DC-738 which eliminates a relatively benign by-product (methanol) was just as ineffective. Either the silicone edge seals permeate moisture at a rate higher than anticipated or there is poor bonding at the various sealant interfaces. Once the moisture has migrated in between the mirror and substrate it becomes trapped. Once trapped, the water cannot

escape since the module cannot "breathe," that is, moisture cannot
freely condense and evaporate during thermal cycling. Consequently, moisture can freeze and the resulting expansion can cause
further delaminations thereby allowing more moisture penetration.
At higher temperatures the trapped water can lead to chemical
degradation. Post-mortem examination of the module revealed that
the adhesive (EC-3549) to polystyrene foam interface was highly
pitted. The corrosion observed virtually replicates the foams
cellular structure, see Figure 1. As a result of the cellular
structure, there are literally hundreds of voids where moisture
can permeate, become trapped and cause mirror corrosion. Again,
the presence of moisture points out the importance of bond
integrity throughout these composite structures which includes
the following interfaces: silver to glass, gray paint and adhesive
to substrate. If there is not intimate contact at each interface a
void or delamination results, creating an area where harmful
entities, i.e., water, can diffuse and initiate degradation.
Non-bonding at these critical interfaces can occur in a variety of
ways, for example, poor adhesive applications, unclean surfaces,
or poor process control.

One of the better heliostat sealants was found to be a butyl
rubber, Adcoseal B-100. This module incorporated a design modification; a 1/16" butyl rubber pad bonded between the mirror and the
polystyrene foam substrate. The module looked extremely good after
12 months accelerated aging. A sealant of this type must be used
with some degree of caution because butyl rubbers are not normally
considered as thermally stable or UV resistant as silicones.

The paper honeycomb substrates sealed with either epoxy-
fiberglass or melamine all showed some form of mirror degradation
after 6 months and increased deterioration at 12 months.
Post-mortem examination of some of these modules dramatizes the
degradation effects atrributed to thermal expansion mismatches in
sandwich heliostat structures. The sine wave ribs were perpendicular to the mirror surface and work against the mirror during
thermal cycling causing debonding in the immediate contact area.
The corrosion pattern replicates the sine wave structure, see
Figure 2. Thus, mirror longevity is just as much a funtion of
heliostat design as is material selection.

Two Gardner mirror control samples (gray pain backing) were
included in the aging study. The mirrors without any added protection showed only a slight amount of pitting after 12 months.
Survival of these mirrors with a minimal amound of damage was
attributed to its ability to "breathe." Another Gardner mirror
control was coated with the adhesive, EC-3549; it showed evidence
of deterioration (spotting, shadows) after 6 months. This corrosion suggests incompatibility between the adhesive and silver.
The EC-3549 is a polyurethane adhesive cured with an amine.
According to a literature source (8), the basicity of an amine in
the presence of moisture can result in silver deterioration.
Unfortunately, it was very difficult to evaluate the effectiveness

Figure 1. Polystyrene foam: adhesive, 3M EC-3549; edge seal, GE 1200 construction sealant; time aged, 12 mos.

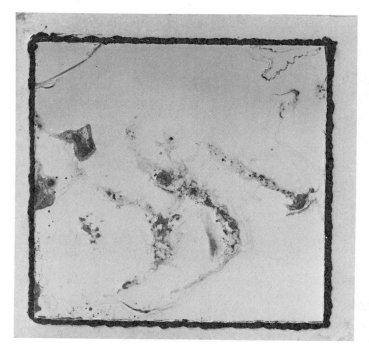

Figure 2. Fiberglass face, sine wave fiberglass body: adhesive, 3M EC-3549; edge seal, GE 1200 construction sealant; time aged, 12 mos.

of the above respective edge seals because material incompatibilities, nebulous bonding and heliostat designs appeared to have
a greater bias on the results than edge seal selection.

Summary and Conclusion

The findings in this aging study resulted in the following
conclusions which should have a bearing on heliostat designs:
1. Edge Seal choices seemed of minor importance compared to
    the overall module design and bonding integrity. Although
    they age well, silicone edge seals appear to bond poorly
    and/or permeate moisture easily.
2. To minimize moisture diffusion pathways either a better
    barrier is needed, that is, an improved design or to
    assure the integrity of the various bond interfaces,
    i.e., edge seals to substrates, silver to glass, etc.
3. Mirror corrosion was manifested by a variety of forms
    such as pitting, spotting, delaminations and discolora-
    tions with no particular pattern. Efforts were not made
    to elucidate corrosion mechanisms, yet we believe the
    first step in the degradation process is a delamination
    or creation of a void area at some critical interface
    where harmful reactants, i.e., water, collect, thereby
    precipitating degradation.
4. Compatibility of various materials in these heliostat
    designs is an important consideration which if properly
    addressed, can add longevity to the system. For example,
    an edge seal which outgasses a relatively benign
    by-product like alcohol (as opposed to acetic acid) would
    be a logical choice as an edge sealant. In general, butyl
    rubber sealants were superior to silicones,
    providing a UV stabilized product is used.

Acknowledgment

Work supported by Department of Energy #DE-AC04-76, DP00789

Literature Cited

1. Burolla, V. P., Roche, S. L., "Silver Deterioration in Second
    Surface Mirrors," SAND79-8276, Jan. 1980.
2. Lind, M. A., Buckwalter, C. Q., Daniel, J. L., Hartman, J. S.,
    Thomas, M. T., Peterson, L. R., "Heliostat Mirror Survey and
    Analysis," Sept. 1979, Pacific Northwest Lab. 3194, UC-62.
3. Second Solar Reflective Materials Workshop, Feb. 12-13, 1980,
    San Francisco, CA, SERI/TP-334-558.
4. Mendelsohn, M. A., Luch, R. M., Yoemon, F. A., Navish, F. W.,
    "Sealants of Solar Collectors," Westinghouse R&D Center, LASL
    Contract.

5.   Allred, R. E., Miller, D. W., Butler, B. C., "Environmental Testing of Solar Reflector Structures," Presentation at 1979 International Solar Energy Society Congress.

6.   Sharma, S. P., Thomas, J. H., Bader, F. E., Electrochem, J., Soc., p. 2002, Dec. 1978.

7.   Allred, R. E., Miller, D. W., Private Communication at Sandia Laboratory.

8.   Hamner, H. G., "Corrosion Data Survey, Metals Section", 5th Edition, 1974 National Association of Corrosion Engineers.

RECEIVED November 22, 1982

Reactivity of Polymers with Mirror Materials

S. K. BRAUMAN, D. B. MacBLANE, and F. R. MAYO
SRI International, Menlo Park, CA 94025

Of eight polymers screened, poly(methyl methacrylate) and possibly poly(vinylidene fluoride) and poly(ethylene terephthalate) show promise as protective coatings for solar mirrors. Polymer-coated mirrors were exposed in a Weather-Ometer and analyzed periodically for mirror and polymer degradation. Failures resulted from physical delamination and chemical reaction (a) at the polymer/mirror interface due to interaction with the degrading polymer or its additives and (b) at the mirror backside due to inadequate protection by the mirror backing or encapsulant.

Organic polymers are being considered as protective coatings on silver or aluminum solar mirrors, either as backing substrates or transmitting superstrates. Although polymers offer the advantage of low initial cost, their long-term stability in the mirror applications is of concern. Environmentally induced polymer/mirror interface reactions could result in a costly loss in the mirror's reflectivity. Considerable information is available on the weatherability of isolated polymers (1), but not when they are in contact with a metal such as silver or aluminum.

In an ongoing program we are studying the reactivity of candidate polymers with mirror materials. This report summarizes progress from initial studies designed to screen the interfacial stabilities of different polymer/metal combinations. The objective of this initial screening is the selection of promising polymer/mirror combinations for subsequent long-term, semi-quantitative environmental evaluation.

Experimental Section

Polymers. The polymers used in this study included ethylene - (18 wt%) vinyl acetate, EVA (Du Pont Elvax 460); ethylene (65

to 75%) - propylene copolymer, EP (Exxon Chemical Vistalon MD
719); polyisobutylene, PIB (Exxon Chemical Vistanex MM L-100);
and poly(methyl methacrylate), PMMA (Rohm and Haas Plexiglas
V811). Polymers obtained and used in film form included Kynar,
a poly(vinylidene fluoride), $PVDF_f$ (Westlake Plastics; 3 mil);
biaxially oriented Kynar, $PVDF_f$ (BIAX) (Pennwalt Corp.; 3.5 mil);
Korad Klear, an acrylate/methacrylate polymer made by multistage
emulsion polymerization, $Korad_f$ (Georgia Pacific, GO212; 3 mil)
(2); Llumar, poly(ethylene terephthalate) (ICI, Melinex O that is
aluminized and sold by Martin Processing Inc., PET_f/Al, 3 mil);
and FEK-244, aluminized poly(methyl methacrylate) with adhesive
backing, $PMMA_f$(3M)/Al/adhesive (3M; 4 mil polymer). Some of the
bulk polymers were compounded with 3 wt% carbon black (Cabot's
Vulcan 3), designated with a subscript c, to reduce ultraviolet
absorption.

Polymer/Mirror Assembly Preparation. Only silver mirrors
were prepared in our laboratories. For the bulk polymers EVA,
PIB, EP, and PMMA, the polymer/mirror assemblies were prepared
by solution (toluene)-casting 2-4 mil (dry) polymer films over a
silver mirror deposited on glass (air surface of 1/16-in. Libby-
Owens-Ford soda-lime float glass). Some of the EVA-coated
mirrors were annealed for 5 minutes at 90 to 100°C. All PMMA-
and PIB-coated mirrors were annealed at 80 to 85°C for 6 minutes.
These combinations are designated polymer/Ag/glass. For the
preformed commercial films $Kynar_f$, $Kynar(BIAX)_f$, and $Korad_f$,
silver was deposited directly on the polymer; the mirror backside
was protected with a backing. These assemblies are designated
$polymer_f$/Ag/backing. For our purpose, no extensive attempt was
made to optimize the reflectance of our mirrors by varying the
fabrication procedures.

All silver mirrors were prepared by electroless deposition
using the commercial solutions and general procedures supplied by
London Laboratories, Ltd. (Woodbridge, Conn.). For Korad film,
however, the aqueous solutions were modified with ethanol to
improve adhesion of the silver to the polymer. An ethanol concentration of 4.8 vol % was used in the sensitizing, silvering,
and reducing solutions. The silver in our mirrors is approximately 600 to 700 Å thick.

Initially, the mirror backs in $polymer_f$/Ag assemblies were
protected by a gasketed, glass cover, sandwich arrangement. Subsequently, we used two commercial mirror backings: a traditional
gray mirror backing (Glidden Coatings and Resins) and an antique
mirror backing paint (PPG 44425 diluted with PPG GV 3003 reducer).

Evaluation. For screening stabilities at interfaces,
polymer/mirror samples were exposed in an Atlas Weather-Ometer
at 70°C and 50% relative humidity, using continuous radiation

from a 6000-watt xenon arc lamp with borosilicate filters. Samples were positioned with the polymer facing the lamp. Exposed samples were removed at intervals for evaluation. First, the reflectance of the intact polymer/mirror assembly was measured. Where possible, the polymer was then removed from the mirror. The separated polymer film was analyzed for changes in tensile properties, molecular weight, and infrared absorptions. The mirror interface was analyzed by scanning electron microscopy.

A Brice-Phoenix Universal Light Scattering Photometer (1000 series with 3200-lumen high pressure mercury light source) was used for measurement of the 90° (45° angle of incidence) specular reflectance of mirrors. For all polymer/Ag/glass assemblies, the reflectance was measured through the glass. For all polymer$_f$/ mirror/backing assemblies the measurements were made through the polymer film. Reflectance was measured at four to eight different locations on each intact mirror sample before the polymer film was removed; the results were averaged. The reflectance of the mirror assemblies is referenced to that of a commercial silver mirror that serves as a common standard.

Mirror components were separated for analysis either by peeling the polymer from the glass or by soaking the entire assembly in water. Typically the metal remains on the polymer film with either procedure. Normally we do not remove such residual metal or even good mirrors before tensile testing of the polymer, but the metal film or mirror backing is first broken up by gently flexing the assembly or metallized polymer film. Elongation and tensile strength at break of the polymer film were measured with an Instron Model TM tensile tester (crosshead speed 2 in. per min; 23°C).

Polymer films were dissolved in appropriate solvents for determination of molecular weight by gel permeation chromatography (GPC). GPC molecular weights are referenced to polystyrene standards, which are suitable for determining relative changes in molecular weights.

Performance Ranking

Using loss in reflectance as an indication of mirror failure and change in tensile properties as an indication of polymer failure, we can rank the polymer/mirror assemblies studied for overall performance. In Table I we rank all the assemblies in terms of poor, intermediate, and good performance. FEK-244 or PMMA(3M)/Al/adhesive is the most durable of the polymer/mirror combinations studied. Poor performers warrant no further study. Intermediate cases show some promise and could possibly be improved with modification of the polymer/mirror assembly. Both physical and chemical failure have been observed in the polymer-

Table I.

RANKING OF THE WEATHERING PERFORMANCE OF
POLYMER/MIRROR ASSEMBLIES

Category	Assembly
Poor	EVA/Ag/glass
	EP/Ag/glass
	EP_c/Ag/glass
	PIB/Ag/glass
	PIB_c/Ag/glass
	$Korad_f$/Ag/Antique
Intermediate	PMMA/Ag/glass
	$PVDF_f$/Ag/sandwich
	$PVDF_f$/Ag/Glidden
	$PVDF_f$(BIAX)/Ag/Glidden
	PET_f/Al
Good	$PMMA_f$(3M)/Al/adhesive/glass

coated mirrors. Our results indicate, however, that both types of failure can often be reduced by proper selection of the polymer/mirror assembly and the additives in the polymer.

Physical Failure: Delamination

Physical failure was evident as delamination of polymer and mirror during weathering of PMMA/Ag/glass. These mirrors remained shiny and clear throughout the exposure period, but they showed some loss in reflectance after 9 days of exposure, when some samples also began to delaminate (See Figure 1). All samples showed some delamination by 12 days when the test was terminated. Analysis of these delaminated samples by attenuated total reflectance IR spectroscopy showed that weathering had no appreciable effect on the polymer surface that was in contact with silver or glass in the samples. Specifically, there was no change in the C=O or C-H peaks, and no OH or C=C developed with exposure. Furthermore, the dissolved film (completely soluble in toluene) showed neither loss in peak molecular weight nor GPC peak broadening with exposure. Although the polymer did not degrade, it exhibited a loss in tensile strength after 9 days of exposure corresponding to the loss in reflectance and onset of delamination (Figure 2; compare this loss in tensile strength to a similar but less extensive loss for $PMMA_f$ in Figure 1).

Because the polymer in PMMA/Ag/glass showed no detectable degradation during weathering, polymer degradation does not contribute significantly to delamination of PMMA. We believe, however, that a physical change in the polymer, possibly indicated by the loss in tensile strength, mechanically disrupts the silver, causing the loss in reflectance and eventual delamination. Stress relaxation and a difference in coefficients of thermal expansion between PMMA and glass are factors that could contribute to a physical change in this rigid polymer. Thus, thinner films in PMMA/Ag/glass weather longer before delamination occurs. Furthermore, delamination apparently can be avoided by eliminating the glass substrate from the mirror assembly. Preliminary experiments in progress indicate that delamination does not occur under similar exposure conditions when the silver is deposited directly onto a preformed polymer film (3M, Acrylar X-2417) to form $PMMA_f$/Ag/backing. Moreover, for the mixed acrylate $Korad_f$ in $Korad_f$/Ag/Antique, delamination was not apparent on weathering.

These results indicate that delamination can be reduced and that PMMA, therefore, shows potential as a coating material for metal mirrors.

Chemical Failure: Polymer/Mirror Interface

Reaction at the polymer/mirror interface, observed only for

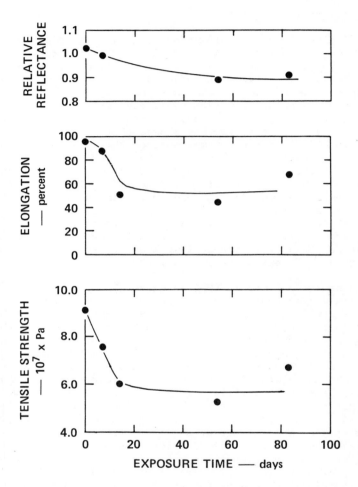

Figure 2.   Weathering of PMMA$_f$ (3M)/Al/Adhesive.

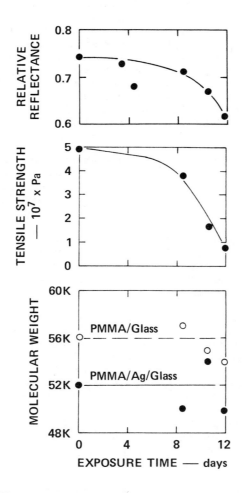

Figure 1.   Weathering of Poly(methyl methacrylate) Systems
● PMMA/Ag/glass
○ PMMA/glass.

silver mirrors coated with EP, EVA, and PIB, was indicated by
varying degrees of loss in reflectance and degradation of the
polymer (EP/Ag/glass >> EVA/Ag/glass > PIB/Ag/glass best); see
Figures 3 to 5. The most noticeable change was yellowing of the
mirrors (indicated by arrows in Figures 3 to 5) that preceded
loss in reflectance. This discoloration is selective, appearing
only where the polymer was in contact with the silver. Furthermore, on prolonged exposure the loss in molecular weight is
slightly greater for the polymer over silver than for the polymer
over glass. Where the exposed polymer could be separated from
the mirror (EVA/Ag/glass), we found that most of the yellow color
remained with the silver and not the polymer. This discoloration
probably is not due to a readily reducible form of silver (e.g.,
oxide, sulfide) because the color could not be removed by
soaking the mirror interface in a reducing solution of dilute
stannous chloride.

Yellowing of the mirrors is light-dependent. When EP/Ag/
glass and PIB/Ag/glass samples were weathered in the dark at 70°C
and 50% relative humidity, their mirrors degraded and discolored
much more slowly than those weathered in the light (e.g., see
Figure 4).

We conclude that we are observing reaction at the polymer-
silver interface that results in reflectance loss. This interface reaction could result from degradation of the bulk polymer
or its additives, initiation at the polymer/mirror interface,
or reaction with the ambient atmosphere.

Interface degradation apparently does not result from any
appreciable reaction of the coated mirror with corrosive gases
permeating in from the atmosphere. Thus, when the degraded
polymer was dissolved off the mirror of a weathered PIB/Ag/glass
sample and the mirror analyzed by X-ray/scanning electron
microscopy, the interface contained no sulfur (0.1% detection
limit) and no more chlorine than did an unweathered control.

Radicals produced in the degrading polymer matrix could
react at the polymer/mirror interface eventually causing mirror
failure. Thus, for all three polymers, yellowing and extensive
degradation of the polymer (evidenced by loss in molecular weight
and where possible - i.e., EVA - loss in tensile properties)
occur simultaneously. Both processes precede any significant loss
in reflectance; see Figures 3 to 5. Of the three polymers (PIB,
EVA, EP) that exhibit this selective yellowing, only PIB/Ag/glass
yellows and degrades without subsequent rapid, major loss in
reflectance. Apparently, degradation of the mirror is retarded
when degradation of the polymer is retarded by shielding the
samples from the light.

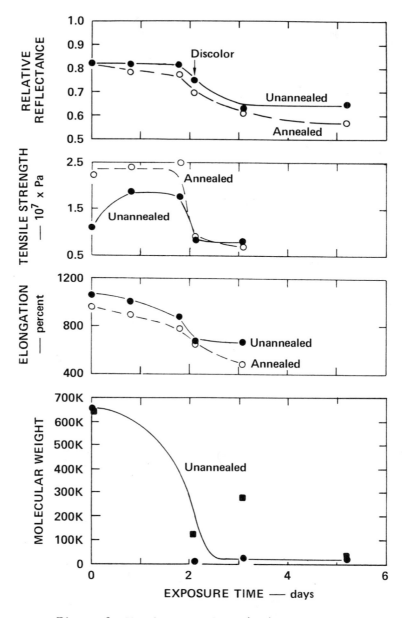

Figure 3. Weathering of EVA/Ag/Glass Systems
● EVA/Ag/glass unannealed
○ EVA/Ag/glass annealed
■ EVA/glass unannealed.

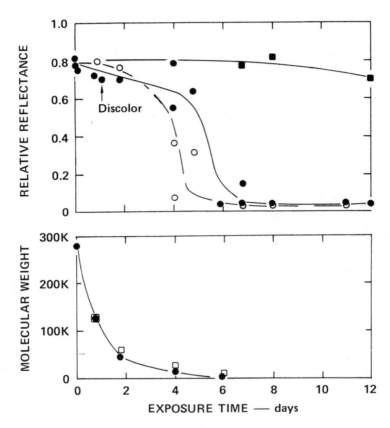

Figure 4. Weathering of Ethylene-Propylene Systems
● EP/Ag/glass
○ EP$_c$/Ag/glass
■ EP/Ag/glass - light blocked
□ EP/glass.

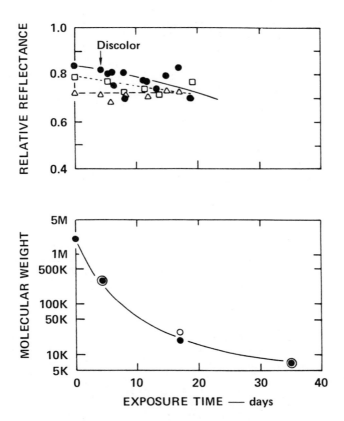

Figure 5. Weathering of Polyisobutylene Systems
● PIB/Ag/glass
☐ PIB$_c$/Ag/glass
△ PIB(-BHT)/Ag/glass
○ PIB/glass.

The antioxidants present in these polymers could also contribute to degradation of the polymer/mirror interface. Thus, ethanol extraction of most of the antioxidant BHT (butylated hydroxy toluene) from PIB prior to mirror assembly fabrication significantly improved the reflectance stability of the mirror upon weathering. Essentially no loss in reflectance occurred on weathering up to 17 days, when the test was terminated; see Figure 5. However, the apparent onset of degradation--flowing of the polymer and yellowing of the film only in contact with silver --was unchanged by the BHT extraction. The three polymers that exhibit polymer-silver interaction and yellowing all contain related hindered phenol-type antioxidants; EVA and PIB contain BHT, and EP contains Irganox 1076. In general, BHT and other hindered phenolic antioxidants are ineffective in the light.

BHT: R = $-CH_3$

Irganox 1076: R = $-CH_2CH_2\underset{\underset{O}{\|}}{C}OC_{18}H_{37}$

To retard photodegradation, we also examined the effect of added carbon black (3 wt%) on the weathering of EP and PIB polymer/mirror combinations. Weatherable carbon-filled polymers might be useful on mirror backs. However, the addition of carbon black showed no benefit; see Figures 4 and 5.

We conclude that EP, EVA, and PIB are not candidate materials for protective coatings in solar reflectors. Even with more suitable antioxidants the stability of EP and EVA appears inadequate. Besides incorporation of a more suitable antioxidant, the use of the more stable PIB would probably require that the material be crosslinked to increase hardness. The usefulness of such a formulated PIB is questionable because of its possible opacity and reactivity with the mirror during crosslinking.

Chemical Failure: Mirror/Backing Interface

Chemical failure of quite a different type was observed for several of the first-surface mirrors polymer$_f$/metal/backing. When interface reaction was apparent, polymer degradation preceded mirror failure. In contrast, for the first-surface mirrors studied (PET$_f$/Al, PVDF$_f$/Ag/sandwich, and Korad$_f$/Ag/Antique) mirror failure precedes polymer degradation. In these cases, we believe mirror failure results from reaction at the inadequately protected backside of the mirrors.

The mirrors PET_f/Al (taped Al-side down to a glass support) and $PVDF_f/Ag/sandwich$ rapidly developed transparent, almost chalky spots on the metal mirrors upon weathering. The appearance of these spots corresponds to the early, moderate loss in reflectance (Figures 6 and 7). Since any loss in tensile properties or molecular weight occurs after the major loss in reflectance, degradation of the reflecting surfaces apparently occurs prior to any major degradation of the polymers. Thus, bulk reaction in the polymer does not account for the mirror degradation. The spotty nature of the mirror degradation, coupled with a general lack of polymer degradation during the mirror degradation, suggests that mirror deterioration results from reaction between the inadequately protected backside of the mirror and the surrounding atmosphere. We suspect that our sandwich arrangement (assembled under nitrogen) did not give an adequate atmospheric seal. Furthermore, for PET_f/Al preexisting tiny pinholes in the aluminum could develop into the transparent, chalky spots that could be a partially hydrated form of Al_2O_3.

If atmospheric degradation of the mirrors is occurring, sealing the backs with a protective coating should improve their weatherability. Thus, some $PVDF_f/Ag$ samples were coated with a Glidden commercial mirror backing ($PVDF_f/Ag/Glidden$). Although the reflectance decreased upon exposure testing (Figure 7) the transparent spots did not develop as early or as extensively on the Glidden-backed mirrors. However, when these fairly rigid samples were handled, the backing showed a tendency to crack. During exposure, the cracks became slightly more defined and the underlying silver became discolored. Similar backing difficulties were also encountered with the biaxially-oriented Kynar sample $PVDF_f(BIAX)/Ag/Glidden$ (Figure 7).

To avoid the above problems, we used the more flexible Antique mirror backing paint (PPG 44425) for the $Korad_f$ samples $Korad_f/Ag/Antique$. However, this backing rapidly embrittled in exposure testing. The embrittlement corresponds to the loss in reflectance for the mirror. The change in tensile properties of the polymer occurs after the loss in mirror reflectance; see Figure 8. Therefore, the degrading polymer is apparently not the cause of the loss in mirror reflectivity. It is possibly, however, that the degrading mirror backing could contribute to deterioration of the mirror either by interacting directly or by providing atmospheric access.

With improved mirror backings, Llumar (PET_f/Al) and Kynar ($PVDF_f/Ag/backing$) might find use as transmitting protective coatings in solar applications. $Korad_f$ appears too unstable to be useful.

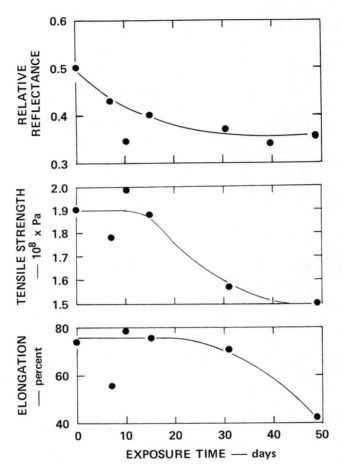

Figure 6. Weathering of PET$_f$/Al (Llumar).

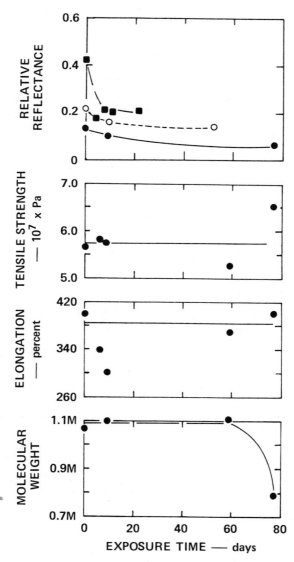

Figure 7. Weathering of Kynar Systems
● PVDF$_f$/Ag/sandwich
○ PVDF$_f$/Ag/Glidden
■ PVDF$_f$(BIAX)/Ag/Glidden.

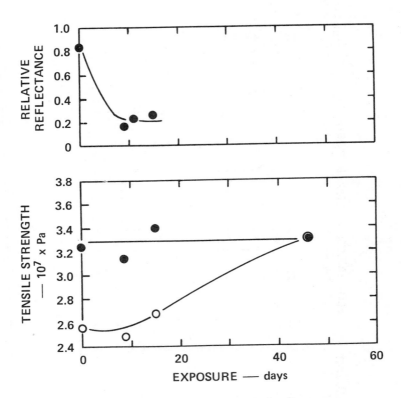

Figure 8. Weathering of Korad$_f$/Ag/Antique
● Maximum tensile strength
○ Tensile strength at break.

Conclusions

Based on our findings, recommendations for polymer coatings for subsequent evaluation can be made. The $PMMA_f$- and possibly $PVDF_f$- and PET_f-containing combinations were the most durable, and they should be studied further. However, the final polymer formulation and polymer/mirror assembly presently cannot be specified.

We have observed mirror failure due to physical delamination and chemical reaction (a) at the mirror/backing interface apparently due to inadequate protection from the atmosphere by the backing and, (b) at the polymer/mirror interface due to reaction with a light-sensitive antioxidant or possibly with the degrading polymer. For both silver and aluminum mirrors, interfacial reactions were only observed at the back surface of the mirror regardless of how the mirror was put down or assembled. We speculate that contact of this surface with the atmosphere prior to coating with test polymer or mirror backing paint could be related to mirror durability. Regardless, improved polymer additives and particularly mirror backings must be found before durable polymer/mirror assemblies can be developed. The backings must be flexible, impermeable and light-stable, and they must not contain solvents that will swell the polymer or disrupt the polymer/mirror interface. Thus, these backings may have to be polymer specific.

Acknowledgment

This work was supported on subcontract No. XP-9-8127-1 by the Solar Energy Research Institute for the Department of Energy.

Literature Cited

1.   Hawkins, W. L.; "Polymer Stabilization"; Hawkins, W. L., Ed; John Wiley & Sons, Inc., New York, 1972; p. 2.
2.   U.S. Patent 3,562,235.

RECEIVED November 22, 1982

Effect of Absorber Concentrations on IR Reflection–Absorbance of Polymer Films on Metallic Substrates

JOHN DAY WEBB, GARY JORGENSEN, PAUL SCHISSEL, and A. W. CZANDERNA

Solar Energy Research Institute, 1617 Cole Boulevard, Golden, CO 80401

A. R. CHUGHTAI and D. M. SMITH

University of Denver, Department of Chemistry, Denver, CO 80208

This paper examines the relationships between the concentration of functional groups and observable infrared reflection–absorbance (IR-RA) spectral changes and applies the findings to bisphenol–A polycarbonate (BPA–PC) on gold substrates. The paper then notes the importance of studying the mechanisms of photochemically and catalytically enhanced degradation at polymer–metal (oxide) interfaces for solar energy collection and gives the rationale for using the IR-RA technique for these studies. An apparatus is described which applies Fourier transform spectroscopy (FT-IR) for the IR-RA studies. An expression is developed showing the independence of the real and imaginary parts of the complex refractive index ($N = n - ik$) at the fundamental oscillator frequency when the frequency distribution of k can be approximated by a Lorentzian distribution. The values of k and n were determined for the polymer between 3200 and 950 cm^{-1} using an iterative calculation and experimentally determined reflectance spectra. From the optical constants determined for each functional group in BPA–PC, the sensitivity of the IR-RA measurement to changes in the functional group concentration was calculated for polymer thicknesses ranging from 10 nm to 10 μm. The optimum sensitivity is generally found at film thickness between 0.1 and 1 μm. At greater thicknesses, IR-RA spectra may be insensitive to changes in functional group concentration, and at thicknesses below 10 nm, the sensitivity is limited by instrumental signal-to-noise ratios.

An understanding of photodegradative reactions of polymers and the extent to which these reactions are influenced by a polymer/ metal (oxide) interface is important for applications involving

the use of polymers to protect metallic surfaces from deterioration in outdoor service. Many such applications exist or are
proposed for solar energy conversion systems [1]. For example,
polymers have been used in attempts to protect reflector surfaces
from exposure to sources of environmental degradation such as
ultraviolet light, temperature, atmospheric gases, moisture, and
abrasive particles, with the objective of maintaining the solar-
weighted reflectance of the unit at a constant value. A polymer
used to encapsulate a photovoltaic device may be in contact with
several different surfaces including a semiconductor, grid metallization, interconnections, antireflection coating, and cover
plate consisting of glass or another polymer. To protect the
substrate in all cases, the bulk properties of the polymeric
coating must be stable with respect to ultraviolet (UV) radiation,
temperature, moisture, and atmospheric gases. However, the polymeric coating may itself contribute to an accelerated attack on
the protected materials if degradative interfacial reactions
change its desirable properties. Delamination at the polymer/
metal interface and/or the concentration of reactive gases on the
"protected surface" could be a likely result of such reactions.
It is known that thermally induced oxidative degradation of certain polymers is accelerated when they are placed in contact with
copper (oxide) surfaces [2,3,4]. What is not known is whether
similar catalytic or synergistic effects result when the polymer
is exposed to ultraviolet radiation while in contact with silver,
aluminum, or other metallic (oxide) surfaces.

Realistic studies of bulk and interfacial degradation of
polymers for outdoor applications must be concerned with the
effects of UV radiation, temperature, temperature cycling, and
atmospheric gases [5]. Experimental methods used to measure these
effects should be sensitive to small chemical changes in the
samples to allow examination of early degradative events and to
perform the experiment in a reasonably short time. The method
should permit these samples to be studied in a configuration that
closely resembles their intended application. The results obtained should be reproducible and comparable to those obtained by
outdoor exposure of similar samples. Finally, the analytical
procedure should not itself contribute to sample deterioration and
should be rapid compared to the events being observed.

Infrared reflection–absorption (IR-RA) spectroscopy was the
primary technique chosen to obtain analytical information on
polymer/metallic samples subjected to degradative conditions since
the method is sensitive and nondestructive [6,7]. The placement
of a specially designed controlled environment exposure chamber
(CEEC) into a Fourier transform infrared spectrophotometer (FT-IR)
makes possible the technique of IR-RA spectroscopy with control of
degradative parameters in situ [8]. Thus, the equipment can
collect low-noise spectra rapidly, enabling continual examination
of the polymer/metal sample during exposure to several degradative
parameters. The CEEC allows simultaneous control of UV exposure

intensity, sample temperature, and the composition of gases surrounding a particular sample. Collection of IR-RA spectra takes place with the sample in situ. The IR-RA, FT-IR method depends on extreme sensitivity to small chemical changes in a given sample so that the study can be brief, and so that only modestly amplified exposures to environmental parameters are necessary to produce detectable degradation.

The purpose of this work is to demonstrate the utility of the method both through theoretical calculations of the relationship between changing polymer functionality and observable IR-RA spectral changes, and through presentation of experimental results obtained using the apparatus mentioned above. The method has been used to study several types of polymer films on various IR-reflective substrates; however, we shall use only our calculations for films of bisphenol-A (BPA) polycarbonate on gold and aluminum substrates to demonstrate the potential of this experimental approach. BPA polycarbonate (BPA-PC) was chosen for initial study since its bulk photochemistry is fairly well known [9,10], thus permitting the validity of this in situ spectroscopic method to be tested. Gold was chosen as a substrate on the assumption that it is a chemically inert IR reflector, thus facilitating studies of the bulk photochemistry of the material. Aluminum is a likely candidate for reflector applications and is more likely than gold to show interfacial activity with respect to the polymeric coating.

The theoretical portion of this paper provides the basis for a definition of optimum sample thickness and polarization state of IR radiation for the study of degradative changes in the major functional groups of BPA-PC. Experimental results, which have been presented elsewhere [8], demonstrate the feasibility of collecting IR-RA spectra on a sample undergoing simultaneous exposure to UV and a flow of gases while maintaining control of the sample temperature. This capability represents an improvement over previously reported methods [11] which depend upon removing the sample from the spectrophotometer for irradiation, resulting in loss of control over the sample temperature, environment, and orientation. Preservation of sample orientation is particularly important if spectroscopic accuracy is to be maintained in the IR-RA experiment.

Theoretical Basis for the IR-RA Experiment

Any technique having utility for the study of reaction kinetics must include the capability of measuring an experimental variable which is related to the concentration of the species of interest as a function of time. The proportionality of IR-absorbance to concentration of an absorbing functionality, which is summarized in Beer's law $A(\nu) = \varepsilon(\nu)bc$, is generally accepted. When the measurement is performed on a sample dispersed at low concentration in a nonabsorbing medium, interfacial reflection effects are low and a linear relationship between absorbance and species concentration is often obtained. A more complex relationship exists between IR-RA values and functional group concentration in an absorbing film on a reflective substrate. Allara et al. [12] showed that the frequencies of band maxima observed in IR-RA spectra do not necessarily correspond to the fundamental vibrational frequencies because of the influences of anomalous dispersion near the fundamental frequencies on the surface reflection measurement.

Theoretical investigations of the dependence of IR-RA intensities and frequencies on the optical constants defining the refractive index $N = n - ik$ of the materials being characterized were reported by Greenler [13,14]. A simplified ray diagram of the IR-RA experiment for a single thin film on a reflective substrate is shown in Fig. 1 [13]. Greenler [13] obtained a set of four complex equations (Fresnel equations) describing the magnetic and electric fields at each interface for a given angle of incidence, assuming parallel or perpendicular polarization. He also presented evidence for the influence of superstrate film thickness and IR incidence angle on the intensities of IR-RA spectral peaks. A similar set of Fresnel equations was utilized by Tomlin [15] for analysis of the reflectance of multilayer stacks of absorbing and/or reflective materials. We have used a computerized solution of the latter set of Fresnel equations to predict IR-RA values for radiation polarized parallel (p), perpendicular (s), or elliptical with respect to the plane of incidence.

Calculation of Reflection-Absorbance. Two different methods for calculating the reflectance-absorbance (RA) of a thin film on a reflective substrate have been reported. Greenler [13] used the relationship

$$RA = (R_o - R)/R_o \tag{1}$$

where R is the measured reflectance of the polymer-coated metal in air at a given frequency and R_o is the measured reflectance of the polymer-coated metal calculated with the assumption that the film is nonabsorbing ($k_2 = 0$). In practice, experimental values of R_o are obtained from the reflection spectrum of an uncoated metal sample similar to that used to produce the laminate. The form expressed in Eq. 1 then becomes analogous to a transmittance measurement. Allara et al. [12] used the form

$$RA = -\log_{10}(R/R_o) \tag{2}$$

which is analogous to a measured absorbance. Equation 1 is obviously the leading term of an expansion of Eq. 2.

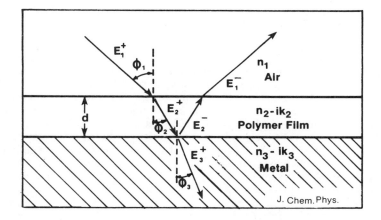

Figure 1. **Ray Diagram of the IR–RA Experiment for a Polymer-Coated Metal.** Adapted from work by Greenler [13]. The subscripts 1, 2 and 3 on the optical constants correspond to the electromagnetic wave in air, polymer film, and metal, respectively.

Relative Concentration of Absorbing Species. The relationship between n and k in $N = n - ik$ is of concern in RA experiments since some of the IR radiation reflected from the polymer-coated metal may have been reflected from the polymer/air interface. Thus, IR-RA measurements of relative changes in concentration (roughly proportional to changes in k) will generally be influenced by associated changes in n with a resultant loss of accuracy. However, there are exceptions to this generalization which may be elucidated by an examination of the mathematical relationship between n and k expressed in the Kramers-Kronig relations [16].

Allara et al. [12] presented a simplified version relating n and k, i.e.,

$$n(\nu_0) = n_\infty + \frac{1}{\pi} P \int_{\nu_1}^{\nu_2} \frac{k(\nu)d\nu}{(\nu - \nu_0)} \tag{3}$$

where n_∞ represents the contribution of all other oscillators in the sample to the value of n and is usually evaluated outside the band of the oscillator under consideration defined by ν_1 and ν_2 but away from other oscillator frequencies and where P denotes the Cauchy principal value. Equation 3 is based on two assumptions: the first is that the vibrational mode having fundamental frequency ν_{max} can be modeled as an independent oscillator. The second assumption is that $\nu \gg (\nu - \nu_0)$ in the range of frequencies included in the absorption band (ν_1 to ν_2), where ν_0 is the frequency within this band at which n is evaluated. The conventions implicit in Eq. 3 are that $\nu_1 < \nu_0 < \nu_2$ and $\nu_1 < \nu_{max} < \nu_2$.

The integral in Eq. 3 may be evaluated by assuming that the k frequency spectrum within the absorption band may be approximated by a Lorentzian distribution:

$$k(\nu) = \frac{1}{\pi} \frac{\Gamma/2}{(\nu - \nu_{max})^2 + (\Gamma/2)^2} \tag{4}$$

where Γ is the full width at half height $k(\nu_{max})/2$ of the distribution. Equation 3 can be integrated to yield Eq. 5 by using the definition of the Cauchy principal value [16] and by inserting Eq. 4 into Eq. 3, i.e.,

$$n(\nu_{max}) = n_\infty + \frac{1}{2\pi^2 A} \left[\ln\left(\frac{(\nu_2 - \nu_{max})^2}{A^2 + (\nu_2 - \nu_{max})^2}\right) - \ln\left(\frac{(\nu_1 - \nu_{max})^2}{A^2 + (\nu_1 - \nu_{max})^2}\right) \right] \tag{5}$$

where the substitutions $A = (\Gamma/2)$ and $\nu_0 = \nu_{max}$ were made prior to the integration. It can be seen from Eq. 5 that $n(\nu_{max}) \cong n_\infty$ for $A \ll (\nu_1 - \nu_{max})^2$ and $A \ll (\nu_2 - \nu_{max})^2$. For an integration interval symmetric about ν_{max}, i.e., $(\nu_2 - \nu_{max})^2 = (\nu_1 - \nu_{max})^2$, Eq. 6 holds:

$$n(\nu_{max}) = n_\infty \qquad (6)$$

For the isolated oscillator that we are considering, the value of n is independent of the value of k at the fundamental oscillation frequency. This is shown for the idealized k − n sets plotted in Fig. 2. As the absorptivity k decreases, the values for n change everywhere except at the fundamental oscillation frequency ν_{max}. This implies that the fraction of the IR beam which is reflected from the polymer/air interface will vary as a function of changes in both k and the associated changes in n at every frequency except the fundamental oscillation frequency. The surface reflection component can make a significant contribution to the total reflected IR signal (vector E_1^- in Fig. 1), especially when unpolarized IR incident radiation is used. Therefore, since an unbiased measurement of changes in functionality concentration (changes in k) is desired, the measurement of absorbance should be performed at a frequency where n is independent of k.

The above treatment shows that for an idealized oscillator, this condition is met at the fundamental frequency, where sensitivity to changes in concentration (or k) is also highest. The fundamental frequencies may be determined experimentally either by obtaining the k spectrum using polarimetry or by obtaining the transmittance spectrum. Data resulting from application of both techniques to BPA-PC will be presented below.

Determination of Optical Constants in the IR for Bisphenol-A Polycarbonate

The purpose of obtaining spectral data on the complex refractive index of polycarbonate polymer was to permit detailed interpretations to be made of the IR-RA spectra collected in situ on metal-backed films of this material. Several of the principal methods for obtaining the optical constants n and k of an isotropic medium have been reviewed by Humphreys-Owen [18]. All of the methods outlined are insensitive to k when k is close to zero, which is the case for frequencies between absorbance bands. For this study, a polarimetric technique (method D in Ref. 18) was chosen to obtain the optical constants of BPA-PC. To apply this method, the ratio of surface reflectances R_p/R_s at two large, but well-separated, angles of incidence (θ_i) was obtained for BPA-PC in the IR. R_p is defined as the reflectance measured for a sample using radiation polarized parallel to the incidence plane and R_s

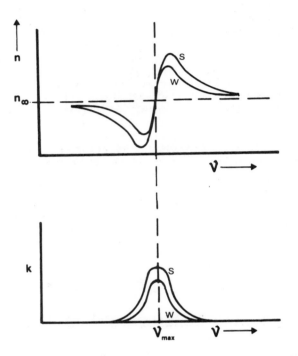

Figure 2. **Plots of n and k for Strong (s) and Weak (w) Absorption**
for Idealized Oscillators. (Adapted from
Nussbaum [17])

is the reflectance measured using light polarized perpendicular to this plane. Measurement of $R_{\theta i}$, which is the ratio of R_p/R_s, at two different values of θ (i = 1 or 2) yields two equations with only n and k unknown. However, the resulting two equations cannot be solved explicitly for n and k [12]. To estimate n and k, an iterative calculation is used that proceeds from trial values for n and k, values for the incidence angles used, and the corresponding measured values for the reflectance ratios, $R_{\theta i}$. Using the initial trial values for n and k, estimates for $\widehat{R}_{\theta_1}$ and R_{θ_2} are calculated at both angles of incidence, and the differences $\Delta\widehat{R}_{\theta_i}$ between the calculated and measured values are computed. Correction factors Δn and Δk are then computed using

$$\Delta R_{\theta_i} = \left(\frac{\partial R_{\theta_i}}{\partial n}\right)_i \Delta n + \left(\frac{\partial R_{\theta_i}}{\partial k}\right)_i \Delta k \tag{7}$$

where i = 1 or 2. The correction factors are added to the current trial values for n and k, and the revised values become the trial values for the next iteration. Convergence is assumed to have occurred when the sum of the absolute values of Δn and Δk falls below 0.005.

According to Allara et al. [12], the ratio I/I_o of detected-to-incident radiation may be related approximately to the absorbance coefficient k by assuming that the absorption band is well isolated. Thus, the ratio becomes

$$I/I_o \cong e^{-4\pi k b \widehat{\nu}} \tag{8}$$

where b is the optical path length (cm) for absorption and $\widehat{\nu}$ is the frequency expressed in wavenumbers. This expression may be combined with the Beer-Lambert absorbance law

$$I/I_o = 10^{-\varepsilon bc} \tag{9}$$

to yield an approximate relationship between k, the decidic molar absorptivity ε, and the molar concentration c of the absorbing species in the solid polycarbonate matrix. By combining Eqs. 8 and 9, k can be expressed as

$$k \cong \frac{2.303 \ \varepsilon c}{4\pi\widehat{\nu}} \tag{10}$$

Molar absorptivities may be calculated from absorption spectra of compounds distributed uniformly at well-defined concentrations in cells of known path length, assuming that Beer's Law is obeyed.

Experimental Apparatus and Procedures

The major pieces of apparatus used to perform the abbreviated exposure experiment consist of an FT-IR spectrophotometer (Nicolet 7199) with a DTGS detector, a controlled environmental chamber (CEEC) with monitoring system and gas supply, and a solar simulator. The polymer-coated mirror sample is mounted inside the CEEC, which is mounted inside the sample compartment of the FT-IR. Details of the instrumentation have been presented elsewhere [8]. A schematic of the CEEC is shown in Fig. 3.

To obtain the reflectance measurements necessary to determine the optical constants of BPA-PC, a specular reflectance accessory (Harrick Scientific Co., Part No. VRA-RMA) was installed in the rear IR beam of the FT-IR. This variable-angle accessory was fitted with a kinematic mounting base and a spring-loaded sample holder to increase reproducibility of sample positioning. The latter change restricted the variability of actual incidence angle to a range of 77.5 to 60 degrees in the high-angle configuration. The incidence angles read from the angular scale on this instrument were calibrated to true values by using a laser reference beam and an aluminum mirror at the sample location. A Perkin-Elmer wire-grid polarizer consisting of gold grid elements on an AgBr substrate was used in conjunction with the specular reflectance attachment. For polymers and other substances that are partially transparent to the incident IR radiation, internal reflections may bias the measurement of surface reflectivity. To reduce this effect while maintaining planarity of the sample surface, one surface of a 3-mm thick sheet of BPA-PC was roughened using a 600-grit emery cloth. The other surface of the sheet was prepared for IR surface analysis by scrubbing with a cotton swab saturated with optical-grade ethanol. Any IR radiation that passes through such a sheet without being absorbed is scattered randomly upon striking the rear surface in addition to being displaced laterally from the optical path to the detector by ~3 mm. Thus, radiation undergoing internal reflection in the thick sheet has a very low probability of reaching the detector and interfering with the measurement of surface reflection.

A reference reflector consisting of aluminum, vacuum evaporated onto a $50 \times 25 \times 3$ mm glass optical flat, was cleaned by centrifugal rinsing with ethanol and chloroform. Single-beam IR reflection spectra of this mirror at two angles of incidence, $76.4°$ and $61.4°$ were collected and stored, each with the polarizer set at zero and $90°$ from the vertical, or with polarizations perpendicular (s) and parallel (p) to the incidence plane, respectively. This procedure was also used to collect the spectra for the BPA-PC sheet. To obtain good signal-to-noise ratios for the spectra collected with parallel polarization, data from 1000 scans of the interferometer were accumulated using a cooled MCT detector to yield an averaged interferogram which was then transformed into a low-noise spectrum. Spectral resolution used was 2 cm^{-1} between

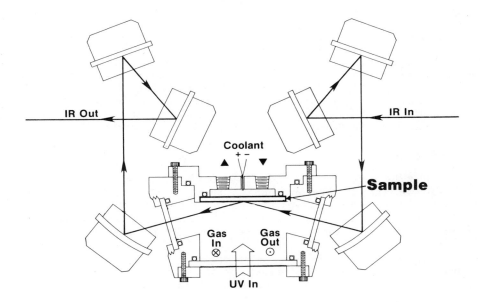

Figure 3. Controlled Environment Chamber (CEEC) Showing IR Transfer Optics

4000 and 400 cm^{-1}. The quantities R_p and R_s were determined for the two incidence angles as a function of frequency by digitally dividing each of the four reflection spectra of the polycarbonate sheet by the appropriate reflection spectrum of the aluminum reflector. The two spectral ratios R_{θ_i} (Eq. 7) were similarly generated from the four reflection spectra, and converged values for n and k were calculated.

To obtain the absorptivity coefficients of the absorbing polymer functionalities for use in predicting values of k, the IR absorption spectrum of the solid polymer was also measured. A solution of BPA-PC was prepared at a concentration of 0.105 M in reagent-grade chloroform. The absorption spectrum of this material was measured over the range 4000 – 400 cm^{-1} at 2 cm^{-1} resolution, using a cell having an effective path length of 0.0133 cm as determined by measurement of interference fringe widths. For qualitative comparison to the k spectra obtained by this method, the IR spectrum of BPA-PC was also measured using KBr discs to reduce interference fringes and solvent absorption effects. The materials were dispersed at ~5% concentration in dry KBr by solvent evaporation, subsequent grinding of the blend, and pressing into discs. Plots of the n and k spectra and the absorbance spectrum for BPA-PC are shown in Figs. 4 and 5, respectively.

Results and Discussion

In this section, the spectral optical constants measured for BPA-PC in the IR are presented first. Using these data, the relationships between IR-RA absorbance values and BPA-PC functionality concentration were determined for a range of applicable film thicknesses using an optical model of the IR-RA experiment.

Optical Constants of Bisphenol-A Polycarbonate in the Infrared.
The spectral values obtained for n and k of BPA-PC between 2000 and 950 cm^{-1} are plotted in Fig. 4. In the 3200–2800 cm^{-1} region, which encompasses the C-H stretching region of the spectrum (not shown), the expected variations in n and k were too small to be detectable above the level of random noise. A constant value of n ≃1.39 was measured within the higher wavenumber region.

From the data presented in Fig. 4, the level of random noise in the determination was estimated at ±0.01 units for n and ±0.05 units for k. The generally decreasing slope of the k spectrum between 2000 and 1300 cm^{-1} may result from a minor systematic error in the measurement. Such an error could be caused by increased scattering of IR radiation by the polymer sheet relative to that produced by the metallized optical flat used as a reflectance standard, or by minor misalignment of the sheet relative to the standard.

An IR absorption spectrum of the polycarbonate polymer used in these determinations is shown in Fig. 5. A comparison can be made

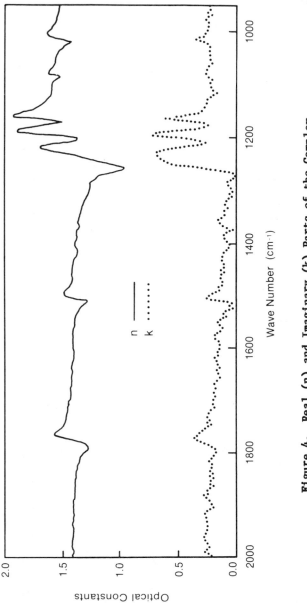

Figure 4. Real (n) and Imaginary (k) Parts of the Complex Refractive Index Calculated From Reflectance Measurements for Bisphenol-A Polycarbonate Film

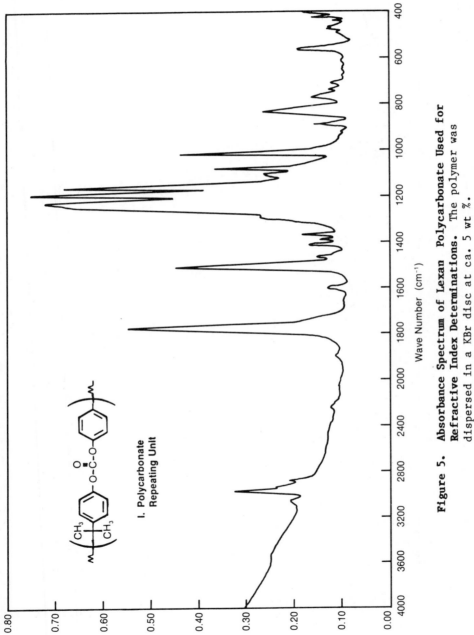

Figure 5. **Absorbance Spectrum of Lexan Polycarbonate Used for Refractive Index Determinations.** The polymer was dispersed in a KBr disc at ca. 5 wt %.

between this spectrum and the k spectrum shown in Fig. 4.   The major absorption bands at 1778, 1505, 1230, 1195, 1165, and 1016 cm^{-1} correspond in both shape and position, but not necessarily in relative intensities to the maxima observed in the k spectrum at 1775, 1505, 1230, 1195, 1165, and 1015 cm^{-1}, respectively.   In Fig. 5, the less intense absorption bands at 2960 cm^{-1} and 1600 cm^{-1} do not have distinguishable counterparts in either the k or the n spectrum of Fig. 4, while the moderately strong absorption band at 1080 cm^{-1} has an associated variation visible in the n spectrum only.   It is important to recall in this context that the intensities of the k spectral peaks are functions of frequency, as well as of absorptivity coefficient and functionality concentration, as demonstrated in the derivation of Eq. 10.

In Table I, a comparison is made between the k values calculated using Eq. 10 and the measured maximum intensities o those peaks in the k spectrum (Fig. 4) for which structural assignments in the polycarbonate repeating unit may be made without ambiguity.   The calculated k values were obtained (Eq. 10) using molar absorptivity coefficients estimated from the IR absorbance spectrum of a BPA-PC solution in chloroform.   Absorptivity coefficients estimated in this manner are in good agreement with published values [19].   Similarly, the correspondence between measured and calculated k values is also fairly good.   The value of k calculated using IR absorbance for the relatively weak methyl band at 2960 cm^{-1} is well below the random noise level of ±0.05 units in the k spectrum determined polarimetrically; thus, it is not

Table I.   Comparison of k Values for Various Polycarbonate Functional Groups Using Two Different Measurement Methods

Functional Group	k Spectrum Maximum (cm^{-1})	ε, Absorptivity Coefficient (ℓ mol^{-1}cm^{-1})	Bulk Concentration (M)	k (from Absorbance Measurement)	k (from Reflectance Measurements)
$-O\overset{O}{\underset{}{C}}$	1230	440	9.45	0.620	0.59
(ring)$-O-$	1195	360	9.45	0.521	0.59
$\overset{O}{\underset{}{C}}$	1775	380	4.72	0.184	0.18
(ring)	1505	144	9.45	0.166	0.19
$-CH_3$	2960	44	9.45	0.025	--

surprising that this band is not distinguishable in the k spectrum. In utilizing Eq. 10 to predict approximate values for k, correct assignment of the absorption bands from which ε is calculated to the various polymer functionalities is obviously important. For example, the two absorption bands at 1230 and 1195 cm^{-1} were assigned to the two distinguishable types of C–O single bonds (inset, Fig. 5) in the polycarbonate repeating unit, on the basis that the intensities of the absorption peaks and those of the peaks in the k spectrum (corrected for frequency difference) (Figs. 4 and 5) are nearly equal. This should be the case for species having similar extinction coefficients present in equal concentrations in a polymer matrix. The intensity of the peak at 1165 cm^{-1} is significantly lower, especially in the k spectrum, and it may result from absorption by another functional group. Gupta [9] assigns this absorption to a C–C single-bond vibration.

For an independent oscillator, the validity of Eq. 6 as applied to various oscillators in the polycarbonate repeating unit may be checked with reference to the n spectrum presented in Fig. 4. Since Eq. 6 was developed with the assumption that the vibrational frequency ν_{max} was well separated from other strong absorption frequencies, an oscillator whose n spectrum conforms to Eq. 6 may be considered to be sufficiently isolated to be spectrally independent of adjacent absorbances.

In Table II, a comparison is made between the values of n measured at ν_{max}, with values of n_∞^- and n_∞^+ estimated from measurements of n at frequencies higher and lower than ν_{max}, such that $dn/d\nu \approx 0$. It is apparent from an examination of Table II that Eq. 6 holds strictly only for the carbonyl vibration at 1775 cm^{-1} and the aromatic vibration at 1505 cm^{-1}. The $n(\nu_{max})$ value for the ester vibration at 1230 cm^{-1} also falls within a range of values defined by n_∞^- and n_∞^+; however, the width of this

Table II. Comparison of n Values Measured at the Fundamental
Vibrational Frequencies to Values Estimated for n

ν_{max}	$n(\nu_{max})$	ν_1	n_∞^-	ν_2	n_∞^+
1015	1.575	950	1.523	1055	1.558
1165	1.694	1110	1.590	1450	1.409
1195	1.652	1110	1.590	1450	1.409
1230	1.449	1110	1.590	1450	1.409
*1505	1.388	1450	1.409	1660	1.415
*1775	1.403	1650	1.415	2000	1.402

*$n(\nu_{max}) \cong n_\infty$

range precludes an unambiguous definition of n_∞ (Fig. 4). We infer that only the carbonyl and aromatic bands in the IR-RA spectrum are sufficiently separated from other absorption bands to enable independent concentration determinations of these polymeric functional groups to be made via RA spectroscopy. Even measurements made at these frequencies may be biased by appearance of nearby absorbances related to the buildup of reaction products during the course of irradiation.

The large variations in n observed at frequencies near the fundamental frequencies and the relationship between n and k expressed in Eq. 3, will clearly influence attempts to measure changes in concentrations of functional groups (i.e., changes in k) using the RA technique. As demonstrated in the derivation of Eq. 6, changes in k can be measured independently of the associated changes in n only at a fundamental vibrational frequency. Since the value of n changes rapidly as a function of frequency near ν_{max}, accurate determinations of ν_{max} are necessary for unbiased measurements of changes in k to be made. Since the influence of surface reflections on positions of the spectral maxima is low at normal incidence, the frequencies of the oscillators (k maxima) may be determined fairly accurately from an IR absorption spectrum of the material to be studied. The assertion of Allara et al. [12] that the positions of the IR-RA band maxima do not necessarily define the vibrational frequencies of the species of interest is reinforced by an examination of the n spectrum in Fig. 4. Therefore, integrating an absorption band to determine changes in concentration, a technique commonly used in transmission-mode experiments, may not be applicable to RA bands. Instead, useful fundamental frequencies of all species to be studied must first be determined by an appropriate transmission-mode experiment or, more accurately, by a determination of the complex refractive indices as a function of frequency. Changes in k may then be related to the changes in the absorption spectrum measured in the RA mode at these frequencies, provided that they do not overlap measurably with the vibrational frequencies of other functional groups (including those of reaction products) and that the absorbances of the materials being measured are low enough to permit good sensitivity to changes in k.

Most Useful Method for Calculating Reflection-Absorbance.
IR-RA intensities calculated using both Eqs. 1 and 2 show that Eq. 2 yields good proportionality to BPA-PC functionality concentration over the widest range of polymer film thickness. Therefore, this equation was utilized in all calculations of IR-RA discussed subsequently.

Effect of p- and s-polarized Incident IR Radiation on IR-RA Intensities. The existence of an optimum polarization state for the study of thin films on reflective substrates can be demonstrated by an examination of the relative magnitudes of the IR-RA

signals predicted for p- and s-polarized IR incident radiation.
The IR incidence angle was not treated as a variable subject to
optimization; the dimensions of the spectrophotometer sample
compartment and other constraints limited the possible range of
incidence angles which could be accommodated in the CEEC. The
chosen design value of 75° was used throughout these calculations.

The optical constants for BPA-PC at the fundamental vibrational frequencies were determined by application of the techniques described above. Values of n_2 and k_2 measured for the
BPA-PC absorption at 1775 cm^{-1}, as well as the optical constants
for air ($n_1 = 1.0$, $k_1 = 0.0$) and for gold ($n_3 = 4.2$,
$k_3 = 27.6$ [19]), were used to predict IR-RA values at 1775 cm^{-1}
for BPA-PC films ranging in thickness from 0.01 to 10 μm on gold
substrates. Calculations were performed for both polarization
states using Eq. 2. Effects of the substrate properties on these
calculations are minor. For example, using the optical constants
of aluminum ($n_3 = 6.8$, $k_3 = 32.0$ [19]) resulted in changes of <1%
in the values obtained.

To compare the magnitudes of the reflection-absorbances
(Eq. 2) obtained using p- and s-polarization, the logarithm of
their ratio is plotted vs. film thickness in Fig. 6 This plot
affirms the conclusion made by Allara et al. [12] and
Greenler [13] that p-polarized light is at least two orders of
magnitude more sensitive than s-polarized light to absorption by
films less than 0.1 μm thick. However, it is interesting to note
that this advantage decreases rapidly for thicker films. In the
particular case illustrated in Fig. 6, s-polarized radiation would
actually yield an RA value about four times greater than would
p-polarized radiation for a 1.25 μm film. The variation in sensitivity for s and p polarization and is a result of the location of
standing wave maxima and nodes in the thicker films.
Greenler [13] showed that the p-polarized electromagnetic vector
has a maximum at the film/reflector interface, while the s-
polarized vector has a node at this interface. The absorption of
s-polarized light in a 1.25 μm film probably results from interactions with carbonyl groups located some distance from the
polymer/metal interface. Calculations show that p-polarized light
yields RA values (Eq. 2) that are more nearly linear with concentration than does s-polarized light, probably because p-polarized
light undergoes less reflection at the polymer-air interface.

The conclusions to be drawn from these results are that, for a
75° incidence angle, p-polarized light will be more sensitive than
s-polarized light to absorbance of films less than 1 μm thick, and
p-polarized light will yield RA values more nearly linear with
respect to concentration than will s-polarized light.

**Sensitivity of the IR-RA Measurement to Changes in
Functional Group Concentration.** The experimental parameter
of polymer film thickness can be treated as a variable
subject to optimization, with the objective of obtaining a

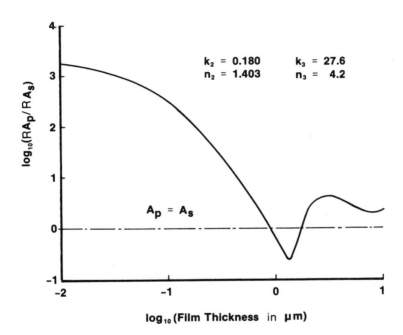

Figure 6. Calculated Magnitudes of Relection–Absorbance Using p-Polarized Light (RA_p) Relative to That Measured Using s-Polarized Light (RA_s) for Film Thicknesses 0.01 to 10 μm. Absorbances were calculated for BPA Polycarbonate Carbonyl Absorption at 1775 cm^{-1}, assuming a gold substrate and 75° incidence Angle.

linear relationship between functionality concentration in the film and measured values of reflection-absorbance. Stronger IR-RA bands of optically thick films exhibit a loss of sensitivity to decreases in absorber concentration, and the weaker IR-RA bands of optically thin films exhibit an unacceptably low S/N ratio. However, an optimum film thickness may exist which will permit simultaneous study of both strongly and weakly absorbing functionalities in a given polymer.

We therefore calculated the dependence of IR-RA intensities obtained using p-polarized light at a 75° incidence angle on reductions in the initial concentration of the methyl, carbonyl, and ester functionalities at 2960, 1775, and 1230 cm^{-1} in BPA-PC films assuming thicknesses of 0.01, 0.1, 1.0, and 10.0 μm. Such reductions in functional group concentrations were modeled to be similar to those caused by degradation in actual films. The n_2 and k_2 values measured at each of the above fundamental frequencies for BPA-PC, and the above optical constants for gold and air were used to calculate the reflection-absorbance of a film having a given thickness deposited on a gold reflector. The initial value of k_2 was then reduced by successive 20% increments to correspond to degradation-related decreases in concentration of the absorbing functionality of the polymer matrix, and the RA values were calculated for the reduced k_2 values. The value of n_2 was held constant since the calculations were performed at a fundamental vibrational frequency, although, strictly speaking, Eq. 6 will be satisfied only for the well-isolated bands centered at 2960 and 1775 cm^{-1}. To facilitate display of these calculated results, the RA values predicted for a functionality at reduced concentration (i.e., reduced k_2) in a film of a given thickness were divided by the respective initial RA values. The resulting ratios are plotted as a function of the logarithm of film thickness in Figs. 7 and 8. From the plot for the moderately strong BPA-PC carbonyl absorption at 1775 cm^{-1} (Fig. 7), a radical departure from concentration linearity is observed for the RA values predicted for a 10 μm film. For such a thick film, the RA technique at 1775 cm^{-1} would be insensitive to reductions in carbonyl concentration because very little radiation will penetrate this film until its optical density has been substantially reduced (by degradation, for example). Films of 1.0 and 0.1 μm thickness are predicted to yield RA values exhibiting much better linearity with respect to carbonyl concentration. Films much thinner than 0.1 μm will yield RA values for carbonyl absorption with unacceptably low S/N ratios. Similar behavior is predicted for the BPA-PC aromatic absorption at 1505 cm^{-1}, which has a k value close to that measured for carbonyl. Noise in this spectral region, which is about ±0.001 absorbance unit for the 100 scans employed for signal averaging, could be reduced by collecting more scans, but time resolution in these studies of reacting systems would be sacrificed. Noise could be reduced more effectively by utilizing a cooled detector or by improving the throughput of the IR transfer

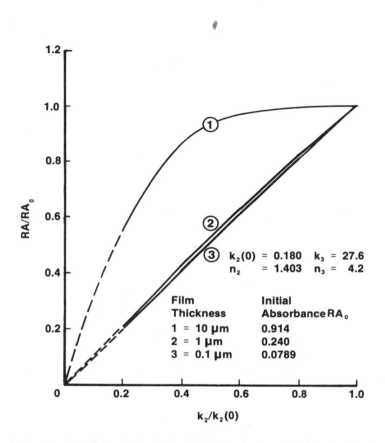

Figure 7. **Sensitivity of Reflection–Absorbance (RA) to Decreases in k at 1775 cm^{-1} of BPA polycarbonate.** The absorbances were calculated for various thicknesses of PC films on gold substrates, assuming p-polarized light at a 75° incidence angle.

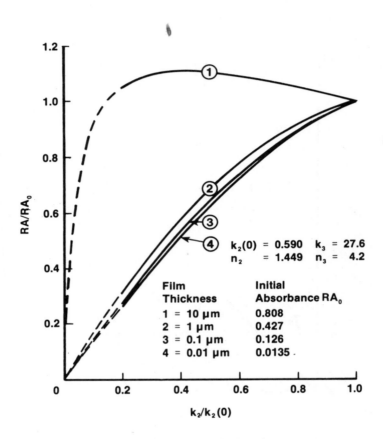

Figure 8. **Sensitivity of Reflection–Absorbance (RA) to Decreases in k (at 1230 cm^{-1}) of BPA Polycarbonate.** The absorbances were calculated for various thicknesses of PC films on gold substrates, assuming p–polarized light at 75° incidence angle.

The plot for the strong carboxylic ester absorption at 1230 cm^{-1} (Fig. 8) shows an interesting variation in the relationship between absorbance and concentration for a 10-μm BPA-PC film. The measured RA values actually increase as ester concentration is decreased because nearly all of the IR signal observed for an undegraded film will be reflected from the film surface. Any light not reflected will be totally attenuated by absorption within the 10 μm film. As the value of k_2 is reduced, the predicted RA values increase because surface reflection decreases and more IR radiation is refracted into the film where it is absorbed. Only if the value of k_2 is reduced to <15% of its original value does sufficient IR radiation pass through the polymer film to permit a roughly linear relationship between RA values and ester concentration.

Calculations (not plotted) for the weak BPA-PC methyl absorption at 2960 cm^{-1} reveal that the predicted IR-RA values for a 1.0 μm film are quite linear with concentration, and deviate only moderately from linearity even for a 10.0 μm film. However, the reflection-absorbance intensity predicted for a 1.0 μm film is only 0.065 units compared to a random-noise level of ±0.0015 absorbance units for measurements made near 2960 cm^{-1}. A reduction of <9% in the methyl functionality concentration in a 1.0 μm film of BPA-PC would not be detectable (S/N > 2) using 100 scans even if p-polarized light were employed.

The sensitivity of the IR-RA technique to reductions in BPA-PC functionality concentration is summarized in Table III. An inspection of Table III reveals that IR-RA measurements sensitive to the widest range of reductions in BPA-PC functional group concentrations are possible on films of about 1.0 μm in thickness. Greater linearity, with little sacrifice in sensitivity, may be obtained from measurements made at the carbonyl and ester frequencies using 0.1 μm films, but sensitivity to reductions in concentration of the methyl functional group will be reduced. Also, any technique that results in increases in signal-to-noise ratio without increasing the measurement time will extend applicability of the method to thinner films with a consequent increase in the linearity of the RA values obtained.

If estimated values of the optical constants for polymer degradation products absorbing at different frequencies than the existing polymer functionalities were available, plots for these species similar to Figs. 7 and 8 could be constructed. Such estimates could be developed from the absorbance spectra of monomeric analogs to these compounds using Eqs. 6 and 10. From these plots, the concentration changes undergone by the polymer functional groups and those of the reaction products could be estimated.

The effect of film thickness on penetration of UV radiation into the films should also be considered. If optimization of film thickness were to be based only on the predictions presented in

Table III, preferential use of 10- m films to study changes in
methyl group concentration might be recommended. However, the
optical model presented here assumes that the concentrations of
the IR-absorbing species are uniform throughout the film. Therefore, if the bulk concentrations predicted by the optical model
are to be valid throughout the film, degradative reactions should
occur uniformly throughout the film. For reactions influenced by
the metallic substrate, this requirement will obviously be met
better by thinner films. Also, for photochemical reactions to
yield degradation products that are distributed uniformly throughout the film, the intensity of UV radiation throughout the film
must also be uniform. This condition likewise will be most nearly
approximated in thin films.

Table III. Sensitivity of the IR-RA Measurement to Reduction in
Initial Concentration of Several Functional Groups in
Bisphenol-A Polycarbonate Films of Various Thicknesses*

Functional Group	Film Thickness, μm			
	0.01	0.1	1.0	10.0
Methyl	Poor	20%	9%	3%
Carbonyl, Aromatic	Poor	3%	1%	Poor
Ester	Poor	6%	3%	Poor

*Percentages represent minimum concentration
changes detectable by an IR-RA signal change
having S/N = 2 using 100 Scans by the DTGS
Detector

In the case of BPA-PC, some of the reaction products have much
higher absorption cross-sections than the undegraded polymer at
the UV wavelengths responsible for photodegradation. Gupta
et al. [9] reported a value of 7300 ℓ mol$^-$cm^{-1} for the absorption
coefficient of phenyl salicylate at 322 nm. The implication is
that, as photodegradation produces UV-absorbing groups on the
polymer chain, buildup of degradation products will eventually be
restricted to the surface of the polymer film, and UV photon flux
at the polymer/metal interface will be limited. The present model
may be extended to account for these effects, but the most
straightforward procedure is to use thin films in these studies.

Acknowledgment

The authors thank M. Lang for contributions in the design of
the transfer optics and for computer programming.
optics used in conjunction with the CEEC.

Literature Cited

1. Schissel, P.; Czanderna, A. W. Sol. Energy Mat. 1980, 3, 225.
2. Allara, D. L.; White C. W. in "Stabilization and Degradation of Polymers," Advances in Chemistry Series; American Chemical Society; Washington, D.C., 1978; Vol. 169, p. 273ff.
3. Jellinek H. H. G.; Kachi H.; Czanderna, A. W.; Miller, A. C. J. Polymer Sci. Polym. Chem. Ed. 1976, 17, 1493.
4. Takahashi T.; Suzuki, K.; Kenichi, A. Kobunshi Kagaku 1966. 3, 1972.
5. Blaga, A.; Yamsaki, R. S. J. Mater. Sci. 1976, 11, 1513.
6. Low, M. J. D. in "Progress in Nuclear Energy"; Elion, H. A.; Stewart, D. C. Eds. Pergamon Press: New York, 1972; Series IX, Vol. 11, p 181ff.
7. Griffiths, P. R. in "Transform Techniques in Chemistry;" Griffiths, P. R. Ed.; Plenum Press: New York, 1978; p. 109ff.
8. Webb, J. D.; Schissel, P.; Czanderna, A. W.; Chughtai A. R.; Smith D. M. Applied Spectrosc. 1981, 35, 598.
9. Gupta, A.; Rembaum, A.; Moacanin, J. Macromol. 1978, 11, 1285.
10. Humphrey, J. S.; Shultz, A. R.; Jaquiss, D. R. G. Macromol. 1978, 6, 305.
11. Lin, S. C.; Bulkin, B. J.; Pearce, E. M. J. Polymer Sci. 1979, 17, 3121.
12. Allara, D. L.; Baca, A.; Pryde, C. A. Macromol. 1978, 11, 1215.
13. Greenler, R. G. J. Chem. Phys. 1966, 44, 310.
14. Greenler, R. G. J. Chem. Phys. 1969, 50, 1963.
15. Tomlin, S. G. Brit. J. Appl. Phys. (J. Phys. D.) 1968, 1, 1677.
16. Stern, F. in "Solid State Physics, Advances in Research and Applications" Seitz F.; Turnbull, D. Eds., Academic Press: New York; Vol. 15, 1963, pp 299-408.
17. Nussbaum, A.; Phillips R. A., "Contemporary Optics for Scientists and Engineers", Prentice Hall: Englewood Cliffs, NJ, 1976; p. 365ff.
18. Humphreys-Owen, S. P. F. Proc. Phys. Soc. 1977, 5, 949.
19. Tompkins, H. G. in "Methods of Surface Analysis"; Czanderna, A. W. Ed.; Elsevier: New York, 1975, Vol. 1, p. 458ff.

RECEIVED February 18, 1983

Adhesives Used in Reflector Modules of Troughs
Effects of Environmental Stress

N. H. CLARK and D. CLEMENTS

Sandia National Laboratories, Albuquerque, NM 87185

V. GRASSO

Budd Company, Fort Washington, PA 19034

Adhesives play a critical role in the
concepts developed for parabolic trough reflector
modules. The need to develop adhesives that are
compatible with silvered glass mirrors, that serve
as structural members and can maintain these
properties over 20 years has lead to a program to
define appropriate candidates.
The program consists of environmentally
stressing complete modules as well as model
systems with stresses of high temperature, high
temperature and RH, and temperature cycling.
Qualitative and quantitative techniques will be
identified that measure changes in the samples.
Changes that can lead to failure will be quanti-
fied if possible. The most successful candidates
will be identified for further study to determine
their lifetimes based on kinetic modeling
techniques.

Sandia National Laboratories has been working with
industry, in particular The Budd Company of Philadelphia and
Parsons of California, to develop state-of-the-art reflector
panels for parabolic trough solar collectors.[1] Typically
parabolic trough concentrators have concentration ratios
(aperture area/receiver area) of between 20 and 50. Figure 1
is a sketch of a parabolic collector showing the component
parts and how a parabolic trough concentrates parallel rays of
solar energy onto a small receiver area at the focus of the
parabola. Figure 2 depicts a stamped sheet metal trough
reflector as fabricated by the Budd Company. Adhesives play a
critical role in the fabrication of parabolic trough reflector
modules. The need to identify adhesives that are compatible
with silvered glass mirrors, that can carry some structural
loads and that can maintain these properties over approximate

0097–6156/83/0220–0169$06.00/0

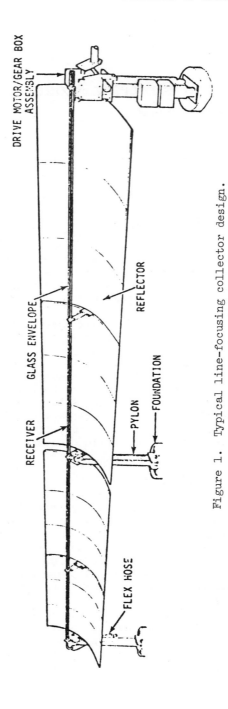

Figure 1.   Typical line-focusing collector design.

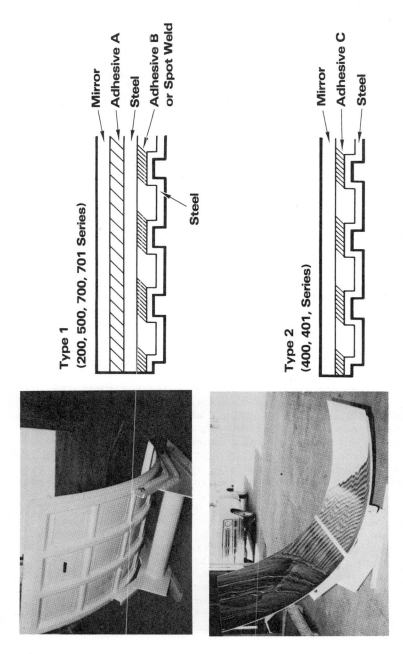

Figure 2. Details of a parabolic reflector: (left) front and backside view of a typical parabolic reflector; and (right) cutaway view of structural details.

lifetimes of 20 years has led to a test matrix that is done in
two phases. In the Phase I, the test matrix was developed to:
  (1) Identify most likely candidate materials for 20-year
       life
  (2) Identify techniques which provide qualitative and/or
       quantitative information that allows identification of
       likely candidate materials
  (3) Identify the most likely failure modes of candidate
       materials
  (4) Identify techniques which provide quantitative
       information on the failure modes identified.
Once failure modes and quantitative measures for them have
been identified, Phase II will commence. Phase II utilizes the
identified failure modes to design a test matrix that will
allow Arrhenius type relations to be developed which may be
used to predict real lifetimes of reflector modules.

Theory
     Traditionally, environmental stress in test systems is
simulated using an environmental chamber in a thermal cycling
mode. Although this method has produced system failures,
interpretation of the failure in terms of cause is very difficult since there are, at a minimum, three variables--(1)
temperature, (2) humidity, and (3) mechanical stress, all of
which may be operating simultaneously.
     Although cyclic environmental chamber test procedures may
suffice for failure processes involving, for example, mechanical stress, kinetic controlled processes dependent upon time
and temperature such as oxidation and diffusion do not lend
themselves to adequate identification and analysis based solely
on number of cycles. Thus Sandia National Laboratories
developed an accelerated aging protocol[2] for environmental
testing which (1) identifies material incompatibilities and
subsequent failure modes in Phase I and (2) proceeds with
kinetic analysis of the Arrhenius type of failure mode processes which allow extrapolation necessary for lifetime prediction of components in Phase II. Thus two phases are necessary
in a complete analysis to accurately predict system lifetimes.
The accelerated aging protocol requires the identification of
the stresses that are most likely to damage the performance of
the component under test. However, data is frequently not
available on the performance of a system under a particular
stress. When this is the case, it becomes necessary to make
predictions of those stresses most likely to cause degradation
and then test to see if the stresses selected are dominant.
     Stresses are herein defined as conditions that will degrade
system performance. For solar systems these stresses can
include temperature, humidity, UV radiation, and temperature
cycling. Once stresses are identified, then tests are designed
to subject the system to each particular stress independently

of the others. Examples of engineering conclusions these tests
can lead to are (1) the stress does not produce a system failure, (2) material incompatibility exists but can be accommodated by design changes or, (3) although the stress results in
failure, an acceptable level of performance has been achieved.
A failure mode is identified when stress results in a failure.
However, such failure may not result in an unacceptable level
of performance. Therefore, once a failure mode has been
identified, a damage parameter, which correlates in an
empirical fashion with the expected failure mode, must be
generated as well to complete Phase I testing.

Following failure mode identification (ie, the material
suspected of significant aging, the stress implicated in its
aging, and the appropriate damage parameter, which quantifies
the expected failure), Phase II kinetic analysis for the predictions of lifetimes is conducted. The usual kind of kinetic
analysis is one that uses Arrhenius principles, but other types
of kinetics may be necessary depending on the empirical form of
the data.

Experimental and Results

Adhesives used in the test matrix included epoxies,
urethanes, and acrylics. (See Appendix I for details of
adhesives chosen.) Two types of epoxies designated as Epoxy X
and Y were used. Both types had been successful in the aerospace industry and Epoxy X had been successful in other solar
programs. Epoxy Y was the only sheet type adhesive included in
the program. Three types of urethanes which were designated as
Urethane A, B, and C were used. Urethanes A and C are both
amide cured urethanes that had successful environmental
histories in automotive applications. Urethane A was highly
flexible and Urethane C had good strength. Urethane C' differs
from Urethane C only in that it has no filler in the base
mixture. Urethane B was chosen since it is a polyol cured
urethane and perhaps less likely to corrode the silver in the
mirrors. Acrylic K was recommended based on its known environmental and strength properties although it had no solar
history. Three types of specimens are used and each will be
described briefly below.

Specimen Type (1)--Full sized 1 m x 2 m troughs are all
similar in construction to the cutaways shown in Figure 2.
Table I shows what adhesives A, B, and C correspond to Budd
type troughs used in this program. One additional type of
trough was used in which the stamped sheet metal back is
replaced by an aluminum honeycomb structure. This is known as
the PAR series. Epoxy Y was used in that structure.

Specimen Type (2)--Sample flat laminates were made by
bonding 10 in. x 23 in. silvered glass mirrors to either

TABLE I
Adhesives Used in Budd Troughs

Series	Adhesive A	Adhesive B
200	Urethane A	Urethane C
500	Epoxy X	Spot Welded
700	Epoxy X	Urethane C
701	Urethane C	Urethane C
400	Adhesive C = Urethane C	

aluminized or painted 10 in. x 23 in. steel with urethanes A, B, C, and C'; Epoxy Y; and Acrylic K.

Specimen Type (3)--Lap shear samples were made by bonding the glass side of 1 in. x 1 in. pieces of silvered glass mirrors to a 1 in. x 6 in. piece of aluminized steel with a common adhesive and then bonding a second 1 in. x 6 in. piece of aluminized steel to the mirrored side with Urethanes A, B, and C.

Three types of stress conditions were chosen as the most likely causes of failure in the trough modules.

1. Thermal cycling--Figure 3 shows cycle used
2. High Temperature--160°F, Low Relative Humidity (RH)--
                        RH $\sim$20 percent
3. High Temperature--160°F, High Relative Humidity (RH)--
                        RH $\sim$80 percent

Controls used included:

1. Outdoor exposure--New Mexico site
2. Indoor--70°F to 80°F, Ambient RH

Full-sized troughs were thermally cycled only. Sample flat laminates were exposed to all the conditions identified above. Lap shears were exposed to all but the outdoor exposure.

Adhesive properties were monitored both directly and indirectly. Direct measurements included changes in shear strength and Tg (glass transition temperature) of adhesive. Indirect measurements include those changes in mirror properties such as reflectance and shape which may occur as changes occur in the nature of the adhesive bond. Techniques used for the above direct and indirect measurement are summarized below.

Indirect Methods

(1) Laser Ray Trace Inspection (LRT)[1]--determines focal length (R in inches) and slope accuracy (S in mradians) of parabolic panels. Basically the LRT system developed by SNLA scans a reflector panel with a laser beam and measures the deviations of the reflected beam from a perfect parabola due to variations in surface contour. Small variations in focal length and local surface contour can be detected.

(2) Ultrasound mapping--detects the presence of voids. An acoustic beam is directed at a sample. If there are voids in

WEEK 1

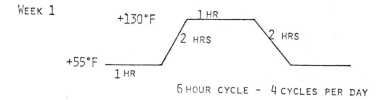

+130°F   1 HR
            2 HRS   2 HRS
+55°F
        1 HR

6 HOUR CYCLE - 4 CYCLES PER DAY

WEEK 2

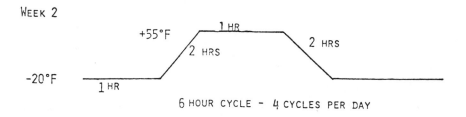

+55°F   1 HR
           2 HRS   2 HRS
-20°F
        1 HR

6 HOUR CYCLE - 4 CYCLES PER DAY

R.H. SHALL BE 50-60%

TO BE REPEATED UNTIL TEST PERIOD COMPLETED.

Figure 3.   Environmental cycle (two-week cycle).

the sample, the beam is modified and data received gives an
indication of voids within the panel. A quantitative technique
developed by SNLA allows numerical determination of the void
areas.

(3) Mapping with a polarizer--measures stress in glass. A
sample is mapped using a polarizer. The technique works on the
principle that unstressed glass does not affect transmission of
polarized light, but stressed glass does. It is hypothesized
that if voids are present, stress is introduced in the glass by
air in void expanding.

(4) Grid Photography--detects changes in mirror surface
which may result from voids, irregular patterns of glass and
poor silvering that will affect trough performance.

(5) Reflectance--measures change in reflectivity of
mirror. If adhesive and mirror are chemically incompatible,
for example, corrosion of mirror may occur which will result in
decreased reflectance.

Direct Methods

(1) Shear strength--measures changes in shear strength by
pulling lap shear samples on an Instron. The load at which
samples pull apart is designated as the failure load.

(2) TG (Glass transition temperature)--measured on
adhesive samples using differential scanning calorimetry
(DSC). Changes in TG most often relate to chemical changes in
the polymer that makes up adhesives.

Results

Results from the specimens used will be organized by the
technique used to measure the change in adhesive properties.
Section A will deal with the results from indirect measurements,
Section B with results from direct measurements.

SECTION A

(Indirect Measurements)

Laser Ray Trace Inspection (LRT)

The LRT data is summarized in Table II. The only change of
note is with Urethane C and since that change indicates
improved slope error, it cannot be related to adhesive failure. The other small changes are not considered significant
enough to warrant any statement about a change in adhesive
properties that might cause them.

Ultrasonic Mapping

Full-Sized Parts. Ultrasonic mapping was done on 1 m x 2 m
troughs containing Urethanes A (200 series) and C (701 series)
as well as Epoxy X (700 series). The results were interpreted
qualitatively and indicated that voids grew with troughs that

Table II

Laser Ray Trace Inspection

1 m x 2 m Troughs Adhesive (Series)	Environmental Condition	Time in Condition	Initial S(mrad)	Final S(mrad)	Initial R(inches)	Final R(inches)
Urethane A (200)	Thermal cycling	2 months	6.2	6.2	19.2	19.2
Urethane C (701)	"	"	4.0	2.5	19.1	19.2
Urethane C (400)	"	"	2.2	2.3	19.1	19.0
Epoxy X (700)	"	"	2.9	2.5	19.1	19.3
Epoxy Y (PAR)	"	"	2.9	N/A	19.4	N/A

used Urethane A, possible growth was noted with Urethane C, and
no growth was shown with Epoxy X.

Sample Flat Laminates. The void growth in flat laminates
was measured quantitatively in most cases. The percent void
area vs time in months in four environmental conditions is
shown in Figure 4. The void growth is accelerated in both
160°F conditions with even greater acceleration with the
addition of high humidity for Urethanes A and Acrylic K.

Grid Photography
     Figure 5 which shows a comparison of an ultrasonic scan and
a grid photograph. If one compares grid photos to ultrasonic
scans, one can see that areas of distortion in the grid photograph correspond to void areas detected by the ultrasonic
scan. Thus grid photography proved to be a technique that was
useful in obtaining qualitative data about the positions of
voids.

Mapping With a Polarizer
     The polarizer was a useful technique in identifying
qualitatively the presence and extent of voids. In general, it
was found that areas over voids had higher stresses in the
glass than other areas. Figure 6 shows a mapping of stress
areas on the corresponding ultrasonic trace of a sample. Thus
void areas correspond to areas of high stress.

Reflectance Measurements
     Reflectance measurements for both small flat laminates and
1 m x 2 m troughs are summarized in Table III. The Acrylic K
adhesive showed degradation in reflectance for small flat
laminates. Urethanes A and C showed degradation for the light
and accommodation. The degradation in reflectances for
Urethane A in the 200 series and C in the 901 series was over
areas that had large voids initially and thus is probably due
to water condensing in void areas not an adhesive/mirror
incompatibility. The decrease in reflectance using the 400
series with Urethane C confirms this theory since the degradation in reflectance took place only over areas that were not in
contact with the adhesive. Thus although changes in
reflectance occurred, they were not related adhesive/mirror
incompatibilities.

SECTION B

(Direct Measurements)

Shear Strength
     Failure loads for the three adhesives tested, Urethanes A,
B, and C, are shown in Figure 7. Urethane A showed virtually
no change in failure load. Urethanes B and C showed changes in

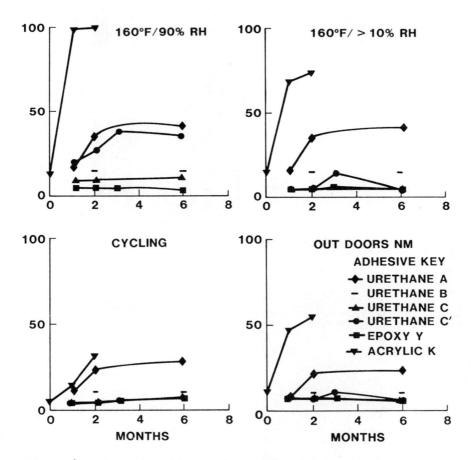

Figure 4. Percent void area in sample laminates as determined by ultrasonic mapping.

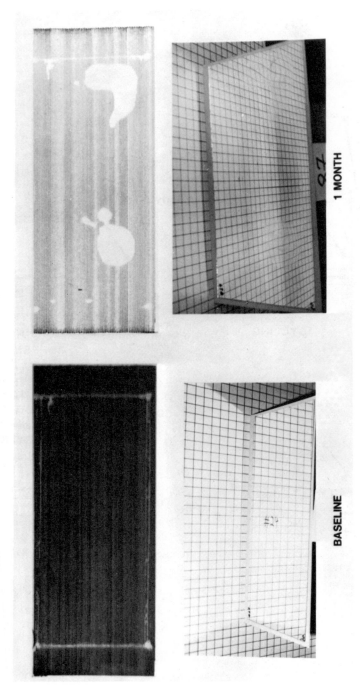

Figure 5.   Comparison of grid photographs and ultrasonic scans.

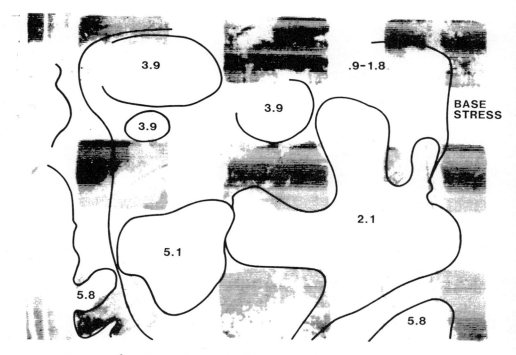

Figure 6. Comparison of ultrasonic trace and stress as measured by a polarizer.

Table III

Reflectance Measurements

Small Flat Laminates

Adhesive	Environmental Condition[1]	Time in Condition	Initial Reflectance (%)	Final Reflectance (%)
Urethane A	A,B,C,D	3 months	95	95
Urethane B	B,C,D	4 months	95	95
Urethane B	A	4 months	95	92
Urethane C	A,B,C,D	4 months	95	95
Epoxy Y	A,B,C,D	3 months	95	95
Acrylic K	A	4 months	95	[2]
	B	4 months	95	66-87[3]
	C	4 months	95	80-93[3]
	D	4 months	95	80

1 m x 2 m Troughs

Urethane A (200)	C	4 months	95	75-94[3]
Urethane C (701)	C	4 months	95	88-94[4]
Epoxy X (700)	C	6 months	95	95
Epoxy Y (PAR)	C	3 months	95	95

[1] A = 106°F/ ∿90% RH, B = 160°F/ ∿10% RH, C = cycling, D = outdoors

[2] System disintegrated to a point at 2 months that no further reflectance measurements where possible

[3] Values varied on different mirrors

[4] High value is over adhesive bead, low value where mirror exposed directly to environment

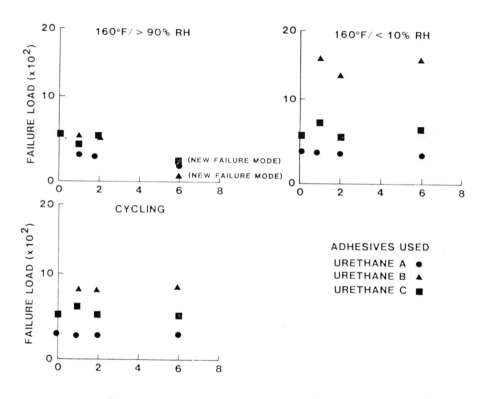

Figure 7. Failure loads for adhesives.

failure load only at the 160°F/   80 percent RH condition. It
is important to note that the failure mode changed from an
adhesive failure to predominantly a paint/copper/silver type
failure.

Glass Transition Temperatures (Tg)

The summary of Tg data is shown in Table IV. The Tg data
was difficult to obtain since measurements were made on very
small adhesives samples taken from lap shear experiments.
Although a change is noted from the original to all samples
taken, this change is likely to be a function of the fact that
larger samples were available for original samples and they
were run at a different time.

Table IV
Summary of Tg Data

Adhesive	Condition*	Time	Original Tg (°C)	After Time Tg (°C)
Urethane A	A,B,C	1,2,6 months	-33°C	-55 ± 2
Urethane B	A,B,C	1,2,6 months	21	15 ± 5
Urethane C	A,B,C	1,2,6 months	-52	-33 ± 5°C

*A = 160°F/ $\sim$ 20% RH, B = 160°F/ $\sim$ 80% RH, C = Cycling

Discussion and Conclusions

In this section the results will be discussed and
conclusions made in the following three areas: (1) Technique
Usefulness, (2) Failure Mode Identification, and (3) Candidate
Identification.

(1) _Technique Usefulness_. For qualitative purposes, grid
photography and the use of a polarizer were identified as the
simplest measurements for determining void areas. However,
voids which occupy relatively small areas are unlikely to be
detected using these techniques. Care must also be taken not
to confuse glass or mirroring imperfections, as determined by
visual inspection, with voids.

For quantitative purposes, ultrasonic mapping found to be
the most useful technique for detecting and quantifying the
growth of voids over time (see Figure 6). Measurements of
shear strength were expected to be the most definitive for
measuring changes in adhesive properties. However, the most
dramatic changes in shear strength shown in Figure 7 for
Urethane A, in 160°F/RH   80, were in general accompanied by a
change in failure mode. The change was from one that involved
the adhesive to one that involved the silver/copper interface
in the mirror. This change in mechanism means that further

work is necessary to define test samples that test only one
failure mode at a time.

Changes in the Laser Ray Trace, the Glass Transition
Temperature (Tg), and in the reflectance did not provide useful
quantitative information.

Although void growth and possibly lap shear data have been
identified as possible quantitative measurements of failure
modes and could be used as damage parameters, additional work
using Urethane C (see candidate identification) and/or one of
the epoxies must be done to ensure the quantitative measurements are measuring only a single mode before Phase II kinetic
analysis is begun.

(2) <u>Failure Mode Identification</u>. Failure modes identified
included increased void area, and change in shear strength of
adhesives. The increase in void area seems to be the most
common cause of failure and occurs in both outdoor samples and
those samples exposed to different environmental stress conditions (see Figure 4). Changes in shear strength of adhesives
were identified as a possible failure mode in one case, but the
ability to use this as a failure mode in Phase II analysis is
limited until appropriate test specimens are designed.

(3) <u>Candidate Identification</u>. Three adhesives were
identified as potential candidates for 20-year life. Both
Epoxies X and Y showed no change in any of the tests used in
this study. One Urethane, Urethane C, which showed only a
possible change in void area in the ultrasonic scan in the
unfilled C' form at 3 months in high temperature condition, is
also a potential candidate. It was expected that epoxies would
do well in these tests, but since they are generally higher
both in initial cost and processing costs, the emergence of a
urethane as a good candidate is promising. Although promising
candidates were identified, none can be guaranteed to last 20
years. Six months of the testing may provide candidates, but
longer tests and work as described in Phase II are necessary to
truly assess 20-year lifetimes.

<u>Literature Cited</u>
1. Champion, R., <u>Proceedings of the Line-Focus Solar Thermal
    Energy Technology Development--A Seminar for Industry</u>,
    SAND80-1666, Sandia National Laboratories, Albuquerque, NM.
2. Gillen, K. and Mead, K., <u>Predicted Life Expectancy and
    Simulating Age of Complex Equipment Using Accelerated Aging
    Techniques</u>, SAND79-1561, Sandia National Laboratories,
    Albuquerque, NM.

APPENDIX I

Information on Adhesives Used

Adhesive	Manufacturer-Designation	Comments
Urethane A	Essex-U-82618 (Base) U-55100 (Curative)	Structural adhesive
Urethane B	3-M 1 x 8A 3504B (Base) 1 x 8A 3504A (Accelerator)	Experimental product
Urethane C	Goodyear - Pliogrip (AX37J110-45, -40, -85)	Experimental product to increase working time
Urethane C'	Goodyear - Pliogrip (6000/6001G)	Product without filter in 6001G to increase flow
Epoxy X	Shell - EPON 828 Quaker - VERSAMIDE 324	Sandia developed product
Epoxy Y	McCann - MA 229	Amide cured epoxy sheet resin

RECEIVED January 24, 1983

Salt-Gradient Solar Ponds and Their Liner Requirements

R. PETER FYNN and TED H. SHORT

Ohio Agricultural Research and Development Center, Wooster, OH 44691

MICHAEL EDESESS

Flow Industries, Inc., 444 Sherman Street, Denver, CO 80203

A case is presented for the use of solar ponds as solar radiation collection and thermal storage devices. The integrity of solar pond brine containment is vital to the efficient operation and longevity of the solar pond. Problems that were encountered while building and operating a solar pond in Wooster, Ohio for space heating of greenhouses are discussed and recommendations made for future liner specifications for use in solar ponds.

A solar pond is a natural or site-built pond, usually one to five meters in depth, which has been altered to retard the loss of absorbed solar energy. The absorbed solar energy in a pond is usually lost because buoyancy of the warmed water causes it to rise to the surface and lose its heat to the atmosphere. The chief solar pond technology - the salt-gradient, or "nonconvective" solar pond - prevents this buoyancy by means of a vertical density gradient achieved with dissolved salt. Because of this, deep water in a solar pond can absorb enough solar radiation to reach high temperatures, up to and even above the boiling point of pure water, and large quantities of heat can be stored in this zone of a solar pond (1).

Two problems which have troubled solar energy conversion technologies and have hindered their commercial development are the lack of inherent energy storage and the consequent inability of solar energy collection to provide firm energy supply without backup, and the materials-intensiveness of solar energy systems and their high initial cost.

Both of these problems are alleviated by solar ponds (2,3). First, the solar pond contains "seasonal storage". It can supply energy weeks, and even months, after absorption of solar radiation, even if no significant solar radiation has been absorbed in the interim. In particular, solar ponds have been

0097–6156/83/0220–0187$06.00/0
© 1983 American Chemical Society

known to provide useful energy after a winter of being covered
with a layer of ice. Second, the solar pond is simple in concept,
easy to build, and has a low requirement for high technology
materials. The primary requirements are for earth-moving, salt,
water, and water and brine transport. On most occasions the solar
pond will need a sturdy membrane liner. This will be the most
technology-intensive requirement of the pond itself. However, in
some cases it may be possible to substitute local clays for the
membrane liner, or to locate the pond on impermeable earth.

The versatility of the solar pond is illustrated by its wide
range of suitable applications. It may serve in rural areas of
less developed countries as a village-centered passive solar
device for pre-cooking, heating water for sanitation and laundry,
and other hot water applications such as for simple village
industries. In more developed areas, heating of washing water as
used in agriculture, and pre-heating of process water can be done.
With heat pumps, solar ponds may provide higher temperature heat
or steam to industrial processes. With absorption chillers, solar
ponds can provide space cooling. In combination with organic
Rankine-cycle engines or thermoelectric devices solar ponds can
generate electricity. With multi-stage distillers or flash
evaporators or other methods, solar ponds can be used to
desalinate or otherwise purify water.

Salt-gradient pond technology

A schematic diagram of a salt-gradient solar pond is shown in
Figure 1. It is a pond in which salt has been dissolved, in high
concentrations near the bottom, and decreasing to low
concentrations near the surface. Sodium chloride is the most
commonly used salt, although there are numerous other
possibilities.

Solar radiation enters the pond, and whatever radiation is
not absorbed in the water on the way down is absorbed on the dark
bottom of the pond. As a result of this heat collection at the
bottom, the deeper water becomes warm.

Higher concentrations of salt prevail in the lower pond
regions than in the upper regions, so the warmer, deep water is at
a higher density than the colder water near the surface. Pure
water, when warmed, becomes less dense. If there was no salt
concentration gradient in the pond there would be continous
convection of the warmed water from the bottom of the pond to the
cooler layers near the top. The higher density created by the
salt prevents this thermal buoyancy convection. Heat transfer to
the surface of the pond occurs primarily by conduction, which is a
slow enough process to enable the lower regions of the pond to
achieve and maintain a high temperature.

In practice, the salt-gradient pond has three layers, as
shown in Figures 1 and 2. In the top layer, the surface
convecting layer, vertical convection takes place due to the

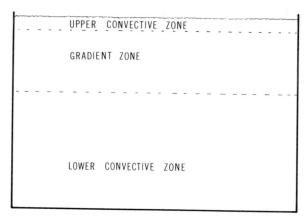

Figure 1.   Cross-section of a salt gradient solar pond.

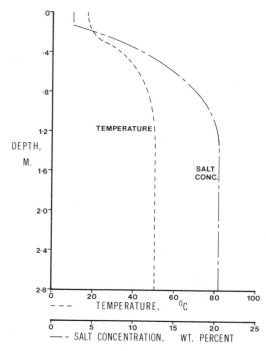

Figure 2.   Temperature and salt concentration in a salt gradient solar pond.

combined effects of wind, evaporation, and diurnal warming. The next layer, which may be approximately one to one-and-one-half meters thick, contains an increasing concentration of salt with increasing depth and is nonconvecting. The bottom layer is a convecting layer which provides most of the thermal storage and facilitates heat extraction. Figure 2 depicts typical temperature and salt concentration profiles as a function of depth in an operating salt-gradient solar pond (4). Because of the massive thermal storage in the salt-gradient pond, daily fluctuations in pond temperature or in its ability to deliver energy are almost insignificant. The one-day temperature loss, in the absence of any insolation whatever, from a typical 3-m deep pond would be about 0.3 degree C. The pond does, however, experience longer term variation in temperature and output. Annual peak temperatures usually occur near the end of summer or in early autumn, and annual low temperatures occur in late winter or early spring.

Experience of the OARDC pond

A salt-gradient solar pond was constructed in 1975 at the the Ohio Agricultural Research and Development Center in Wooster, Ohio (5). The first liner, a chlorinated polyethylene liner, 30 mil in thickness, failed on filling of the pond. It was replaced by a second liner of the same material which failed two years later in 1977. The third liner, Shelter-Rite XR-5, has been in use since August 1977 and continues in service.

The pond liner and its installation

The main problems experienced with the laminated CPE liners installed at the OARDC solar pond in Wooster were as follows:

Patches. These were glued on with a supposedly failsafe adhesive but they fell off the vertical side walls of the liner.

The Liner Surface. This was cracked in some places, even though maximum pond temperatures had only been 44 deg C. (This did not lead to an immediate leak, but was indicative of a material breakdown in that area which could have led to a leak later.)

The Laminate. The liner laminate had been extruded between the fabric of the reinforcing scrim by the water pressure, leaving holes in the liner.

Bubbles. Bubbles formed between the laminations of the liner due to air trapped between the laminations expanding with the heat.

<u>Failure</u>. The liners physically failed when put under stress, even though the temperature of the pond brine was not particularly high.

In view of these difficulties, it is strongly recommended that prospective builders cooperate with the liner material manufacturer, the liner fabricator and the liner installer from the initial pond design stage. Solar ponds must be leakproof. If solar ponds are to be accepted, salt seepage must be absolutely prevented. The liner must be impermeable. The objective of a solar pond is the collection of solar energy. This energy is stored as warm or hot brine and will be lost to the soil if the brine leaks out of the pond. Further, wet soil has a much higher thermal conductivity than dry soil, leading to higher thermal losses out of the pond bottom. A leaky solar pond which has wet soil under the pond will reduce pond performance dramatically. The maximum potential temperature will not be achieved, due to the continuous heat losses. Groundwater contamination by salt affecting ground water in wells and aquifers is also an undesirable consequence of pond leakage.

Liner specifications

A solar pond liner fabrication specification and installation procedure should include the following points: - (These are by no means exhaustive.)

<u>Color</u>. Black is generally most desirable because of its resistance to ultra violet radiation. However, the liner colors will not affect pond performance since the deposition of dust and biological material on the walls and bottom of the liner gradually change its color to a dark grey/green.

<u>Temperature</u>. From a local ambient minimum (-30 deg C) along the top of the pond to a maximum of 100 deg C at the bottom. A solar pond can be designed to achieve different maximum temperatures depending on the application. High temperatures are not necessarily synonymous with high performance. For example for winter space heating in Wooster, Ohio a deep pond would be used (approximately 4 meters), and the maximum temperature would only be 55 to 60 deg C. However, a large quantity of heat would be stored in the pond. Electricity generation in the Southwest United States would require much shallower ponds, (approximately 2 meters) that may achieve temperatures in excess of 100 deg C at times.

<u>Weather</u>. Ability to withstand ultraviolet light while under thermal and chemical stresses. The most vulnerable area is the material around the pond surface exposed to the atmosphere.

Chemicals. Ability to withstand saturated brine (NaCl) solution at up to 100 deg C, with free chlorine and ions of copper, iron, aluminum, sodium, chloride and sulphate present in the solution.

Microorganisms. The liner must remain impervious to attack from algae and other water- or soil-borne organisms.

Rodents. The liner and foundation should resist rodent attack. Rodents will burrow alongside a warm solar pond during the winter.

Wicking. The liner fabric should not wick or delaminate (if laminated) if a raw edge is exposed to saturated brine at up to 100 deg C. (Standards need to be set on this).

Welds. Welds should be at least 5 cm. in width and as strong as the parent material.

Permeability. The liner should be impermeable to hot brine. A permeability of 10^{-10}cm/sec should be acceptable.

Durability. Mechanical specifications need to be established. Specifically, the material must be highly resistant to abrasion, puncture, and tearing and must maintain its flexibility over its life.

Loading. The liner must not serve as a load bearing structure. However, as any part of the liner may come into this category for a short period of time, the integrity of the liner structure must be absolute, and the liner and its seams must have adequate strength and elasticity to withstand such stresses.

Solar pond construction

In the construction of a solar pond, the following points must be taken into account.

Liner Support. Salt water is dense, and some ponds have to be deep (4 m). The pressure at the bottom of a salt pond is considerable, and liner movement has been seen in all the ponds in Ohio because of this pressure. The correct backfill with adequate compaction is essential. A compaction of at least 85% to 90% of Proctor density should be used.

Drainage. Avoid drainage of any kind near the outside of the pond walls. Rain water, particularly, must drain away from the pond perimeter. If the pond is constructed in a high rainfall area, footer drains should be installed to keep runoff from infiltrating down around the walls of the pond.

Overflows. Any pond needs an automatic overflow device for excess rain and will require a system for heat extraction. These devices should not be brought through the pond wall or bottom, perforating the liner, but should be of a siphon type, coming over the top edge of the pond.

Brine Disposal. With present solar pond designs, a means of temporarily holding or disposing of brine of varying salt concentrations needs to be provided to minimize environmental impact.

Leaks. A leak detection and containment system is necessary. Having detected a leak, a means of locating it and repairing the liner has to be defined. Pumping the brine into holding reservoirs is possible, but heat will be lost, and the salt gradient will need to be re-established. A means of repairing a liner in situ without a lot of pumping needs to be formulated.

Sand. Sand should not be used as a fill material on walls and corners. It will slump under the liner on sloping walls putting undue stress on the liner. If a small leak develops, sand will erode, reducing liner support and exacerbating the situation.

Conclusions

Future builders of solar ponds must make sure that their pond sub-base is adequately compacted and prepared for accepting a liner.

Collaboration between the pond builder, the liner manufacturer, the liner fabricator and the liner installer is absolutely essential.

Methods for leak detection, location and repair need to be formulated.

A correct pond design requires proper sizing, a quality liner, a well built foundation, overflow provision and heat extraction and pond gradient maintenance equipment. Each component is necessary for the achievement of trouble-free solar pond operation. Capital saved on a cheap liner and cheap foundation, can prove to be a very expensive mistake.

The solar pond is one of the most promising energy technologies for large scale collection and storage of solar energy. The purchase and installation of a liner is one of the major capital costs incurred in building solar ponds. Ponds can be hundreds of acres in size. A difference of cents per square meter can make or break projects of such size. Further work is needed to develop an inexpensive, reliable liner with low permeability and high chemical resistance to sodium chloride at temperatures up to the boiling point of water. The advent of such a liner would bring the solar pond firmly into the market place as a viable alternative to present, non renewable energy source, methods for generating electricity and providing space heating.

Literature Cited

1. Weeks, D.D.; Bryant, H.C. Proc. AS/ISES., 1981, p768.
2. Bryant, R.S.; Bowser R.P.; L.J. Wittenburg. Proc. ISES.,
    1979, p80
3. Nielsen, C. E. "Nonconvective salt gradient solar ponds."
    Marcel Dekker, New York, 1980.
4. Fynn, R. P.; Badger, P. C.; Short, T. H.; Sciarini, M. J.
    Proc. AS/ISES., 1980, p38
5. Fynn, R. P.; Short, T. H.; Shah, S. A. heating. Proc. ASAE.
    Energy Symposium, 1980, p531.

RECEIVED November 22, 1982

Flexible Membrane Linings for Salt-Gradient Solar Ponds

RALPH M. WOODLEY

Burke Rubber Company, 2250 South 10th Street, San Jose, CA 95112

Hypalon, chlorosulfonated polyethylene, exhibits very low weight gain in brine solutions, even at elevated temperature. Although Hypalon has excellent outdoor weatherability and is widely used as a pond liner, it has not been recommended for continuous high-temperature (90 °C and up) service. As a flexible membrane liner, Hypalon is used as an uncured compound, to give easy and reliable seams both in the factory and field. Maximum service temperature for continuous exposure is 50 °C for potable water. However, Hypalon has been used successfully to line a salt gradient solar pond, withstanding heat on a continuous basis up to the boiling point (109 °C) without degradation. Laboratory testing confirms that the weight gain at 100 °C drops dramatically from distilled water through gradually increasing brine concentrations. A return to distilled water from brine resumes the weight pickup, indicating that the hot brine does not cure the sheet. A new low-swell industrial grade performs even better. This combination of easy seaming and repairs, outstanding weathering--fully exposed, and performance in high-temperature, high-concentration brine solutions makes Hypalon a prime candidate for lining salt-gradient solar ponds.

Utilization of solar energy to meet a portion of the needs of today's high technology, energy-intensive society, has become a national priority. Most solar energy systems collect solar energy for immediate use, but do not have the capability of storing solar energy as well. This means that the solar energy is generated on an intermittent basis - working while the sun is shining brightly, but not working at night, or as effectively on rainy or

0097–6156/83/0220–0195$06.00/0

cloudy days. This usually means that the solar energy
source is used with a conventional backup - essentially
duplicating the required capacity.

Imagine, if you will, a solar energy collector that
simultaneously stores the collected energy for use at
anytime - day or night, rain or shine, summer or winter.
This collector/storage system has no moving parts, uses
the cheapest materials available, and requires very little
maintenance.

Wishful thinking? No - such a system does exist, and
is being developed commercially. The most complicated
thing about the system is its name - Non-convective Salt
Gradient Solar Pond. (Figure 1)

It is simply a pond, three to five meters deep, of
almost any shape and volume. It contains water plus a
water soluble material whose density increases with
concentration. There are three layers - the bottom
being a uniform, dense solution that stores the collected
heat. The center layer is a non-convecting composite of
constantly diminishing concentration (and density)
layers working up toward the surface. The top layer is
essentially fresh water, and is kept as thin as possible.
The sun's rays penetrate the upper and middle layers,
warming the dense solution at the pond bottom. This
bottom layer expands and convects the warmer solution
upward toward the surface. This convection is stopped
by the gradient layer, which remains less dense than
the heated solution below. The heat buildup continues
at the pond bottom as long as the solar insolation
continues, reaching temperatures as high as 100°C or
higher (limited only by the boiling point of the
solution). Usable heat is obtained by circulating the
bottom storage layer solution through some type of heat
exchanger, then returning it to the storage layer for
reheating. (Figure 2)

The solar pond can be located anywhere, but efficiency
dictates an area of relatively high solar insolation.
Economics dictates proximity to a cheap source of solution
material (usually a salt). To minimize heat losses, the
pond should be as close as possible to the point of heat
use.

The Need for Pond Linings

Initial projects of major size have been proposed
for the Great Salt Lake, The Salton Sea and Owens Lake
in the United States (as well as at the Dead Sea in
Israel). Cheap salt is readily available at these
sites, and there is little concern about salt contamination of the local ground water. These projects will

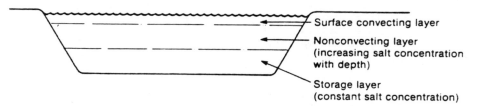

Figure 1.   Salt gradient solar pond.   (Reproduced with permission from Ref. 1.   Copyright 1980, Solar Energy Research Institute.)

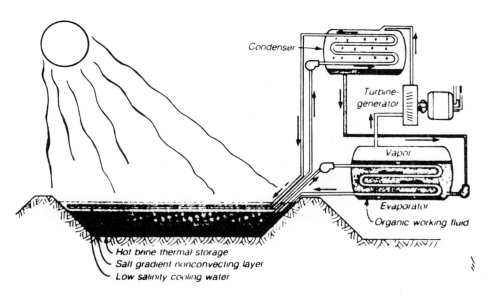

Figure 2.   Solar electrical power.   (Reproduced with permission from Ref. 5.   Copyright 1981, McGraw-Hill.)

probably not be lined as there is no economic justifi-
cation or environmental necessity to mandate a liner.
    Only if those environmentally non-sensitive sites
near naturally occurring salty lakes develop methane
or other gas bubbles, may a membrane liner be needed
regardless. Test results indicate that bubbles upset
the gradient layer. However, many smaller installations
of salt gradient solar ponds will be in areas with
permeable soils, or where the leakage of salt brine into
the ground and water table would be an environmental
disaster. Impermeable liners will be a "must" in these
areas.

The Requirements for Impermeable Liners

    Salt gradient solar ponds present special problems
not frequently encountered in combination. They include:
  1. Chemical resistance to salt brine – ranging
      from saturated solutions diluting down to
      essentially fresh water.
  2. Heat resistance – serviceable from freezing
      or below, to over 100°C.
  3. Weatherability – resistant to UV and Ozone and
      capable of continuous outdoor exposure.
  4. Durability – cost effective performance for
      twenty (20) years or more.
  5. Reliability - pin-hole free construction,
      resistance to mechanical damage, delamination
      and blistering.
  6. Repairability - easy to seam with seams stronger
      than the parent sheet. Homogenous seams that
      do not require special equipment or skilled
      operators for field repairs.
The location of a sizable volume of salt brine in
most areas presents a potential for serious contamination.
Leakage of brine from a damaged liner can "poison" the
soil and the ground water over a wide area, if not
promptly repaired. Proper selection of the impermeable
liner is essential to wide-spread use of salt gradient
solar ponds.

Candidate Materials for Impermeable Membranes

    Many of the lining materials commonly used for
liquid storage cannot be used in salt gradient solar
ponds. Compacted soils, native clays, soil additives
or soil cement, swelling clays such as Bentonite, are
not impermeable to high temperature saturated salt brine
solutions. Only the flexible membrane lining materials
offer the potential for impermeability (zero leakage)

to high temperature concentrated salt brine solutions over long periods of storage time.

Flexible membrane liners for ponds, pits, lagoons, canals and reservoirs were introduced following World War II. Many plastics and rubbers have been used. Thermoplastics are readily and reliably seamed, but are usually buried for protection against UV. Vulcanized or cured rubbers have good weathering for exposed service, but are not readily seamed once they are cured. Thermoplastic rubbers offer good weathering in addition to easy seaming. They are usually supported or reinforced with an open-weave scrim encapsulated between two or more plies of rubber to improve their "green" strength. Examples of each group would include:

> Thermoplastics
> Polyethylene
> Polyvinyl Chloride
> Vulcanized Rubbers
> Butyl
> EPDM
> Thermoplastic Rubbers
> Chlorinated Polyethylene
> Chlorosulfonated Polyethylene

There are many alloys, blends and modifications of the basic materials, as well as new polymers that are constantly being introduced. Time and testing of the various candidates will identify those materials that will perform best.

In 1970, DuPont introduced Hypalon 45, a thermoplastic grade of chlorosulfonated polyethylene synthetic rubber, for applications as a pond liner material. It was one of the first thermoplastic elastomers retaining easy seamability both in the fabrication plant and in the field, and possessing outstanding resistance to outdoor weathering.

Liner field performance under the actual conditions of use is the most reliable indicator of suitability. Salt gradient solar pond technology has largely been developed within the last five (5) years. Very few commercial installations exist even today, so that little comprehensive information about flexible membrane liners has been developed.

However, there is a closely related end use that offers extensive commercial background and experience.

The initial use of hypalon as a liner for a salt gradient solar pond came about strictly as an accident. For many years, the manufacturers and distributors of propane and butane liquified gas, have used huge underground salt deposits to provide storage for liquid gas buildup during the summer. The gas is withdrawn during

the winter season of peak demand to supplement production
capacity. These storage caverns are called "salt domes"
and are also used for the strategic crude petroleum
reserve storage program. (Figure 3)

Salt domes are formed by drilling into a salt
deposit, then flushing fresh water down the hole, which
dissolves the salt, creating a cavity. When the proper
size and shape is formed, the top is sealed around the
access pipe. The liquified gas (or crude oil) is then
pumped into the cavern, displacing the saturated brine.
The salt itself is impermeable to the gas or oil, and
the underground storage cavern will hold the liquified
gas. Cheap storage in an environmentally secure location
is the end result. (Figure 4)

To retrieve the stored product, the process is
reversed. Saturated brine is pumped into the cavern
and the displaced liquified gas is withdrawn. Saturated
brine must be used, as fresh water only makes the hole
bigger, rather than merely displacing the product.
Therefore, a working salt dome must provide (gallon for
gallon) surface storage for saturated brine. These brine
pits are slowly filled with brine as it is displaced by
the liquified gas during the summer, and slowly emptied
during the winter to retrieve the product. These brine
pits are usually lined with a flexible membrane liner to
prevent wide-spread salt contamination of the ground and
ground-water, as well as to prevent loss of the saturated
brine, needed for recovery.

Review of Installations Using Flexible Membrane Liners

Hypalon has been a material of choice for years as a
brine pit liner - providing excellent resistance to the
concentrated brines and weathering well, in spite of the
wide level fluctuations of brine stored from Winter to
Summer. Normally the brine is recirculated to maintain
a saturated solution available for displacement.
Occasionally, however, an unexpected rainfall can deposit
a layer of fresh water on top of the brine. An example
follows:

In early 1976, we installed a Hypalon membrane liner
for the storage of saturated brine at an Arizona gas
terminal. The terminal is located above a salt dome
used to store liquified butane and propane during the
summer months, which is withdrawn during the peak use
of the winter months. The pond measures approximately
405 x 580 x 40 feet deep and holds saturated salt brine
to displace the liquified gas for withdrawal.

During the first summer of operation, the

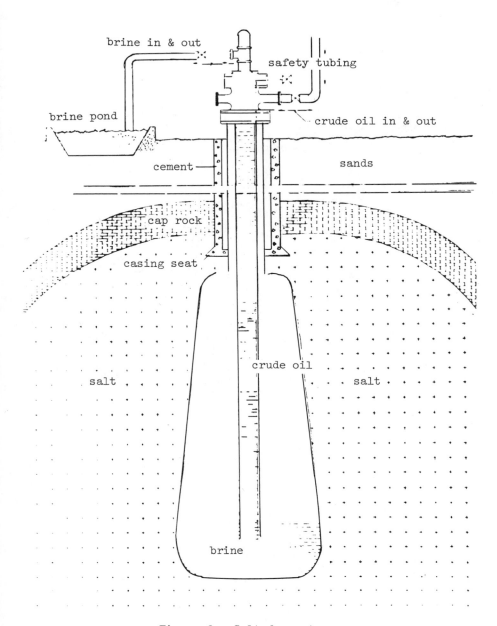

Figure 3.   Salt dome storage.

Figure 4. Major salt deposits in the United States and Canada. (Reproduced from Ref. 1. Copyright 1980, Solar Energy Research Institute.)

circulating pump broke down and was out of service for over 6 weeks. A rare summer rain deposited a layer of fresh water on top of the brine. Upon start-up of the repaired pump, the housing grew so hot that the operator dragged a thermometer across the bottom of the pond, measuring 190°F. Inadvertently, they had formed a salt gradient layer preventing convection and allowing the bottom temperatures to build.

While it was not possible to inspect the bottom at the time, we installed an identical pond adjacent to the first one the following spring. The first pond was drained to install an equalization pipe between the two ponds. The Hypalon, although slightly puckered from the heat, was in excellent physical shape, and we were able to bond a shroud over the equalizer pipe without difficulty, using Hypalon adhesive. (If the material had been completely cured by the hot brine, this would not have been possible).

Both ponds are still in active service, and have performed well for over 5 years. The terminal is located in the middle of farming lands, where salt leakage would have had serious effects. No contamination problems have occurred involving the Hypalon-lined brine ponds.

The Hypalon liner had performed well under the temperature conditions accidentally imposed upon it, which were far above the recommended operating range.

At about the same time, one of the first salt gradient solar ponds was being constructed and put into operation at the University of New Mexico in Albuquerque. Constructed in late 1975, it used a 5 ply Hypalon liner of 45 mil thickness to line a small, circular pond of approximately 60,000 gallons capacity. During the summer of 1977, it exceeded 90°C for the first time. It peaked above 90°C again in 1978 and 1979, and reached 109°C in July, 1980. The pond is still operational and being monitored, so we have not been able to inspect the Hypalon liner - but it is still functioning without leakage after six years of service, and there are no apparent signs of any degradation. (Figure 5)

Test Data on Hypalon Performance in Salt Brines

In an effort to confirm the performance of Hypalon liners in concentrated brine solutions at high temperatures, immersion tests were set up. Potable grade Hypalon, which was used in the brine pits as well as in the New Mexico salt gradient pond was compared to a new grade of Hypalon - industrial grade, which shows less water swell at elevated temperatures. Both materials were tested in distilled water as a control and in 20%

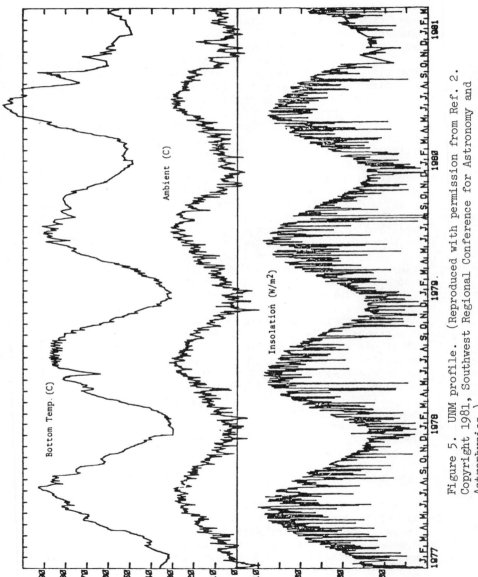

Figure 5. UNM profile. (Reproduced with permission from Ref. 2. Copyright 1981, Southwest Regional Conference for Astronomy and Astrophysics.)

NaCl solution for 30 days @ 212°F. Results confirm that
Hypalon shows much lower weight gains in brine than in
distilled water even at elevated temperatures (212°F).
(Figure 6) Performance of industrial grade Hypalon is
superior (less weight gain) to that of potable grade,
and industrial grade is recommended for salt gradient
ponds, as potability is not a consideration. (Figure 7)

Additional testing at varying salt brine concentrations indicates that any salt concentration above 1-1/2%
exhibits the low swell effects on Hypalon. Only the
surface layers of a salt gradient pond, which remain
cool, do not inhibit absorption. A low brine concentration at high temperature cannot occur in a salt gradient
pond. (Figure 8)

The "pickling" effects of high temperature brine
on Hypalon are not permanent or irreversible. Samples
showing low weight gain after 30 days @ 212°F in 20%
NaCl solutions, showed typical increases when transferred
to distilled water at 212°F. Also, the fact that the
Hypalon exposed to hot brine over long periods can still
be seamed or patched, indicates that curing of the
Hypalon does not take place. (Figure 9)

The Future for Flexible Membrane Liners in Salt Gradient Solar Ponds

The non-convective salt gradient solar pond will not
vaporize metals, flash steam or power your car or
watch. Its advantages are significant, however:
1. Simplicity - no moving parts, except for heat
    exchanger.
2. Collection plus storage - constant output
    possible.
3. Inexpensive materials - salt, water, lined
    hole-in-the-ground.
4. Provides heat for hot air, hot water, power.

Its drawbacks must be considered, as well:
1. Not portable.
2. Large dedicated area required - rural.
3. Location - near raw materials.
4. Pollution potential - in many areas.

The use of an impermeable flexible membrane liner
will allow salt gradient solar ponds to be located near
the energy user, by preventing salt contamination of
the soil and ground water.

The lessons learned in the pond, pit, lagoon
and reservoir field over the past 10-25 years must
be understood and applied. Design and construction

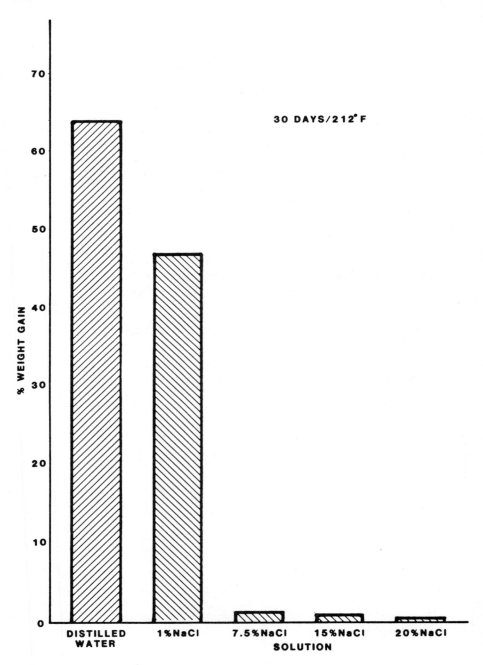

Figure 6.   Percent brine in industrial Hypalon.

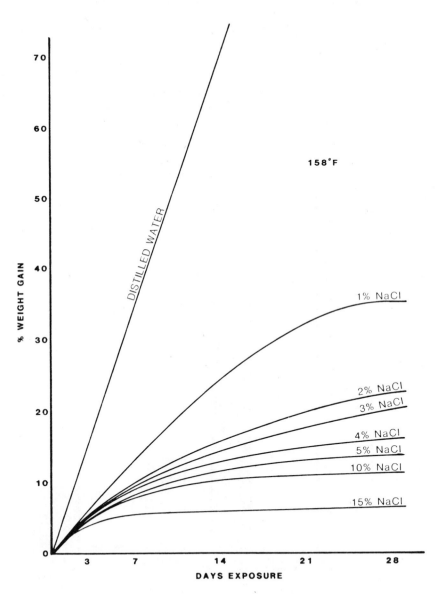

Figure 7.   Percent brine in potable Hypalon.

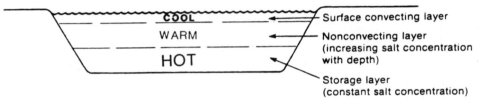

Figure 8.   Temperature gradient.   (Reproduced with permission from
Ref. 1.   Copyright 1980, Solar Energy Research Institute.)

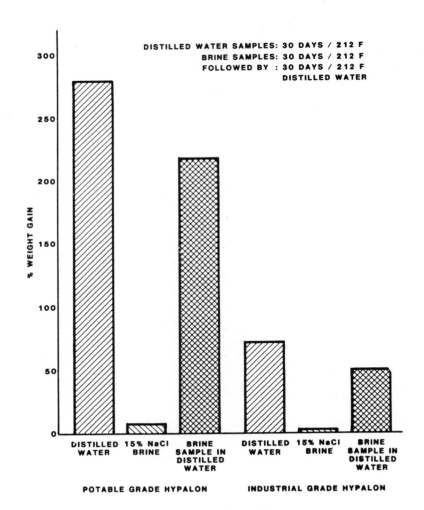

Figure 9.   Brine/water effects.

of the pond, including soils compaction, slope
stability, soil testing and other pre-installation
considerations are mandatory. It must be remembered
that the membrane lining is not designed as a structural member. Its impermeability is only effective
if the thin sheet is seamed correctly, installed
correctly and is not ripped, punctured or otherwise
overstressed during installation or operation.

Lining materials, such as Hypalon, have the
characteristics needed for long service life in the
proper setting.

The environmental risks may be reduced if current
research on the use of fertilizer solutions in place
of salt solutions for non-convective gradient type
solar ponds is successful. Even though the raw material
costs are higher, reduced environmental effects of a
major leak in a farming area offer a great incentive
for further development.

Testing is currently underway at The University
of New Mexico utilizing a solution of potassium
nitrate as the brine. Although more expensive than
sodium chloride, the solutions of potassium nitrate
tend to be self-stratifying, making it easier to
develop and maintain the gradient layer. Also, if
the solution was released into the ground through
an accidental rupture of the liner, no serious contamination of surrounding farm lands would occur.

Without reliable membrane liners, only the areas
of brackish water or brine will be acceptable for
salt gradient ponds.

In summary, salt gradient solar ponds offer
great promise as a means of collecting and storing
the sun's energy. Although not a high-temperature
collector, the salt gradient pond can furnish heat
and/or power, utilizing a Rankine Cycle engine with
a suitable low-boiling organic liquid. The solar
pond provides both the hot brine for the evaporation
and the cool surface water for the condenser.

The selection of a lining material is most
important. Leaks cause loss of brine, loss of heat,
loss of insulation, and are potential environmental
pollutants. The well documented history of the use
of chlorosulfonated polyethylene for long term
outdoor storage of saturated brine solutions makes
it an excellent candidate.

Literature Cited

1. Jayadev, T.S.; Edesess, M., "Solar Ponds",
   Solar Energy Research Institute SERI/TR-731-587,
   April, 1980.

2. Weeks, D.D.; Long, S.M., Emery, R.E. and
   Bryant, H.C., "What Happens When a Solar Pond
   Boils?" Southwest Regional Conference for
   Astonomy and Astrophysics.

3. Zangandro, F., "Observation and Analysis of a
   Full-Scale Experimental Salt Gradient Solar
   Pond", Doctorate Thesis, University of New
   Mexico, May, 1979.

4. Bryant, H. C., personal communication.

5. Power, Vol. 125, No. 6, McGraw-Hill, 1981.

RECEIVED November 22, 1982

Plastic Pipe Requirements for Ground-Coupled Heat Pumps

PHILIP D. METZ

Brookhaven National Laboratory, Solar and Renewables Division,
Upton, NY 11973

This paper describes the temperature, pressure, fluid compatibility and mechanical strength requirements which must be met by plastic pipe materials for various ground coupled heat pump applications. Existing suitable materials are cited as well as areas where improvements are required.

A ground coupled heat pump is a heat pump which uses the earth as a heat source or sink, often by circulating a fluid through serpentine buried pipes or vertical heat exchangers. Ground coupled heat pumps can operate with much higher efficiency than air source heat pumps because the ground is warmer in winter and cooler in summer than the ambient air, is insensitive to short-term weather fluctuations, and removes the need for a defrost cycle. Additionally, it is feasible to design ground coupled heat pump systems which require no auxiliary heating, even in very cold climates. The design of these systems has been discussed elsewhere.(1)

Ground coupled heat pumps can also be used to provide backup heat for solar heating systems, especially solar heat pump systems. The use of ground coupled heat storage tanks in solar assisted heat pump systems has been studied extensively.(2) While long-term inground heat storage is not possible for residential size solar systems, it may be feasible for larger systems in which the storage surface/volume ratio is smaller. This paper considers only stand-alone ground-coupled heat pumps which can be considered solar in the sense that the heat they draw was provided by the sun, and is available in winter due to the thermal properties of the earth.

0097–6156/83/0220–0211$06.00/0

Ground coupled heat pumps may now be feasible, largely due to the availability of durable and inexpensive plastic pipe. Pipe materials such as polyethylene, PVC, and polybutylene have been found to be adequate in some instances. However, no material standards or recommended design practices exist to facilitate the optimization of the performance, cost and reliability of these systems.

Temperature Requirements

The range of fluid temperatures anticipated for ground coupled heat pump systems is approximately -10 to $+50^{\circ}C$. The latent heat of freezing of the ground source keeps fluid temperatures from dropping much below the freezing temperature of the ground, even in very cold climates. It is not feasible to allow ground sink temperature to rise much above $40^{\circ}C$ as heat pump cooling efficiency decreases as sink temperature rises. Systems in which space cooling is not required will not encounter fluid temperatures above about $25^{\circ}C$.

Pressure Requirements

Liquid source heat pumps typically require a fluid flow rate of roughly 5.4×10^{-2} L/S per kW (3 GPM per ton) of heating capacity. In order to minimize circulation pump power, pipe (which ordinarily would be considered oversized for these flow rates) of 4 to 6 cm inner diameter is used. Earth coil lengths are typically 26 m per kW (300 ft per ton) of heating capacity. Thus, system operating pressures for a typical 11 kW (3 ton) heating load do not exceed $7-14 \times 10^{4}$ N/m^{2} (10-20 psi), lower than most existing schedules for plastic pipe. However, the system must also withstand water line pressure, i.e. 7×10^{5} N/m^{2} (100 psi).

Fluid Compatibility

In systems where subfreezing temperatures are not expected, water can be used as a heat transfer fluid (corrosion inhibitors may be necessary). All common existing plastic pipes are compatible with water. In systems where freezing is expected (roughly the northern half of the United States), antifreeze is required. Solutions which have been used include ethylene glycol – water, propylene glycol – water, and $CaCl_2$ – water. Organic antifreezes such as glycols are expensive and may enhance strees cracking in some pipe materials such as in the lower density polyethylenes, particularly at higher temperatures. The extent of this problem over long times is not known. Ethylene glycol is poisonous, and therefore a danger to underground water supplies. Inorganic antifreeze solutions, such as $CaCl_2$ – water, are corrosive and require inhibitors. Otherwise, because these solutions are ionic, galvanic corrosion between dissimilar metals not even in physical contact is possible.

Mechanical Strength Requirements

Plastic pipe for ground coupled heat pumps must withstand installation at depths of approximately 1-2 m for horizontal systems or in vertical shafts as deep as 100 m in vertical systems without extraordinary excavation precautions, as well as ground settling after installation, and thermal stresses during operation without failure.

Pipe Materials

Polyethylene, PVC, and more recently polybutylene have been used for pipes in ground coupled heat pump systems. The use of rigid pipe materials such as PVC has diminished due to the labor involved in connecting joints and the risk of leaks. Low density polyethylene pipe has been used widely in heating-only (i.e., low temperature fluid) systems in Sweden (3) with an ethylene glycol-water solution. This pipe is unsuitable for higher temperatures. Medium density (PE 2306) polyethylene pipe has been used in experiments in the United States at temperatures between -10 and 50°C using water and ethylene glycol-water solutions.(1) This pipe is pressure rated at about 25°C, and has no rating at higher temperatures. Some longitudinal cracks caused by mechanical breakage were found in pipe after heating to about 50°C. High density polyethylene and polybutylene have better pressure resistance and durability at elevated temperatures.

Conclusion

Suitable pipe materials for ground coupled heat pumps may already exist. However, no material standards, recommended design practices, installation procedures or safety precautions have been developed. The operating characteristics of these systems are now known well enough for these standards to be developed. This will facilitate the optimizations of the performance, cost and reliability of these systems.

Acknowledgments

Work performed under the auspices of the U.S. Department of Energy under Contract No. DE-AC02-76CH00016.

Literature Cited

1. Metz, P. D., Design, Operation and Performance of a Ground Coupled Heat Pump System in a Cold Climate, Proc. 1981 Intersoc. Energy Conv. Conf., ASME, Atlanta, August 9–14, 1981, BNL 29625.
2. Metz, P. D., The Use of Ground Coupled Tanks in Solar Assisted Heat Pump Systems, I. Comparison of Experimental and Computer Model Results, Proc. ASME Solar Energy Div. Fourth Ann. Tech. Conf., ASME, Albuquerque, April 1982, BNL 30913.
3. Mogensen, P., Experiences from Earth Heat Pumps in Sweden, Proc. Nordic Symposium on Earth Heat Pump Systems, Goteborg, Sweden, October 15–16, 1979, Chalmers Univ. of Technology.

RECEIVED November 22, 1982

POLYMER PHOTODEGRADATION IN
SOLAR APPLICATIONS

New Approach to the Prediction of Photooxidation of Plastics in Solar Applications

J. E. GUILLET, A. C. SOMERSALL, and J. W. GORDON

University of Toronto, Department of Chemistry, Toronto, M5S 1A1 Canada

A critical feature of the designs for solar arrays is that the components must maintain their integrity and function for an extended period of time in an outdoor environment. There is little experience on plastic cover and capsulant materials which could be used to predict the long-term stability and performance of these materials, and no scientific basis on which one could make reliable extended lifetime predictions. We have, therefore, developed a computer model which can generate realistic concentration versus time profiles of all chemical species formed by photooxidation using as input data a choice set of elementary reactions with corresponding rates and initial conditions. Attempts are being made to interpret some potential failure mechanisms in the solar applications in terms of predicted chemical changes such as scission, crosslinking and oxidation over time. Photooxidation of an ethylene-vinyl acetate copolymer was examined as a particular case.

Any attempt to develop a technology for producing low-cost, long-life photovoltaic modules and arrays must come to terms with the weathering effects experienced by the materials exposed to the sun's ultraviolet, oxygen, moisture and the stresses imposed by continuous thermal cycling, among other things. Polymeric substances which could find application as convenient protective covers, pottants/adhesives and backcovers undergo slow, complex photooxidation which changes the chemical and physical properties of these materials. Absorption of the ultraviolet in the tail of the solar spectrum can cause the breaking of chemical bonds, resulting in embrittlement and increased permeability, or to crosslinking which can produce shrinking and cracks. In addition, oxidation often leads to discoloration and reduced transparency, and to the formation of polar groups which could affect

0097–6156/83/0220–0217$06.00/0

electrical properties. These changes obviously limit the usefulness of
solar devices which, for economic reasons, must perform satisfactorily
for up to 20 years in rather severe environments.

Much work has already been done to unravel the complexities of
photooxidative processes and to develop some highly effective light (UV)
stabilizers which can delay the onset of polymer embrittlement. A good
review has been presented by Carlsson et al.(1). Some polymer systems have also been exploited to fabricate plastics with controlled lifetimes in the short range(2, 3) . However, very little is known about the
ultimate changes that occur in polymeric substances after very long
periods such as the 20-year regime appropriate for economic photovoltaic power plants.

There is no good way to predict the rates of the chemical and/or
physical changes which occur from accelerated tests. In part, the problem has been that there is no adequate laboratory method to measure
these effects over such extended times. Furthermore, accelerated
tests (in which materials are exposed to higher intensity UV sources in
controlled atmospheres) are of limited value in predicting rates since
there is often no reciprocity between intensity and time of exposure.

An alternative approach developed in our laboratories has been to
simulate the process of photooxidation with a computer model which could
be verified with experimental data from accelerated and outdoor exposures. Extrapolation of the numerical solutions for species concentrations over time should provide some understanding and control of the
degradation process over extended lifetimes.

Computer Model

The model consists of a set of reactions (Table I) for the basic reaction sequence based on the now well established mechanism of hydrocarbon oxidation. Rate parameters have been assigned to these fundamental equations, based on our best estimate from the literature(4)
(Table I). A problem with such a simulation study is that the predicted
rates will be only as reliable as the rate parameter data base employed.
We anticipate continuous refinement of this data base in further work.
In addition, a number of techniques has been developed for performing
the sensitivity analysis required to appreciate the relative importance
of particular reactions and rate data(5) .

Numerical methods for the solution of the resulting large set of
coupled differential equations have been developed. The method originally due to Gear(6) has been applied by Allara and Edelson for the
pyrolysis of alkanes(7) and later by others(8, 9) for similar processes.
Smog formation and detailed small-molecule photochemistry have also
been studied by these numerical simulations(10) . Semi-quantitative

Table I. Elementary Reactions in Polymer Photooxidation
and Corresponding Rates[a,b]

Reaction	Rate constant
RO_2 + RH $\longrightarrow$ ROOH + RO_2	0.1×10^{-2}
RO_2 + RO_2 $\longrightarrow$ ROH + Ketone + $^1O_2$	0.2×10^{8}
RO_2 + ROH $\longrightarrow$ ROOH + Ketone + HOO	0.5×10^{-1}
HOO + RH $\longrightarrow$ HOOH + RO_2	0.5×10^{-2}
HOO + RO_2 $\longrightarrow$ ROOH + $^1O_2$	0.1×10^{10}
RO_2 + Ketone $\longrightarrow$ ROOH + Peroxy CO	0.5×10^{-2}
RO_2 + ROOH $\longrightarrow$ ROOH + Ketone + OH	0.5×10^{-1}
RO_2 + SMROH $\longrightarrow$ ROOH + Aldehyde + HOO	0.5×10^{-2}
RO_2 + Aldehyde $\longrightarrow$ ROOH + SMRCO	0.1×10^{3}
OH + RH $\longrightarrow$ RO_2 + Water	0.3×10^{9}
Ketone $\xrightarrow{h\nu}$ KET*	0.3×10^{-5}
SMKetone $\xrightarrow{h\nu}$ KET*	0.3×10^{-5}
KET* $\longrightarrow$ $SMRO_2$ + SMRCO	0.5×10^{7}
SMRCO $\longrightarrow$ $SMRO_2$ + CO	0.5×10^{6}
KET* $\longrightarrow$ Alkene + SMKetone	0.5×10^{8}
KET* + O_2 $\longrightarrow$ Ketone + $^1O_2$	0.1×10^{10}
KET* + ROOH $\longrightarrow$ Ketone + RO + OH	0.1×10^{8}
KET* $\longrightarrow$ Ketone	0.1×10^{10}
$^1O_2$ $\longrightarrow$ O_2	0.6×10^{5}

Continued on next page

Table I (continued)

Reaction	Rate constant
$^1O_2$ + Alkene $\longrightarrow$ ROOH	0.1×10^4
$SMRO_2$ + RH $\longrightarrow$ SMROOH + RO_2	0.1×10^{-2}
SMROOH $\longrightarrow$ SMRO + OH	0.3×10^{-4}
SMRO + RH $\longrightarrow$ SMROH	0.1×10^6
SMRCO + O_2 $\longrightarrow$ SMRCOOO	0.4×10^{10}
SMRCOOO + RH $\longrightarrow$ SMRCOOOH + RO_2	0.1×10^{-1}
SMRCOOOH $\xrightarrow{h\nu}$ $SMRCO_2$ + OH	0.1×10^{-3}
ROOH $\xrightarrow{h\nu}$ RO + OH	0.3×10^{-4}
RO $\longrightarrow$ $SMRO_2$ + Aldehyde	0.1×10^6
RO + RH $\longrightarrow$ RO_2 + ROH	0.1×10^6
$SMRCO_2$ + RH $\longrightarrow$ Acid + RO_2	0.1×10^6
RO_2 + RO_2 $\longrightarrow$ ROOR	0.1×10^5

$\underline{a}$ $[O_2] = 10^{-3}$ M (constant); SMProduct = Product from chain cleavage.

$\underline{b}$ This set of reactions and rate constant values is by no means final. We have used it only to demonstrate the approach and to test the usefulness of the method by adjusting a few of the important variables. Much more work is required to identify the most appropriate input data block.

prediction has been possible in these cases. However, we report here
the first attempts to simulate photooxidation kinetics in the case of polymers. Our modified program calculates by stepwise integration in real
time the varying concentrations of chemical species formed during photooxidation. To validate our numerical procedure we employed the data
base given for the cesium flare system and generated curves identical to
those of Edelson(11) for the same system (Figure 1). The excellent
agreement between predicted and actual rate curves shows that the program itself (irrespective of the data base) performs in a satisfactory
manner.

Mechanism of Photooxidation

As a starting point for polymer photooxidation we looked at a formal linear low-molecular-weight alkane (RH). In principle, this should
be similar to amorphous high density polyethylene where short-range
diffusion rates in reaction centres should approach that of viscous liquids
In practice, many polymers will show only chemical changes in the hydro
carbon moiety since bond breakage will commonly take place initially in
the more labile C−H and C−C bonds. Initiation will take place following
UV absorption by ketone or hydroperoxide groups or even fortuitously by
some C−H bond cleavage. The possibilities of energy transfer among
different groups have also been included in the model. Propagation takes
place via the formation of peroxy radicals followed by hydrogen atom
abstraction from the backbone and repeated fast oxygen addition. Peroxy
radical chain carriers terminate by disproportionation to form alcohols
and ketones. Further photolysis of ketone products leads to another autocatalytic chain.

Initiation

$$RH \xrightarrow{\quad ? \quad} R\cdot \xrightarrow[\text{fast}]{+O_2} RO_2\cdot$$

$$\text{Ketone} \xrightarrow[\sim 10^{-5}]{h\nu} KET^* \xrightarrow[\substack{\phi \sim 0.02 \\ k = 10^7}]{\text{Norrish I}} SMR\cdot + SMRCO\cdot$$

$$ROOH \xrightarrow[\sim 10^{-4}]{h\nu} ROOH^* \xrightarrow[\phi = 1]{\text{very fast}} RO\cdot + OH\cdot$$

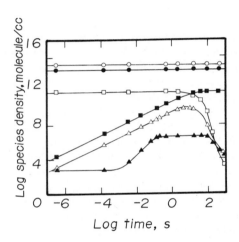

Figure 1. Our numerical solutions for concentration versus time profiles for the cesium flare (cf. ref. 10): (○) N_2, (●) O_2, (□) Cs,
(■) CsO_2, (△) Cs^+, e^-, (▲) O_2^-.

Propagation

Peroxide chain

$$\boxed{RO_2 \cdot} + RH \xrightarrow[k = 10^{-3}]{} ROOH + R \cdot \xrightarrow[\text{fast}]{+O_2} \boxed{RO_2 \cdot}$$

$$ROOH \longrightarrow RO \cdot + OH \cdot$$

$$\begin{matrix} RO \cdot \\ OH \cdot \end{matrix} + RH \xrightarrow[\substack{k \sim 10^5 \\ k \sim 10^9}]{} \begin{matrix} ROH \\ HOH \end{matrix} + R \cdot \xrightarrow[\text{fast}]{+O_2} \boxed{RO_2 \cdot}$$

Ketone chain

$$\text{Ketone} \xrightarrow{h\nu} KET^* \xrightarrow[\substack{\phi \sim 0.2 \\ k \sim 10^8}]{\text{Norrish II}} SMKetone + Alkene$$

Norrish I
$\phi \sim 0.02$

$$SMR \cdot + SMRCO \cdot \qquad\qquad KET^*$$

Termination

$$RO_2 \cdot + RO_2 \cdot \xrightarrow[k \sim 10^7]{} ROH + Ketone + {}^1O_2$$

$$RO_2 \cdot + RO_2 \cdot \xrightarrow[k \sim 10^4]{} ROOR + {}^1O_2$$

$$RO_2 \cdot + \begin{matrix} ROOH \\ ROH \\ Ketone \\ Aldehyde \\ etc. \end{matrix} \xrightarrow[k \sim 10^{-2} - 10^{-3}]{} ROOH + Other\ products$$

Preliminary Results

Early computer modelling results have shown that the processes of photooxidation under the conditions of our scheme would involve a long induction period, of up to several years in the pure hydrocarbon, followed by a fairly rapid deterioration (Figure 2). Initiation is effected

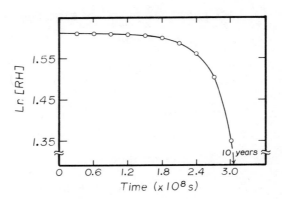

Figure 2. Photooxidation of linear alkane (RH) versus time.

fortuitously in the program by assigning a low rate constant for R−H cleavage or using low initial concentrations (ca. 10^{-5} M) for either ketone or hydroperoxide moieties. The principal products of photooxidation are ketones, alcohols, water and alkenes, with smaller quantities of aldehydes, acids, carbon monoxide, etc. (Table II).

The rates of formation of the major species show exponential behavior after the induction period is over (Figure 3). This conforms to the rate behavior for polyethylene observed experimentally in accelerated tests (Figure 4). An increase in the light intensity (reflected in a chosen systematic increase in all the photochemical absorption rates in the program) shows a systematic change in the exponential formation of ketone products (Figure 5) while the induction period is shortened (Figure 6). Increase in termination rate shortens the kinetic chain length and reduces the formation of product ketone. In this case, the exponential formation of ketone remains the same but is shifted in time to lengthen the induction period (Figure 7).

The autocatalytic process is initiated by a narrow concentration range of ketone initiator. At low levels ($<10^{-7}$ M) the induction period is excessively long for typical computer times and at high levels ($>10^{-3}$ M) the mutual termination of peroxy radicals is too fast to permit significant H-abstraction from the substrate.

These preliminary results are consistent with all of our experience to date on the photooxidation of ketone-containing polymers(12).

Conclusions

Much work remains to be done in refinement of the model to allow for the inclusion of substituent groups, the reactivity of secondary and tertiary C−H bonds, the significance of diffusion, the influences of temperature cycling and dark reactions, and the impact of additives. In conclusion, we believe that these modelling studies which can simulate real systems (given reliable input data), represent a novel approach to the general understanding of polymer photooxidation phenomena which should lead to a new understanding of the study of controlled lifetimes for polymers and for the development of procedures which would allow the prediction of performance of plastics for solar applications.

Table II. Final Concentration Array.
Time of Photooxidation, 10 Years

Species label	Initial conc., M	Final conc., M
RO_2	--	0.25×10^{-7}
RH	5.0	3.8
ROOH	--	0.77×10^{-4}
ROH	--	0.49
Ketone	10^{-5}	0.29
$^1O_2$	--	0.23×10^{-12}
HOO	--	0.25×10^{-10}
HOOH	--	0.21×10^{-4}
Peroxy CO	--	0.89×10^{-3}
OH	--	0.10×10^{-16}
SMROH	--	0.19
Aldehyde	--	0.19×10^{-3}
SMRCO	--	0.10×10^{-14}
H_2O	--	0.40
KET*	--	0.88×10^{-15}
SMKetone	--	0.14×10^{-1}
$SMRO_2$	--	0.14×10^{-5}
CO	--	0.19×10^{-1}
Alkene	--	0.16×10^1
ROOR	--	0.22×10^{-3}
RO	--	0.48×10^{-14}
SMROOH	--	0.18×10^{-3}
SMRO	--	0.14×10^{-13}
SMRCOOO	--	0.11×10^{-6}
SMRCOOOH	--	0.43×10^{-4}
$SMRCO_2$	--	0.11×10^{-13}
Acid	--	0.15

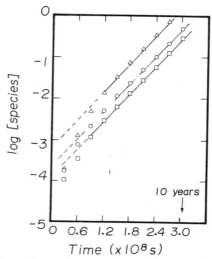

Figure 3. Variation of major product concentrations during photooxidation: ($\triangle$) alkene, ($\bigcirc$) ROH, ($\square$) ketone.

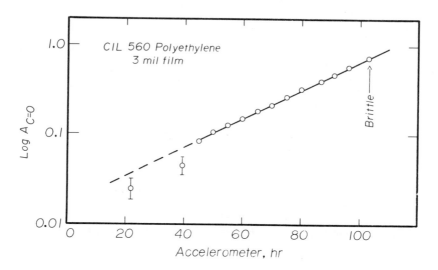

Figure 4. Photooxidation of polyethylene.

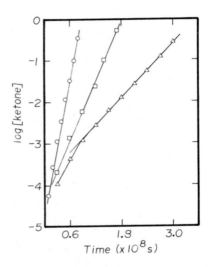

Figure 5. Effect of intensity on product formation during photooxidation: (O) N x 5, slope 5.5 x 10^{-8}, (□) N x 2, slope 2.4 x 10^{-8}, (Δ) N, slope 1.1 x 10^{-8}.

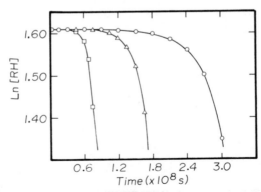

Figure 6. Photooxidation as a function of intensity of light: (□) N x 5, (Δ) N x 2, (O) N.

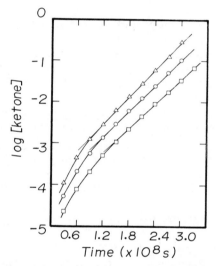

Figure 7. Effect of termination rate on product formation during photo-oxidation: (△) D, (○) D x 2, (□) D x 5.

Acknowledgments

This research is supported by a grant from the Jet Propulsion Laboratory, Pasadena, California, and is part of a Solar Energy Project, administered through the U.S. National Aeronautics and Space Administration.

Literature Cited

1.   Carlsson, D. J.; Garton, A.; Wiles, D. M. "Developments in Polymer Stabilisation", vol. 1, G. Scott, Ed., Applied Science, Barking, 1980.
2.   Guillet, J. E. "Polymers and Ecological Problems", J. Guillet, Ed., Plenum, New York, 1973.
3.   Scott, G. J. Polym. Sci., Polym. Symposia 1976, 57, 357.
4.   We acknowledge useful discussions on this subject with Dr. Keith Ingold of the National Research Council of Canada. Any omissions or errors are ours.
5.   Edelson, D.; Allara, D. L. Int. J. Chem. Kinet. 1980, 12, 605.
6.   Gear, C. W. Commun. ACM 1971, 14, 176.
7.   Allara, D. L.; Edelson, D. Int. J. Chem. Kinet. 1975, 7, 479.
8.   Sundaram, K. M.; Froment, G. F., Ind. Eng. Chem. Fundamen. 1978, 17(3), 174.
9.   Olson, D. B.; Tanzawa, T.; Gardiner, W. C. Jr. Int. J. Chem. Kinet. 1979, 11, 23.
10. Ebert, K. H.; Ederer, H. J.; Isbarn, G. Angew. Chem. 1980, 19, 333.
11. Edelson, D. J. Chem. Ed. 1975, 52, 642.
12. Guillet, J. E. Pure Appl. Chem. 1980, 52, 285.

RECEIVED February 18, 1983

Effects of Photodegradation on the Sorption and Transport of Water in Polymers

C. E. ROGERS

Case Western Reserve University, Department of Macromolecular Science,
Cleveland, OH 44106

The use of polymeric materials as protective coatings
for solar energy systems must be assessed in terms of
prospective changes in properties induced by long term
exposure under conditions favoring photooxidative degradation. Changes in water sorption due to progressive changes in polymer composition by degradation can
be estimated from knowledge of the dependence of water
uptake on the functional group composition of the polymer. The effects of cyclic variations in temperature
and humidity, combined with progressive photodegradation as a function of material thickness, can then be
estimated to determine the probability of serious dis-
ruptions in material continuity and related properties.
A model treatment has been developed to simulate the
phenomena for purposes of design lifetime performance
prediction.

The use of polymeric materials in applications which involve
exposure to solar radiation, oxygen, and varying temperatures is
common. Assessment of the stability of polymeric materials under
such conditions is a primary design factor for the prediction of
lifetime and performance. Despite considerable work by many
investigators, the feasibility and validity of preditions of long
term performance based on short term tests generally remains
questionable at best. The interdependence of material and
environmental factors, changing progressively with exposure induced degradation, confounds any simple analysis and predictive
model scheme. The situation requires an enhanced understanding
of the temporal and spatial variations in polymer composition,
structure, and morphology due to degradation coupled with an
understanding of the consequent changes in properties including
interactions with environmental agents such as water which affect
their sorption and transport behavior. These considerations are
especially germane for polymeric materials which are used as protective coatings or encapsulants for electrical or electronic

0097–6156/83/0220–0231$06.00/0

devices in which the material is subject to delamination, cracking, or other failure mode as a result of the combined action of
degradation and moisture attack.

In this present case, we will consider the nature of modes
of sorption and permeation of water in polymers in terms of degradation induced changes in polymer composition. These changes
can accentuate the effects of temperature-humidity cycling on
water sorption and diffusion to cause major changes in properties with progressive degradation.

Moisture Sorption

The equilibrium sorbed moisture content is judged to be a
primary parameter affecting the selection and design of protective materials. Those polymers with high sorbed water contents,
or those which can be anticipated to achieve high water contents
upon photooxidative degradation, would not be suitable candidate
materials. Such materials would have an adverse effect upon substrate (e.g., solar cell) operational capabilities and, in addition, would not be expected to maintain adequate adhesive bonding
to the substrate. Dimensional changes in the material due to
swelling would be a major design complication.

The magnitude of equilibrium water sorption in the initial
materials can be assessed by direct measurement and/or by the application of established theories of polymer solutions. The
change in sorption due to the formation of polar degradation products may be estimated by consideration of the stoichiometry between sorbed water and polar substituent groups established in
this study by reference to literature data (1,2) as confirmed by
experimental results. The data in Table I of moles of sorbed
water per mole of polar groups shows that there are two classes
of behavior. There is nearly a one-to-one relationship between
water uptake and the molar concentration of the hydrogen-bonding
donating groups: hydroxyl, carboxyl, peptide, (and free amide).
There is much less sorption on nitrile, carbonyl, ester, and
ether groups.

Knowledge of the course of formation of oxidation products
allows an estimation of the expected change in water sorption.
One example (3) is the formation of carbonyl and hydroxyl groups
during the accelerated weathering of polyethylene under UV exposure. A corresponding increase in water sorption occurs due to
the later formation of hydroxyl groups as degradation products.
The variations of sorption mode with time, concentration, temperature, and other experimental variables has been shown (4) to
lead to significant changes in polymer properties.

A calculation of the predicted change in equilibrium sorbed
water content due to changes in polymer structure caused by
photodegradation can be made on various levels of sophistication.
A reasonably rigorous approach would need to consider effects due
to chemical group changes (e.g., formation of carbonyl with corre-

Table I: Water-Polymer Functional Group Ratios

Functional Group	Moles Water/Mole Group
(Donating)	
Hydroxyl	0.93
Peptide	1.1
Carboxylic Acid	0.78
(Accepting)	
Nitrile	0.22
Ketone	0.20
Butyral	0.13
Alkyl Ester	0.15
Aryl Ester	0.15
Methacrylate Ester	0.11
Ether	0.06

sponding loss of the precursor reactant group), loss of volatile
products with concurrent decrease in residual sample mass, and
structural/morphological changes, especially void and crosslink
formation and perturbation of crystalline regions.

In this present case, we will neglect any effects on sorption due to gross changes in structural/morphological aspects of
the polymer. The occurrence of such changes due to photodegradation would be expected to seriously affect other polymer properties of consequence for the application such as adhesion and
impact strength. This, of itself, would suffice to eliminate the
material from consideration for the application.

The effect on sorption of progressive changes in chemical
group composition can be expressed quite simply by a weighted
solubility coefficient:

$$S = \phi_o S_o + \phi_1 S_1 + \phi_2 S_2 + \ldots + \phi_i S_i + \ldots$$

where ϕ_i is the mole fraction (or volume fraction) of the composition groups in the polymer (ϕ_o is the original polymer; ϕ_1, ϕ_2,
$\ldots$ are for degradation products) and S_i are the corresponding
characteristic solubilities.

Values of S_i may be estimated by consideration of solubility
parameter concepts coupled with polymer solution theories such as
the simple Flory-Huggins expression. A more direct approach
would utilize the experimental data given in Table I or corresponding experimental data for sorption in polymers with composition groups characteristic of the original polymer and the primary degradation products. A representative set of data (5),
given in Table II, can be used to estimate the change in water
sorption in polyethylene at constant temperature and relative
humidity as the concentration of hydroxyl groups increases with
progressive photodegradation, as shown in Figure 1.

Determination of the dependence of ϕ_i on exposure time would
then suffice for a prediction of the change in water sorption
with degradation. The effects of changes in temperature and in
relative humidity also can be predicted by incorporating the
functional dependence of solubility on those parameters, as discussed below, into the expression above. This method can be expected to hold with reasonable validity up to moderate extents of
degradation beyond which other changes in physical properties
would become overwhelming considerations.

Moisture Diffusion and Permeation

Prediction of moisture diffusion parameters in polymers
undergoing photooxidative degradation is considered to be of
somewhat lesser significance than sorption predictions for three
reasons. First, it seems probable that the occurrence of anomalous sorption-diffusion behavior may be expected in most polymer

Table II: Water Transport Parameters
25°C
60% R.H.

Polymer	$\underline{P} \times 10^9$	$D(c=o) \times 10^9$	$S = \underline{P}/D$
Polyethylene	9	230	0.039
Poly(vinyl) alcohol)	9.6	1.25	7.68

Units:

$\underline{P}$ $ccSTPcm^2sec^{-1}/cmHgcm^{-1}$

D cm^2sec^{-1}

S $ccSTP/cmHgcm^3$

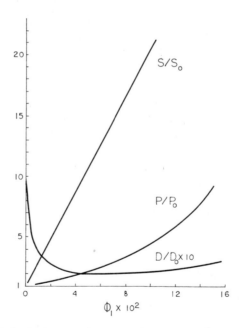

Figure 1.   Predicted change in relative water transport parameters with progressive conversion of $-CH_2CH_2-$ to $-CH_2CHOH-$ (volume fraction ϕ_1) due to degradation.

materials as they undergo progressive degradation (3). The
nature and magnitude of such effects, related to structural re-
laxations and/or specific sorption-immobilization modes, will depend upon the polymeric material and the exposure conditions.
Second, the probability of the formation of crazes, cracks, and
other gross defect structures to include voiding and delamination, will affect the overall moisture transport to underlying
substrates much more profoundly than bulk diffusion, per se.
Third, it can be anticipated that in the absence of appreciable
degradation causing either of the above two circumstances, the
long term exposure to the environment for any polymer of reasonable application thickness will result in static sorbed moisture
content at the polymer-substrate interphase within a time which
is short relative to the desired exposure lifetime. The value of
the nominal diffusion parameter is necessary for prediction of
the equilibrium sorption level under cyclic temperature-humidity
conditions, as discussed below.

An estimate of the change in water permeability upon degradation can be combined with the estimate of solubility, as given
above to obtain an estimate of the change in the diffusion coefficient. One method for the prediction of permeability constants
which has been shown to be reasonably successful is the Perma-
chor method as developed by Salame (6,7). For the case of water
permeation (7), the expression is

$$\underline{P} = (2.5 \times 10^{-7})\exp(-60.5\Pi/RT)$$

where $\underline{P}$ is the permeability constant in units of gms cm/sec cm^2
cmHg, $\overline{R}$ is the gas law constant, T is the temperature, and Π is
the polymer's "Permachour". The value of Π is calculated by adding individual segmental values for the atoms and groupings in
the polymer repeat unit in much the same way (and with some
correspondence in theoretical basis) as the calculation of solubility parameters (δ) based on group contributions. A Table of
values of Π for water permeability has been established (7).
This empirical treatment, when compared with the standard Arrhenius expression for temperature dependence of permeation, states
that the preexponential term is effectively a constant for water
in all polymers (2.5×10^{-7}). The "Permachour" correlates the
apparent activation energy for permeation as $E_P = 60.5\Pi$. This
tacit theoretical relationship is not completely satisfactory, it
can be circumvented or revised to derive more rigorous expressions (8), but the ease of application and established success of
the present form can justify its use in this present case.

An effective value of Π can be calculated for a degrading
polymer by a weighting expression similar to the one for the
solubility coefficient:

$$\Pi = \phi_0\Pi_0 + \phi_1\Pi_1 + \phi_2\Pi_2 + \ldots + \phi_i\Pi_i + \ldots$$

where ϕ_i are the mole fractions of compositional groups and Π_i are the characteristic Permachour values.

As an example, referring to the Table of Π-values (7) we note the following for the groups indicated:

$$-CH_2CH_2- \text{ (density 0.92)} \qquad\qquad \Pi = 101$$

$$-CH_2\underset{O}{\overset{}{C}}- \qquad\qquad \Pi = 100$$

$$-CH_2\underset{OH}{\overset{}{CH}}- \qquad\qquad \Pi = -43$$

The minor difference in Π between the polyethylene repeat unit and carbonyl group repeat unit and the major difference for the hydroxyl group repeat unit is consistent with both selective sorption behavior (Table I) and observed permeation behavior (e.g., Table II and all references). Thus, knowledge of the dependence of ϕ_1 for hydroxyl degradation product on exposure time should lead to prediction of permeation behavior.

For direct water contact at 23°C the expression for $\underline{P}$ reduces to:

$$\underline{P} = (2.5 \times 10^{-7}) \exp(-0.102\Pi)$$

For the case in example, when only one degradation product is of consequence in affecting permeation, the expression for the relative change in permeation is:

$$\underline{P}/\underline{P}_o = \exp[(0.102)(\phi_1)(\Pi_o - \Pi_1)]$$

$$= \exp[(0.102)(\phi_1)(144)]$$

The result of this calculation and the relative change in D from the relationship $D = \underline{P}/S$ is illustrated in Figure 1. We note a nonlinear increase in $\underline{P}$ with increasing hydroxyl content and a concurrent very pronounced decrease in D. This type of behavior is to be expected in consideration of the increase in cohesive energy density of the polymer, with increasing hydroxyl content, which acts to restrict polymer segmental chain motion thereby reducing the ease of diffusion of small molecules through the polymer matrix. The relative decrease in D is more than offset by large increase in solubility. The product of D and S leads to a moderate increase in $\underline{P}$ with increasing ϕ_1. These predictions are generally confirmed by experiment (3).

It must be realized that there is another factor that should be considered for a reasonable prediction, namely, the spatial dependence of degradation products through the polymer

sample thickness. Such a distribution can be anticipated as
the result both of light attenuation and oxygen diffusion control of the photooxidative degradation process. The desorption
rate and gradient of volatile degradation products also may be
a factor affecting water transport behavior (transient sorption
sites and plasticization effects). The effects of a gradient in
polymer composition on transport has been established (9–11) and
could be incorporated into the predictive scheme given here.
The simplest procedure is to consider the sample as effectively
being two films in series (3,12) with the top film of thickness
ℓ_1 undergoing degradation and the bottom film of thickness ℓ_2
remaining undegraded; $\ell_1 + \ell_2 = \ell$ = the total film thickness.
The net permeation, $\underline{P}'$, is then given by the standard relationship:

$$\frac{\ell}{\underline{P}'} = \frac{\ell_1}{\underline{P}} + \frac{\ell_2}{\underline{P}_o}$$

with $\underline{P}$ and $\underline{P}_o$ as the permeabilities of the two film layers. In
absence of transport data on films of measured ϕ_1 and ℓ_1, we do
not make a direct comparison of prediction and experimental
data.

Effects of Temperature–Humidity Cycling

A very important consideration in the selection and design
of protective materials is the magnitude of sorbed moisture content as a function of encapsulant thickness. This moisture content is expected to vary as a function of distance into the
coating as the coating is exposed to cyclic temperature and humidity conditions anticipated in exposure environments. The consequent changes in anisotropic swelling as a function of distance and time could lead to the onset or enhancement of various
modes of unfavorable deformation or other physical modifications
in the encapsulant material.

A prediction of the course of sorption–desorption phenomena
in encapsulant materials related to cyclic variations in the ambient environment during photooxidative degradation must consider a number of contributing factors and the interplay between
those factors. The "normal" dependence of diffusion and sorption parameters can be expressed by the following simple relationships:

Temperature dependence

$$D = D_o \exp(-E_D/RT)$$

$$S = S_o \exp(-\Delta H_s/RT)$$

Concentration dependence

$$D = D(o)\exp(Ac) \approx D(o)(1 + Ac)$$

$$S = S(o)\exp(Bp) \approx S(o)(1 + Bp)$$

Degradation dependence

$$D = D(t=o)[1 - F \text{ (exposure time)}]$$

$$S = S(t=o)[1 + F' \text{ (exposure time)}]$$

The estimation also should consider the effects of:

Radiation attentuation changes due to density, color, and other changes.

Leaching of reaction products by liquid water contact.

Loss of leachable and volatile products leading to material shrinkage.

Crosslinking (and other changes in polymer molecular weight and its distribution).

Swelling-induced polymer chain orientation normal to the coating surface which decreases diffusion rates and modifies sorption equilibrium.

Cooperative degradation reactions involving sorbed water and/or swelling induced stresses and/or deformations.

The restraints on isotropic swelling due to coating adhesion to the substrate.

The number of factors to be considered for a quantitative prediction of the effects of temperature-humidity cycling requires a mathematical model with sufficient parameters to describe most of the above variables. This necessitates the use of computer analog or other computative methods. For the present case, we will only describe a qualitative model representation based on knowledge of the effects of general "normal" dependencies of sorption and diffusion parameters on the major anticipated factors. A more complete analysis will be considered pending the accumulation of knowledge regarding the dependence of other parameters and factors in selected degrading material systems. A basis for judgment as to the dependence of sorption and transport on the course of degradation are derived from the data and trands observed in this study, from the data of Kimball

and Munir (13), and from other, less specific experimental and
literature sources.

The simple predictive method assumes the solution of equations for sorbed concentration as a function of thickness and
time, c(x,t), as given in standard monographs on diffusion (14).
For a model prediction it is necessary to choose a representative set of temperature and humidity cycles. The choice made
for present purposes is a uniform sinusoidal temperature cycle
and the condition of constant water content. This leads to a
temperature-related variation in relative humidity in which the
relative humidity increases with decreasing temperature and vice
versa.

The qualitative prediction then considers the effects of
passing from low to medium to high to medium to low to medium
(etc.) temperatures to establish a quasi-equilibrium sorbed content at the coating-substrate interface with a fluctuating sorbed concentration in the surface regions. This results in the
sorbed concentration undergoing sorption and desorption in surface regions with a gradual buildup in the interior regions of
the coating to achieve eventual equilibrium in terms of a "kinetic" temperature-humidity averaged value.

The range of surface sorption and desorption and the change
in interface sorption to be expected changes when the sorption
and diffusion parameters are constants (D(o), S(o)), when they
exhibit concentration-dependence (5)(D(c), S(p)), when the diffusion parameter decreases and the sorption parameter increases
as a result of degradation (D(o), S(o)), degrade), and the effect of degradation with a concentration-dependence of sorption
and diffusion (D(c), S(p), degrade).

The latter case, a decrease in D with increasing concentration-dependence coupled with an increase in S magnitude and concentration-dependence as a function of progressive degradative
is deemed to be a very probable expectation for many encapsulant
materials. This can result in the formation of a surface layer
which undergoes wide variations in swelling under cyclic ambient
conditions. The result of such variations in anisotropic swelling can easily enhance various failure modes. This is considered to be a significant factor for selection, design, and prediction of encapsulant material performance.

Typical values useful for predictive purposes are given in
the literature for water diffusion, its apparent activation
energy, and sorbed concentration as a function of relative vapor
pressure. The data in Table III include values calculated for
diffusion parameters at 5°C and 30°C as representative values
for nominal temperature extremes for an environmental exposure
cycle. It is seen that values may easily vary from polymer to
polymer by four or more orders of magnitude.

The values of (Dt/ℓ^2) for 30°C and 5°C are those calculated
for a coating 1 cm thick exposed to an average cycle time of 12
hours. Comparison of those values with standard plots of relative uptake ratios (14) shows that polysilicone easily achieves

Table III: Calculated Diffusion Parameters for Water
             in Polymer Coatings

Polymer	Silicone Rubber	Polyethylene	Nylon
E_D(kcal)	3	14	6.5
D(30°C)	7.0×10^{-5}	2.3×10^{-7}	1×10^{-9}
D(5°C)	4.5×10^{-5}	4.6×10^{-8}	–
$(Dt/\ell^2)_{30}$	3.0	0.01	4×10^{-5}
$(Dt/\ell^2)_5$	1.9	2×10^{-3}	–
$\ell(Dt/\ell^2 = 0.1)_{30}$	5.5 cm	0.3 cm	0.02 cm

sorption and desorption equilibrium throughout the sample.
Polyethylene sorbs to about 10% of equilibrium and then desorbs
to about 5% of nominal equilibrium. Sorption in Nylon is re-
strained to surface layers.

A calculation can be made of the film thickness required
to achieve an advance of the sorbed moisture only to the sub-
strate-coating interface (in one initial sorption cycle process)
which corresponds to the condition that $Dt/\ell^2 = 0.1$ at 30°C for
D values as given in Table III and for t equal to 12 hours.
That condition would require a silicone rubber film of 5.5 cm
thickness, and a Nylon film of only 0.021 cm thickness. These
values reflect the dependence of sorption rate on a concentra-
tion-independent diffusion coefficient.

As stated previously, more refined predictions require
both a more extensive data base and a more sophisticated (computer) mathematical model. Such a procedure is considered within the range of feasibility.

Acknowledgments

Support of this research program by a contract from the
Jet Propulsion Laboratory as part of the Low-Cost Solar Array
Project (Flat-Plate Solar Array Project) is gratefully acknowledged. We thank Drs. A. Gupta and J. Moacanin for their discussions and cooperation.

Literature Cited

1. J. Crank and G. S. Park, eds., Diffusion in Polymers, Aca-
    demic Press, New York, 1968.
2. A. D. McLaren and J. W. Rowan, J. Polym. Sci., 7, 289(1951).
3. A. Dudek and C. E. Rogers, to be published.
4. A. Sfirakis and C. E. Rogers, Polym. Engr. Sci., 20, 294
    (1980).
5. C. E. Rogers and D. Machin, "The Concentration Dependence of
    Diffusion Coefficients in Polymer-Penetrant Systems", CRC
    Critical Rev. in Macromol. Sci., April 1972, pp. 245-313.
6. M. Salame and J. Pinsky, Mod. Pack., 36, 153 (1962).
7. M. Salame, 67th AlChE Meeting, Feb. 1970.
8. C. E. Rogers, "Prediction of Polymer Permeability", CSL Con-
    ference on Chemical Defense Research, November 1981, to be
    published.
9. S. Sternberg and C. E. Rogers, J. Appl. Polym. Sci., 12,
    1017 (1968).
10. A. Peterlin, J. Appl. Polym. Sci., 15, 3127 (1971).
11. J. H. Petropoulos, J. Polym. Sci. Phys. Ed., 12, 35 (1974).
12. D. Benachour and C. E. Rogers, ACS Symp. Ser., 151, 263
    (1981).
13. W. H. Kimball and Z. A. Munir, Polym. Engr. Sci., 18, 230
    (1978).
14. J. Crank, The Mathematics of Diffusion, 2nd Ed., Clarendon
    Press, Oxford, 1975.

RECEIVED December 27, 1982

UV Microscopy of Morphology and Oxidation in Polymers

P. D. CALVERT, N. C. BILLINGHAM, J. B. KNIGHT, and A. UZUNER

University of Sussex, School of Chemistry and Molecular Sciences, Brighton, England

Ultraviolet and fluorescent microscopy has been applied to a variety of polymer systems to investigate changes of morphology and composition on the scale of 0.25 μm upwards. Studies are briefly described on the behaviour of stabilisers in polypropylene, diffusion of additives in polymers, spherulite morphology, polyolefin oxidation, inhomogeneities in epoxy resins and polymer blends.

The ultraviolet microscope, operating in the region 230 nm to 280 nm, was developed by Kohler (1) with the intention of taking advantage of the increased resolving power theoretically associated with shorter wavelengths. The increased resolution actually yielded little new information but the microscope did show unexpected contrast effects in biological samples which were completely transparent in visible light. It was later shown that these effects are due to the strong absorption of uv by nucleic acids and this observation quickly led to extensive use of uv microscopy to study the distribution of nucleic acids within cells. Techniques were developed for making quantitative and spectroscopic analyses of the species present and for working with living cells by allowing only brief exposures to the damaging uv light.

The development of the electron microscope has meant that there is little advantage in using uv light to obtain increased resolving power, as compared to the enormous increase allowed by electron illumination. Rather, most uses have been to make qualitative or quantitative concentration observations on systems where one component is strongly uv absorbing. In principle, similar measurements could be made with a wide range of coloured substances using a normal visible light microscope. However, in all forms of light microscopy the depth of focus is limited, particularly as the magnification is increased. The result is that very thin samples are required for successful light microscopy so that only absorbing species with high extinction coefficients will yield acceptable contrast. The main advantages of uv illumination are thus the greater range of absorbing compounds with a high extinction coefficient and the number of uv absorbing substances which are of interest in their own right. A further advantage of the uv microscope is that it can be used to excite and observe fluorescing compounds. These can offer greater sensitivity as fluorescence is observed against a dark background, but the range of suitable substances is more limited and quantitative analysis presents more problems.

Most commercially important synthetic polymers have no strong uv absorption in the easily accessible range from 250 nm to 400 nm. Hence useful application of the uv microscope will depend on there being added uv absorbing molecules or attached side groups whose concentration varies within the polymer. Since the only systems which obviously fall into this category are polymers containing uv stabilisers this has, until recently, been the only application in polymer science. It is our belief that the potential range of applications is very much wider than this, in that uv absorbers and fluorescers can be selectively bound to specific chemical entities in the polymer or will preferentially interact with or dissolve in parts of the structure. In this way these molecules can be used as stains and probes of the morphology of the polymer on the scale from 0.25 μm upwards, in a manner very similar to that in which the biologist uses stains to develop contrast in tissue specimens. Further, in so far as uv absorbers resemble other small molecules of interest, such as drugs and pesticides, they can be used to study the transport of such molecules in polymers.

Outside of polymers uv microscopy has principally been used by biologists to study the distribution of nucleic acids in whole cells. This work has been reviewed by Freed (2). The technique has also been used to measure lignin concentrations in wood (3) and plants and to study coal, oil and peat. Kam and co-workers were able to measure lysozyme concentrations in solution around growing crystals of this enzyme and so to determine the importance of diffusion in limiting the crystal growth (4). Sodium and potassium are slightly transparent in the uv and uv microscopy has been used to observe segregation during solidification of Na-K alloys (5).

Each of these applications is specific to a particular problem so that uv microscopy should be seen as a technique with some specialised applications rather than being generally useful. Also each new application requires the development of new methods of sample preparation and this can be time consuming. In this paper we describe the apparatus and review those areas of polymer science where we have found that this method gives valuable new information. We have recently completed a detailed review of uv microscopy (6) and therefore concentrate on recent advances.

Apparatus

Figure 1 shows, schematically, the equipment which we use for uv microscopy of polymers. It comprises a normal Zeiss Universal microscope fitted with uv-transparent quartz optics, front-surface mirror beam switches, a 150w Xenon arc lamp and a uv-sensitive TV viewing system. A waveform monitor displays the intensity distribution on any single line of the TV image so allowing determination of concentrations of uv absorbing species. Wavelength selection is by interference, liquid or colour glass filters. A 10 cm cell filled with distilled water acts as a heat filter. Images can be directly recorded on normal photographic film and analysed by microdensitometry.

Concentration measurements may be made after calibration with samples of known thickness and uniform absorber concentration. Other sources of contrast such as diffraction effects can be detected by comparing uv and visible light images. Quantitative measurements can also be made in fluorescence though care must be taken to avoid saturation which occurs at high concentrations where most of the incoming uv is absorbed.

At some cost to the uniformity of the field of illumination it is possible to use the microscope in conjunction with a hot stage (Mettler FP2) and directly record concentration changes during crystal growth and melting.

Additives in Crystalline Polymers

Commercial crystalline polymers contain a variety of impurity species and additives, most of which are excluded from the crystalline regions as the spherulites grow. Light stabilisers such as substituted benzophenones and benzotriazoles are frequently added to polyolefins in concentrations of 0.1 to 0.5%. Since these absorb strongly around 320 nm while the polymer is transparent down to 200 nm this system is ideal for uv microscopy. Curson (7) and Frank and Lehner (8) have looked at polypropylene containing uv absorbers and showed that the additive was concentrated close to the spherulite boundaries.

Ryan studied the rejection process during crystallisation (9). A wave of rejected additive builds up ahead of the growing spherulite as shown in Figures 2 and 3. The shape of this wave depends on the diffusion coefficient of the additive and the growth rate of the spherulite and has been compared with computer simulations of the rejection.

Additive concentration gradients are induced by the crystallisation process but would be expected to relax on subsequent annealing. In fact the gradients do relax to some extent but a stable gradient of uv absorption remains with the spherulite centre less absorbing than the boundary (10). This cannot represent a concentration change of the additive within the amorphous regions of the polymer so it must be due to a variation in the

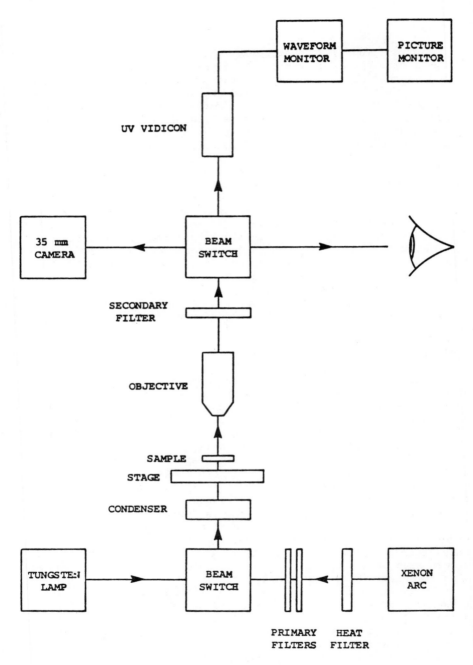

Figure 1. Block diagram of the UV microscope.

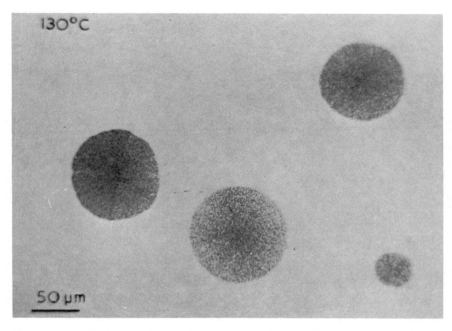

Figure 2. Polypropylene film containing 0.1% Uvitex OB, viewed in fluorescence during crystallization at 130 $^{\circ}$C.

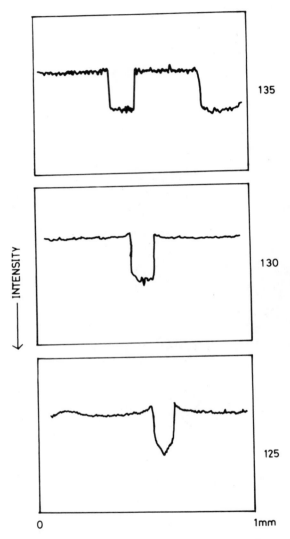

Figure 3. Additive distributions for polypropylene spherulites containing 0.5% Uvitex OB crystallizing at 125, 130, and 135 $^{\circ}$C. Waveform monitor trace of UV image.

amorphous content of the spherulite. Thus the crystallinity of a spherulite must decrease from the centre to the boundary and can be measured along a radius in annealed samples by uv microscopy. This principle can be widened to say that uv absorbing stains can be used to probe for local density variations within polymers provided that these density variations lead to solubility changes for the stain in the polymer.

Atactic Impurities in Crystalline Polymer

The most likely cause of the crystallinity variations in polypropylene spherulites is the accumulation of rejected atactic and low molecular weight impurities. This view is supported by the observation that adding increasing amounts of atactic material to polypropylene purified by octane extraction, leads to changes in the crystallinity distribution (Figure 4).

Atactic polypropylene is not uv absorbing and so cannot be seen directly. We therefore prepared some stained material by reacting atactic polymer with N,N-dimethylamino sulphonyl azide (dansyl azide) which covalently bonds the fluorescent dansyl groups to the polymer at levels of about 1 per hundred monomer units. By incorporating this material into isotactic polypropylene we can follow the rejection of the atactic material using the uv microscope. We can reverse the relationship and stain the isotactic material in which case we see bright fluorescing spherulite centres with darker atactic-rich boundaries as shown in Figure 5.

We are now analysing the rejection behaviour of the atactic material along similar lines to those used for the uv absorbing additives to determine the mobility of the atactic fractions within the polymer.

Diffusion and Loss of Additives

The useful lifetime of a polyolefin corresponds essentially to the end of the oxidation induction time. In stabilised materials this may be determined either by the time at which the stabiliser has been chemically consumed or that at which it has been lost from the sample by evaporation or extraction. We have developed a model to describe these additive loss processes but it requires a knowledge of the diffusion rates and solubilities of additives in polymer (11, 12).

The most common method of measuring diffusion in polymers uses radiolabelled additive. Typically a thin film of the labelled additive is put on the surface of the polymer and the increase in activity at the opposite surface is monitored as the additive diffuses into the polymer. We have used uv microscopy to follow the diffusion of additives and of atactic polypropylene into polypropylene. Samples are exposed to a saturated solution of the additive in a non-swelling solvent such as glycerol then

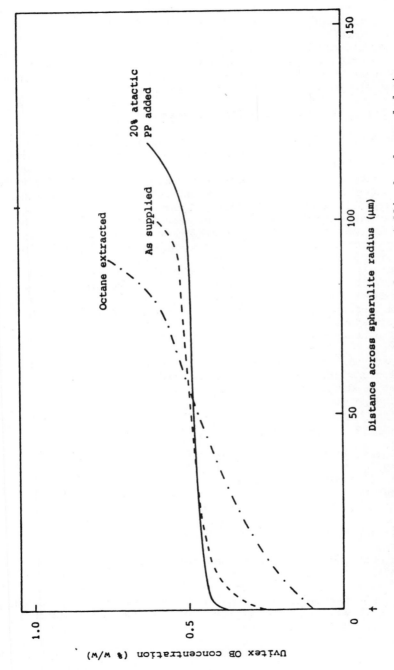

Figure 4. Distribution of Uvitex OB in samples crystallized and annealed at 125 °C. Observed at room temperature.

a

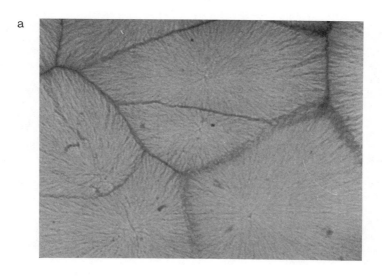

b

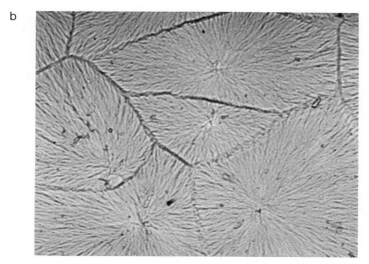

Figure 5. Fluorescently labelled polypropylene containing 40% by
weight of unlabelled atactic polymer, crystallized and annealed at
140 °C. (a) Fluorescence mode; and (b) visible transmission. Dark
boundaries in fluorescence picture are atactic rich.

sectioned and the additive profile within the polymer is determined from the uv monitor display. After calibration the diffusion rate and the saturation solubility can be directly measured.
This has the advantage over other methods that it is far faster
since the diffusion distance need only be about 50 μm. Thus under
equivalent conditions a diffusion coefficient can be determined
in 2 hours which took 2 days by the tracer method.

The microscope requirements are that the illumination must
either be narrow in wavelength or correspond roughly with the
shape of the absorption maximum of the additive as shown in
Figure 6. Otherwise the Lambert-Beer law may not hold. Figure
7 shows a section of a film which has taken up additive and the
corresponding waveform monitor trace. Figure 8 shows the fitted
and observed diffusion profiles for such a sample. Table I gives
data for the diffusion of octoxybenzophenone in polypropylene
compared with the data of other workers.

Localization of Thermal and Photo-oxidation

Most studies of oxidation in polymers implicitly assume that
the process is homogeneous and little or no reference is made to
possible non-uniformities. However in solid polymers there are a
number of reasons why oxidation may be uneven and the brittle
cracking, which is the main undesirable consequence of degradation, is localised by its very nature. It is known that oxidation
is localised to sample surfaces in conditions where the kinetics
are so fast that oxygen diffusion is limiting (12) or where the
polymer is in contact with a catalyst such as copper (13). After
eliminating effects of this sort there is still reason to think
that oxidation may be localised. It is well established that
impurities in polymers can catalyse oxidation and rejection processes of the kind described here are expected to lead to concentration of these catalytic impurities in spherulite boundaries,
where they may cause localized degradation.

As followed by oxygen uptake or development of carbonyl
groups, oxidation in polyolefins is an autoaccelerative process,
due to the ability of the hydroperoxides initially produced to
decompose with the generation of new free radicals, a decomposition which is catalysed by both heat and solar uv. The fact that
it is autoaccelerating means that it may also become localised.
Whether this happens will depend on the branching rate of the
chain reaction and the diffusion rate of the oxidation products
within the polymer. If one area of a sample becomes more oxidized
than the surrounding regions, oxidation can then proceed more
rapidly there and the oxidation products diffusing outwards will
spread the "disease" to the surroundings.

There are a number of pieces of evidence which would support
the idea that oxidation is localised. Thermally oxidising samples
often do show brown spots of degradation which grow with time. If
samples are oxidised to embrittlement the resultant molecular

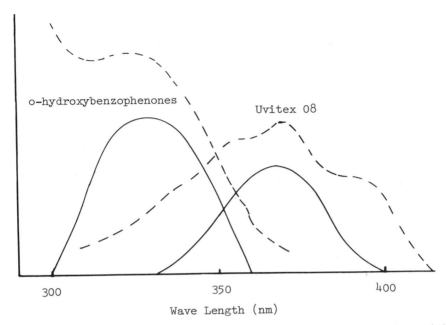

Figure 6. Absorption spectra of diffusing additives compared with illuminating spectra. (----), Absorption spectrum; and (——), illumination spectrum.

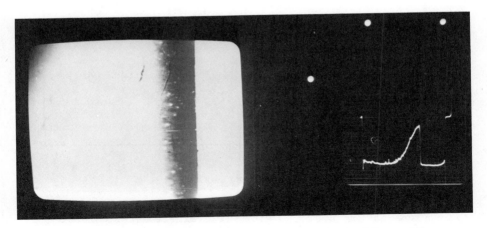

Figure 7. UV image of absorber diffusing into a film, with waveform monitor trace of intensity.

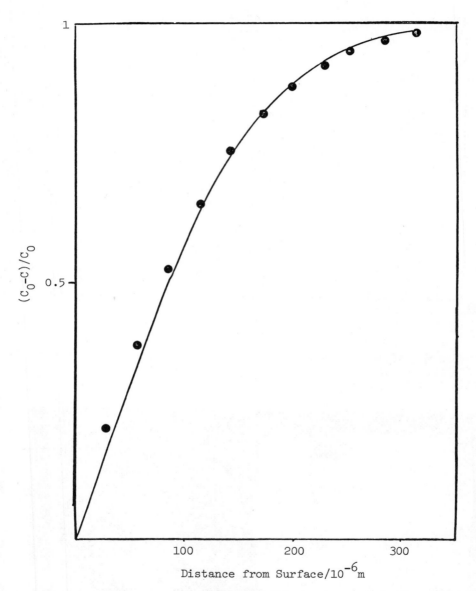

Figure 8. Concentration distribution for Uvitex OB diffused into polypropylene at 125 $^\circ$C for 20 min. ($\bullet$), Experimental points; and (———), fitted curve for D = 6.5 μm s^{-1}.

TABLE I

Diffusion Coefficients of 2-Hydroxy-4-Octoxybenzophenone (UV 531) in Isotactic Polypropylene			
Temperature °C	D, cm^2 sec^{-1}		
	Our Data	Data of Johnson and Westlake[a]	Cicchetti, et al[b]
30	1.76×10^{-11}		
40	4.48×10^{-11}		
44	1.29×10^{-10}	1.7×10^{-10}	
50	3.88×10^{-10}	3.39×10^{-10}	
60	1.23×10^{-9}	1.02×10^{-9}	
75	6.47×10^{-9}	4.7×10^{-9}	8.63×10^{-10}
90	1.98×10^{-8}		2.99×10^{-9}
Activation Energy (E_d) kcal/mole	25.09	23.6	20.8

[a] Measured at 44-75°C, J. Appl. Poly. Sci. (1975) 19, 1745.
[b] Measured at 80-110°C, Eur. Poly. J. (1967) 3, 473.

weight decrease is not sufficient to explain either the brittle-
ness or the fact that remoulding results in the regain of much of
the toughness of the polymer (14).

With this in mind we applied uv microscopy to search for
localisation of oxidation in polypropylene.

Methods

The carbonyl absorption at around 280 nm can be used to see
oxidation in polypropylene directly but this is only possible in
heavily oxidised samples which are also too brittle to be sectioned for microscopy. It is possible in this way to look only at
samples which have been directly crystallised from the melt as
thin films. In order to work with samples with more normal oxidation levels we must stain the oxygen containing groups to make
them uv absorbing. In principle it should be possible to separately stain carbonyl or peroxide groups but we have concentrated
on 2,4-dinitrophenylhydrazine (DNPH) as a stain for carbonyl
groups. Dansyl hydrazine was also tried as a fluorescent stain
but was limited in its ability to penetrate the sample so that
only the surface was stained.

The DNPH staining was done by immersing 10 µm sections in 1%
DNPH in isopropanol-5% conc HCl at 60°C for 24 hours (15). The
section was then extracted with fresh isopropanol at 60°C for 24
hours to remove excess DNPH. Ir and uv spectroscopy of this
staining reaction in 100 µm films showed that there was no further
hydrazone formation after this time with about 66% of the carbonyl
groups having reacted with DNPH. The remainder are probably
esters and acids. During the reaction the hydroperoxides appear
to be converted to carbonyl groups. Quite a number of other
species are present and side reactions do occur. However the most
important limitation of this method is that atactic and low molecular weight species are extracted from the film by the isopropanol. Up to 50% of the carbonyl content is removed from heavily
oxidised polypropylene by prolonged extraction with isopropanol
at 60°C. Methanol is a better solvent in this respect but is less
satisfactory for the staining reaction due to more rapid acetal or
ketal formation. No staining could be seen from aqueous acid
solutions of DNPH. The net effect of the staining process is that
the extinction coefficient at 352 nm of stained oxidised polypropylene is 20-fold greater than that at 280 nm for the unstained
material, a 20-fold sensitivity gain. Oxidation can be observed
at about 1/4 of the induction time in unstabilised material. In
our samples the induction times for thermal oxidation were about 4
hours at 120°C or 12 hours at 100°C.

Observations

From our studies of additive rejection we believed that similar rejection of impurities, including oxidation products, during

crystallisation would lead to high subsequent oxidation rates at spherulite boundaries. Staining of samples which were crystallised and then thermally oxidised revealed localised centres of oxidation but ones which bore no relationship to the spherulite morphology (Figure 9). Within no spherulite was there any gradient in UV absorption between the center and the boundary. If these oxidized sections were flexed the cracks ran apparently at random. Similar samples which were crystallised and oxidised as thin films tended to crack along spherulite radii. If the sample was first oxidised to about the induction time and then crystallised there was again no evident relationship between oxidation and morphology. However the staining process had clearly extracted material from the spherulite boundaries leaving them as channels in the film. If the unstained sections were flexed they frequently cracked along the spherulite boundaries.

We conclude from this that crystallisation of pre-oxidised polymer does lead to segregation of partly oxidised, isopropanol soluble material to the spherulite boundaries which weakens them. At low levels of oxidation where the isopropanol extractable material is only a small fraction of the total carbonyl content, this segregation does not lead to increased oxidation rates locally. At higher oxidation levels much of the oxidised material is extracted so that we cannot follow the process.

Billingham and Manke (16) have investigated the effects of oxidation on the isotactic and atactic components of polypropylene by extraction, GPC and staining. They found that at low levels of oxidation the 2-3% of polymer which was extracted by boiling heptane was responsible for 30% of the oxidation as measured by carbonyl content. This difference would not be expected on the basis of the stereochemical difference between atactic and isotactic polymer or on the basis of the ready access of oxygen to the amorphous regions containing the atactic material. It is most likely to stem from a greater density of chain defects such as unsaturation in the atactic polymer.

Although there is no strong morphological effect on oxidation kinetics there is a high degree of localisation of oxidation as seen in Figure 9. In sections across compression molded films this is seen as a "wood-grain" parallel to the surface. By studying films made by compression moulding under very low pressures it can be seen that this effect arises from variations between the original granules of powdered polymer, figure (10), such that at low oxidation levels a few granules are heavily oxidised while the rest are unaffected. Those particles which were heavily oxidised also contained clusters of sub-microscopic dots observable in visible light microscopy. Similar clusters of dots could also be seen in some individual powder particles when molten at 170°C, suggesting that the dots are infusible particles rather than voids or bubbles. Up to 10% of particles of unprocessed commercial diluent-phase polypropylene contained these dots. Scanning electron microscopy with EDAX, carried out by Mr A.Cobbold of ICI

Figure 9. Section of polypropylene viewed at 350 nm after oxidation
and staining with DNPH. Bar is 50 μm.

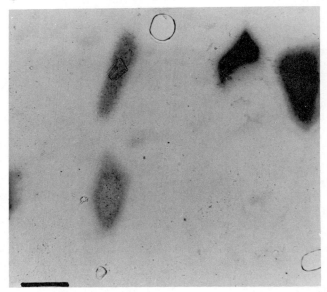

Figure 10. Section of lightly pressed polypropylene partially
oxidized and stained with DNPH. Viewed at 350 nm. Bar is 100 μm.

identified these dots as particles of less than 0.5 μm containing
Ti, Al and Cl. Thus they are almost certainly catalyst residues.
     Different diluent-phase polypropylenes showed similar
behaviour but the proportion of affected powder particles varies.
Clearly these local variations would tend to disappear in the
processing of the polymer to produce stabilised pellets. Variations in oxidation rates of individual particles were also found
with gas-phase polymerized propylenes (Figure 11), in this case
the catalyst residues could not be seen.
     Figure 12 shows a stained section of an oxidized pellet of
gas-phase polypropylene cut normal to the extrusion direction.
The oxidation appears to be spreading out from centres of high
oxidation into the relatively unoxidised surroundings. This suggests that oxidation is enhanced by the diffusion of reaction
intermediates from heavily oxidised regions, analogous to a bad
apple spreading rot through a barrel.
     Most of our studies were on thermally oxidised material. In
photo-oxidation similar localised oxidation occurs which is un-
related to the spherulitic morphology. The catalyst residue
effect was not specifically investigated but is expected to be the
same.

Inhomogeneities in Thermosetting Resins

     Most thermosetting resins have a highly exothermic curing
reaction which is consequently autoaccelerative. It seems
possible that local hot-spots, in curing more rapidly than the
surroundings, could deplete the region of hardener such that the
final fully-cured resin would vary locally in degree of crosslinking. Accordingly we set out to apply uv microscopy to search
for this and other sources of inhomogeneity in epoxy resins.
     There are a number of possible approaches. The simplest is
to use a density marker, a compound which does not react with the
resin but will be distributed in such a way as to reflect its
local solubility within the structure. This solubility should be
higher in the less cross-linked regions. The marker can either
be diffused into the cured resin or blended in before curing. A
second approach is to use a reactive stain which attaches to
unreacted resin or catalyst remaining after curing. In view of
the interest in water diffusion in resins it would also be
valuable to have a reagent which could mark the ingress of water.

Methods

     The resin system was DGEBA (Epikote 828) cured with a molar
equivalent of triethylenetetramine (TETA).
     As non-reactive stains we used a variety of uv absorbing and
fluorescing compounds including the benzophenones and optical
brighteners used in our experiments on polypropylene. The most
successful reactive stain was dinitrofluorobenzene (DNFB, Sanger's)

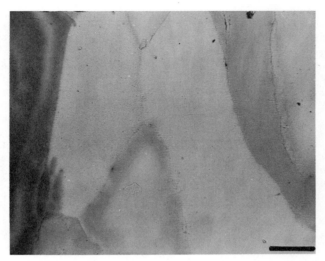

Figure 11. Sample as in Figure 10 except that the polymer was from
a gas-phase process. Bar is 100 μm.

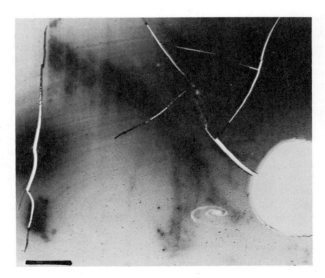

Figure 12. Section from an extruded pellet of gas-phase polypropylene after oxidation and staining. Bar is 100 μm.

reagent) which is extensively used in protein chemistry. This reacts with primary and secondary amines (17) and so stains unreacted groups on TETA.

The main problem is to obtain good sections of fully cured resin. The epoxy can be microtomed easily when partially cured but after full curing it becomes too brittle. Accordingly to study cured samples the resin was allowed to partially cure at room temperature, sectioned and the sections cured at 120°C for 2 hours mounted in a special holder to keep them flat. The DNFB staining was performed by immersing the sections in a 2% solution of DNFB in boiling ethanol for two hours. The unreacted stain was removed by boiling the section in dimethylformamide for 48 hours.

Observations

Non-reactive stains diffuse into cured resins uniformly with no sign of local concentration variation. When incorporated into the epoxy resin before curing these stains did tend to form streaks and to concentrate in haloes around bubbles which form during curing.

The reactive stain, DNFB, showed large bands of unreacted amine to be present in the cured resin, Figure 13. The visible light micrographs show little contrast except for diffraction effects around bubbles and faint parallel lines due to corrugations in the sections.

These samples were mixed thoroughly by hand stirring the amine into the resin with a glass rod for 5 minutes. Similar zones of unreacted amine were found when the mixing was carried out at 60°C. In order to achieve uniformity it was necessary to mix the components in a high shear mixer-emulsifier for 15 minutes or to blend them as a 50% solution in dichloromethane which was then evaporated. In these cases the sections were uniform except for a halo of stained amine around the bubbles. The absorption intensities in these sections were too great to permit determination of the concentration of unreacted amine but we expect to obtain quantitative information by using narrow band illumination in the blue region where the absorption is tailing off.

We have also observed low levels of cure close to the surface of resin cured in 1 cm tubes in a water bath at 60°C. This reflects the fact that the perimeter remains at 60°C while the centre is heated and further cured by the exotherm. Post-curing removes this difference.

Figure 14 shows the diffusion of dinitrophenol in aqueous solution into resin filled with 5% short glass fibres. The stain clearly follows the water up the fibre-resin interface.

Thus well-mixed, fully cured epoxy resins appear uniform on a scale of 1 μm upwards. However, very vigorous mixing of the components is needed to achieve this uniformity and in many normal circumstances cured epoxy resins will be inhomogeneous.

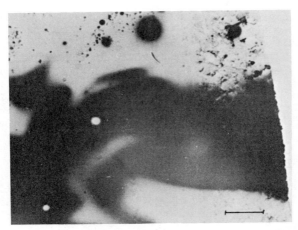

Figure 13. UV picture of epoxy resin cured with TETA hardener at 50 °C. Dark streaks are DNFB staining of unreacted amine groups. Bar is 100 µm.

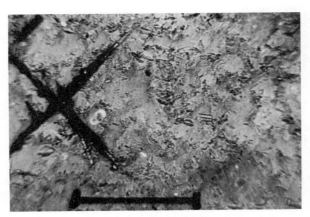

Figure 14. UV picture of cured TETA/epoxy containing a small number of glass fibers. Sample was immersed in an aqueous solution of dinitrophenol for 24 hrs. before examination. Bar is 300 µm.

Conclusions

Uv microscopy has wide applicability to the study of the degradation and reactions of solid polymers. However each new use does require a substantial effort for the development of suitable techniques. There are pitfalls in interpreting the results and it is preferable to combine uv, visible and fluorescent microscopy to avoid misinterpretations.

Acknowledgments

We wish to acknowledge the work of past and present members of our research group in developing the techniques described here and for allowing us to quote their unpublished results. Thanks are due particularly to T.G. Ryan, J.B. Knight and A. Uzuner. We also thank the Science Research Council for the award of a grant to allow the purchase of the uv microscope and Dr T. Henman and Mr A. Cobbold of ICI Petrochemicals and Plastics Division for their help with electron microscopy.

Literature Cited

1. Kohler A.Z.; Wiss Microskopie, (1904), 21, 129, 275.
2. Freed, J.J.; in 'Physical Techniques in Biological Research' 2nd Ed, Vol.IIIc, Ed. A.W. Pollister, Acad Press, New York (1969).
3. Wood, J.R.; Goring, D.A.I, J. Microscopy (1979) 100, 105.
4. Kam, Z.; Shore H.B., Feher G, J. Mol. Biol (1978), 123, 539.
5. Forty, A.J.; and Woodruff, D.P.; Tech Metals Res, (1968), 2 97.
6. Billingham, N.C. and Calvert, P.D.; Dev Polymer Character, (1982) 3 Ch.6.
7. Curson, A.D.; Proc. Roy. Microsc. Soc., (1972), 7, 96.
8. Frank, H.P. and Lehner, H.; J. Polymer Sci. Symp. (1970), 31, 193.
9. Calvert, P.D. and Ryan, T.G.; Polymer, (1978), 19, 611.
10. Calvert, P.D. and Ryan, T.G.; Polymer, (1982), 23, 877.
11. Calvert, P.D. and Billingham, N.C.; J. Appl. Polymer Sci, (1979), 24, 357.
12. Billingham, N.C.; and Calvert, P.D.; Dev. Polymer Degradation, (1980), 3 Ch.5.
13. Allara, D.L., White, C.W, A.C.S. Adv Chem. Ser, (1978), 169, 273.
14. Adams, J.H.; J. Polymer Sci. Chem, (1970), 8, 1077.
15. Kato, K.J.; Appl. Polymer Sci, (1974) 18, 2449, 3087; (1975) 19, 951; (1976) 20, 2451; (1977) 21, 2735.
16. Manke, A.S.; D.Phil Thesis, University of Sussex, 1981.
17. McIntire, F.C.; Clements; L.M, Sproull M, Anal Chem, (1953), 25, 1757.

RECEIVED November 22, 1982

Novel Diagnostic Techniques for Early Detection of Photooxidation in Polymers

RANTY H. LIANG, DANIEL R. COULTER, CATHY DAO,
and AMITAVA GUPTA

California Institute of Technology, Jet Propulsion Laboratory,
Pasadena, CA 91109

A Laser Photoacoustic Technique (LPAT) has
been developed to detect and monitor outdoor photooxidation in Ethylene Methylacrylate copolymer
(EMA). LPAT has been used to demonstrate that the
Controlled Environmental Reactor (CER), an accelerated testing chamber that was developed at JPL, is a
valid accelerated simulator of the real-time outdoor
photooxidation with respect to the rate of formation
of the hydroxyl functional group.

Polymers for solar energy applications have to meet rigorous
goals in terms of material and fabrication costs. In addition,
solar applications require that these low-cost polymers meet
stringent performance criteria over long periods (> 20 years) of
outdoor deployment. Therefore, development of reliable models for
life prediction of polymeric components is essential, if projec-
tions of life cycle cost of solar energy conversion devices are to
be meaningful. These models are based on understanding of degradation mechanisms of the materials. Validation of the models,
however, is often difficult and time consuming, and as a result,
has been attempted almost exclusively on results of accelerated
tests of material specimens. In order for accelerated test data
to be valid it must be derived from experiments in which test conditions preserve the basic mechanisms of degradation. However,
development of valid accelerated test methodology is by far the
most controversial aspect in validating theoretical models.

Ideally the best way to validate accelerated testing methodology would be to compare accelerated data with those obtained
under real-time outdoor exposure. Typical real-time outdoor testing data, however, show that observable changes in material properties (degradations) have induction periods, thus making validation of accelerated test procedures by this means a very time consuming process. Furthermore, primary degradation mechanisms,
which may, in general, be elucidated only during the very early
stages of aging, may be masked as a result of the inability to detect degradation during the induction period.

0097–6156/83/0220–0265$06.00/0
© 1983 American Chemical Society

At JPL, a three-fold approach is taken in developing lifetime prediction models of polymeric materials. It involves:

1)   Determination of degradation mechanisms of polymeric materials.

2)   Development of accelerated testing hardwares and procedures which duplicate degradation observed under outdoor conditions.

3)   Development of sensitive diagnostic techniques which can be used to detect early real-time outdoor degradation.

Recently, we have demonstrated that the Laser Photoacoustic Technique (LPAT) can be used to monitor incipient photooxidation in polymers aged outdoors. We have also used this technique to show that the Controlled Environmental Reactor (CER), an accelerated testing chamber that was developed at JPL, appears to be a valid accelerated simulator of the real-time outdoor photooxidation with respect to the rate of formation of the hydroxyl functional group.

In LPAT[1], a sample is placed inside a specially designed cell containing a sensitive microphone. The sample is then illuminated with chopped laser raidation as illustrated in Figure 1. Light absorbed by the sample is converted in part into heat by non-radiative de-excitation processes within the sample. The resulting periodic heat flow from the sample to the surroundings creates pressure fluctuations in the cell. These pressure fluctuations are then detected by the microphone as a signal which is phase coherent at the chopping frequency. The magnitude of the resulting photoacoustic signal is directly related to the amount of light absorbed by the sample. Since only the absorbed light is converted to sound, light scattering, which is a very serious problem when dealing with many solid materials by conventional spectroscopy, presents no difficulties in LPAT. Moreover, LPAT is extremely sensitive in detecting small amounts of absorption as compared to conventional absorption spectroscopy. For instance, Tam and Patel[2] demonstrated that laser photoacoustic spectroscopy can be used to accurately measure extremely weak absorption of water in the visible region, and we have obtained sensitivity improvement of 2 orders of magnitude by LPAT as compared to FT-IR for small absorbances in polymers.

The Controlled Environmental Reactor (CER) was developed because several key design requirements which control the closeness of correlation of accelerated test data with data obtained in field are not met by available commercial weatherometers. This may explain why correlation between weatherometer and field data is often so poor. Details of design and operational procedures of the CER have been reported before[3] and will be described only briefly in this paper.

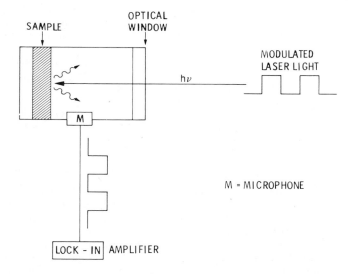

Figure 1.   Schematic diagram of photoacoustic set up.

Experimental

1) Controlled Environmental Reactor (CER)

A 550 watt Conrad/Hanovia medium pressure A.C. mercury lamp surrounded by a 1 cm thick pyrex water jacket, for cooling and infrared absorption, provided the irradiance, while an electrical heater and fan system was used to adjust the sample temperature. The lamp was allowed to operate from its standard power supply with no attempt at regulation of the irradiance. Lamp voltage and current were the only parameters monitored to characterize its performance. The exposure region was a perforated aluminum cylinder 34 cm in diameter and 23 cm high.

A 5 cm thick thermal blanket surrounded the assembly and allowed control of the temperature of a typical sample between 30 to 60°C. A water nozzle was employed to simulate rain and fog.

A strip chart single channel recorder with an adjustable set point, using voltage from one of the copper constantan thermocouples installed on a test module was used as an on-off temperature controller. Gross adjustment of the temperature was achieved by blocking the fan inlet and air outlet. Fine tuning was accomplished by adjusting the heater current. A cyclical variation of ± 1°C was achieved using the various adjustments.

2) Actinometry

The 200 to 400 nm spectral irradiance inside the CER was measured using a Gamma Scientific Spectroradiometer. A selenium photovoltaic cell and Corning 7-45 ultraviolet filter was used to monitor the UV irradiance. A 1 x 2 cm silicon solar cell was used to measure the near-IR irradiance. CER photon flux was also calibrated by using 0-nitrobenzaldehyde (0-NBA) as an actinometer[4]. Outdoor photon flux was measured by dispersing 0-NBA in thin films (25 μm) of polymethyl methacrylate. These films were then exposed at the outdoor site behind a neutral density filter and were examined on a weekly basis. Outdoor weekly UV photon flux was calculated based on the conversion rate of the 0-NBA.

3) Sample Preparation

Samples of Ethylene/Methylacrylate copolymer (EMA) were obtained from Gulf Oil Company (TD938) and reprecipitated from hot cyclohexane. The purified samples were then compression molded into thin films (25-50 μm). FT-IR spectra were recorded on a Digilab FT-IR Spectrophotometer Model FTS 15.

4) Laser Photoacoustic Technique (LPAT)

LPAT was carried out by using a hydrogen fluoride (HF) chemical laser operating at 2.83 μm as the excitation source and a

condensor microphone as the detector. Calcium floride which does
not absorb 2.83 μm light was used as the window material. Samples
of EMA were aged inside the CER as well as under real-time outdoor
exposure at Pasadena, CA. Photoacoustic signals from the aged
EMA films were compared with those obtained from a control EMA
film. The difference in the microphone signals is assumed to be
due to the formation of hydroxyl groups as a result of aging.
Since photooxidation requires the access of oxygen, formation of
the majority of hydroxyl group is expected to take place on the
polymer surface and thus polymer film thickness was not taken into
consideration in calculating hydroxyl concentration. However, the
background photoacoustic signal between control and aged EMA
films which arises from absorption throughout the bulk was calibrated with respect to their thickness. Calibration was also carried out to correlate the microphone signal to the change in absorbance.

Linearity of the photoacoustic signal with respect to the
amount of light absorbed was also verified in a separate experiment in which a CO_2 laser operating at 10.6 μm was used as the excitation source to illuminate thin films of PMMA. Figure 2 and 3
illustrate that the photoacoustic signal increases linearly with
respect to both laser power and percentage of light aborbed up to
~ 300 milliwatts.

Results and Discussions

EMA like many other polymers, undergoes photooxidation when
it is exposed to solar ultra-violet light in the presence of oxygen[5]. This is evidenced by the formation of hydroxyl and hydroperoxyl groups as detected by FT-IR. Figure 4 illustrates the FT-
IR absorption spectra of EMA before and after 200 hours of accelerated aging inside the CER. An increase in absorbance at 3530
cm^{-1} is unmistakable. Figure 5 shows the FT-IR absorption signal
of EMA at 3530 cm^{-1} as a function of accelerated aging time inside
the CER. Equivalent outdoor aging time calibrated by actinometry
is also shown in Figure 5. The fact that after 10 hours of accelerated aging (equivalent to 65 days of outdoor exposure), the –OH
and –OOH absorption peaks barely begin to appear in the FT-IR
spectrum leads to the conclusion that techniques other than FT-IR
are needed, in order to monitor early photooxidation.

Figure 3 is a plot of photoacoustic signal from LPAT as a
function of percent of light absorbed. The slope of the plot
yields the sensitivity of LPAT which is 2.5×10^{-3} % absorption/
μV. This can be translated into a detection limit of 10^{-5} absor-
bance/μV, at least 2 orders of magnitude better than FT-IR.

Figure 6 is a log-log plot of hydroxyl concentration in EMA
as a function of accelerated and real-time aging. Both FT-IR and
LPAT data are illustrated in Figure 6. Whereas FT-IR could not
detect any hydroxyl or hydroperoxyl formation with less than 10

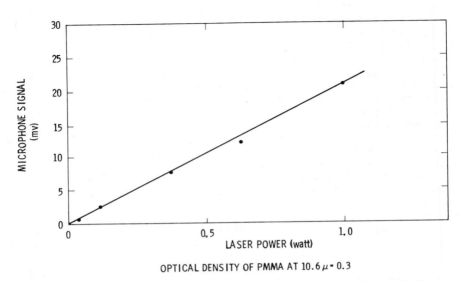

OPTICAL DENSITY OF PMMA AT $10.6\mu = 0.3$

Figure 2.   Photoacoustic signal of PMMA as a function of CO_2
            laser power.

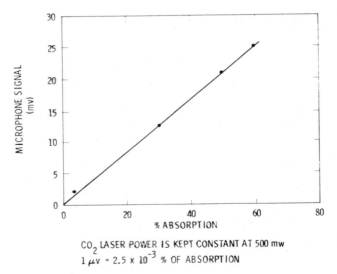

CO_2 LASER POWER IS KEPT CONSTANT AT 500 mw
$1 \mu v = 2.5 \times 10^{-3}$ % OF ABSORPTION

Figure 3.   Photoacoustic signal of PMMA as a function of
            percentage absorption.

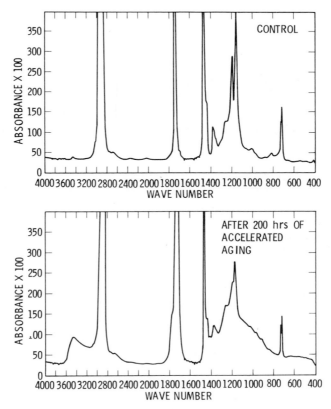

Figure 4. FTIR spectra of ethylene methyl acrylate (EMA).

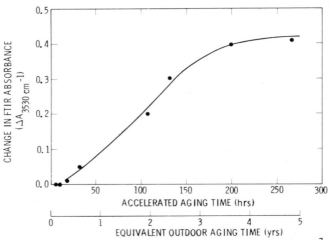

Figure 5.   FTIR absorption signal of EMA at 3530 cm^{-1} as a
            function of accelerated aging time in the CER.

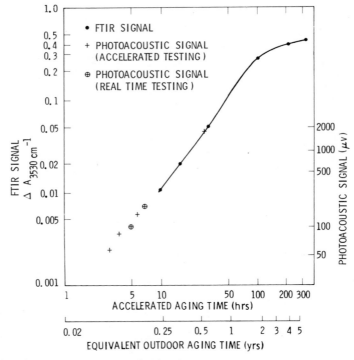

Figure 6.   Formation of (OH) as a function of accelerated and
            real time aging.

hours of accelerated aging, LPAT readily detects signal at an earlier stage. Two samples of EMA films which were aged at the JPL outdoor site also showed formation of hydroxyl group by LPAT after 45 and 55 days of outdoor exposure. These real-time data points correlate well with data which were obtained under accelerated testing conditions as illustrated in Figure 6. This indicates that up to 55 days of outdoor exposure, the accelerated testing chamber (CER) appears to simulate the outdoor degradation with respect to photooxidation.

In the future, attempts will be made to refine LPAT and it will be applied to other systems. The improved sensitivity of LPAT as compared to FT-IR is extremely valuable in detecting early photooxidation of formulated polymeric systems (e.g., systems with protective additives) whose induction period may be considerably longer than unformulated EMA. Longer outdoor real-time exposure will also be carried out in order to validate the accelerated testing procedures in the CER.

Acknowledgments

This research is supported by the Flat-Plat Solar Array Project sponsored by the Department of Energy by agreement with the National Aeronautics and Space Administration. The authors would also like to thank Dr. Susan D. Allen for her assistance in carrying out some of the photoacoustic measurements.

Literature Cited

1. For Review and Ref. see A. Rosencwaig; Anal. Chem. **47** (6) 1973, 595 A.

2. A.C. Tam and C.K.N. Patel, Applied Optics, **18**, p. 3348 (1979).

3. E. Laue and A. Gupta, "Reactor for Simulation and Acceleration Solar Ultraviolet Damage", JPL Publication 79-92, DoE/JPL 1012-31, 5101-135, 1979.

4. A. Gupta, A. Yavrouian, S. Di Stefano, C.D. Merritt, G.W. Scott, Macromolecules, 1980, 13, 821. G.W. Cowell, J.N. Pitts, Jr., JACS 1968, 90, 1106.

5. B. Ranby and J.F. Rabek, "Photodegradation, Photo-oxidation and Photostabilization of Polymers", Wiley-Interscience, 1975 and ref. therein.

RECEIVED January 21, 1983

Photodegradation of Poly(*n*-butyl acrylate)

H. R. DICKINSON,[1] C. E. ROGERS, and R. SIMHA
Case Western Reserve University, Department of Macromolecular Science,
Cleveland, OH 44106

The photodegradation of poly(n-butyl acrylate) results
in the loss of the n-butyl ester side group to yield
crosslinks, carboxylic acid groups, keto and aldo groups,
and volatile products including butanol and butene.
The degradation follows an apparent first order reaction
with an initial rate (up to 2000 hours exposure in a
QUV apparatus) that is faster than the subsequent rate.
An assessment of the kinetic data in terms of proposed
reaction mechanisms and concurrent changes in properties
such as dynamic mechanical behavior can serve as a preliminary basis for evaluation of the material's ability
to retain useful properties for time periods consistent
with certain design requirements for solar energy system applications.

The chemical and mechanical stability of poly(n-butyl
acrylate)(PnBA) to weathering, especially to solar radiation, is
of interest for possible use of this material as an encapsulant/
pottant for silicon cell solar energy arrays. This application
requires that the material retain an acceptable level of its
desirable properties, such as transparency, elastic modulus,
etc., over several years of exposure to intermittent moisture,
temperatures ranging from -10 to 50°C, solar radiation, and
other norms and extremes of exposure conditions. Knowledge of
the dependence of changes in properties and composition of the
material on exposure conditions is a requisite for establishing
reasonable estimates of its prospective performance lifetime
characteristics.

Degradation of poly(alkyl acrylates) and poly(alkyl metha-
crylates) has been the subject of several studies (e.g., 1-5).

[1] Current address: Ayerst Laboratories, Inc., Pharmaceutical Research and Development, Rouses Point, NY 12979.

Morimoto and Suzuki (1), in their investigation of the kinetics
of chain scission and crosslinking during photolysis of PnBA,
observed only an increase in molecular weight; the presence of
chain scission was not apparent. McGill and Ackerman (2), using
a high pressure mercury lamp, found that approximately 80% gel
fraction was obtained after twenty minutes of exposure. An increase in the weight to number average molecular weight ratio
was observed during the early stages of degradation. In a study
of the effects of γ-radiation on a series of poly(alkyl acrylates), Burlant, et. al. (3) found that the extent of degradation
was independent of polymer structure and reaction temperature.
The concurrent temperature dependence of the crosslinking reaction, below and above the polymer softening point, indicates
that the segmental mobility of the radicals is an important factor in the crosslinking step. A comparable consideration should
be applicable to photodegradation processes.
     More recently, Gupta and coworkers (4,5), using the same
PnBA samples as were used in the present study, investigated
photodegradation at 293°K and 77°K with radiation of 253.7 nm
and 310 nm. They identified certain principal photoproducts and
measured their quantum yields of formation. The data obtained
from a number of different experimental methods were used to
formulate a mechanism of photolysis of PnBA which has a close
relationship to the results of this present study.

EXPERIMENTAL

Sample Preparation: The PnBA used was provided by the Jet Propulsion Laboratory. The material was prepared by thermal polymerization of nBA by refluxing the monomer in cyclohexane under
high purity nitrogen for periods up to 48 hours (5). This method was chosen to avoid contamination of the polymer by trace
amounts of initiator which might affect the photooxidation kinetics. Molecular weight, as measured by HPLC, was about
8×10^5. Other measurements gave a $\overline{M}_w$ of 940,000 and a $\overline{M}_n$ of
620,000. The polymer was clear, transparent, and very tacky.
The glass transition temperature in the literature is 219°K.
     Samples of PnBA were dissolved in dichloromethane,
filtered, and cast into thin films on various substrates.
Quartz substrates were used for UV-visible spectroscopic analysis, as well as for gel content and weight loss measurements.
Salt (NaCl) flats were used as substrates for FTIR analysis.
The films were dried slowly to remove solvent followed by vacuum
drying for at least 24 hours.

Radiation Procedures: The samples were irradiated at 40°C with
either a QUV apparatus (Q-Panel Co., Cleveland, Ohio) which uses
mercury fluorescent sunlamps (Westinghouse FS-40) with an emission maximum at 313 nm, or a more intense (by a factor of about
40) medium pressure mercury lamp (Hanovia) with an emission
maximum at 366 nm. All irradiations were pyrex glass filtered.

For the torsional braid analyses, a fiberglass braid was
coated with polymer, dried, then exposed to the medium pressure
mercury lamp for 0, 42 and 160 hours. Since the fiberglass
braid is not transparent, it was rotated at one revolution per
minute about 1 inch from the water jacketed lamp.

Radiation flux at the sample plane was measured using o-
nitrobenzaldehyde (oNBA) as an actinometer. The films were held
parallel to the plane of the QUV lamps at 7 cm from the outer
edge of the nearest lamp. The films were contained in a light-
proof collimator box equipped with a photographic shutter. The
concentration of photosensitive o-nitrobenzaldehyde in the actinometer films is calculated from the optical density at 320 nm
and the molar extinction coefficient found by Pitts et al (6).

The total intensity of light emitted from the FS-40 lamps
was estimated to be 2.1×10^{-9} moles of photons/cm^{2}sec. The intensity at 5 cm, measured by electronic spectroradiometer at the
Q-Panel Company, is 18 µwatts corresponding to 4.7×10^{-11} moles
of photons/cm^{2}sec.

Sample Characterization: The procedures for the various analytical methods used for sample characterization in the study are
described below. The procedures generally followed established
practice with modifications to account for loss of material and
spectral band broadening with progressive degradation.

Fourier transform infrared (FTIR) spectra were measured
using a Digilab FTS-14 spectrometer with the Real Time Disk
Operating System (RDOS). The samples were scanned 256 times at
a resolution of one point every four wavenumbers with double
precision.

The changes in concentration of PnBA groups and reaction
products were determined from the changes in area of selected
assigned bands characteristic of the groups. The group band
assignments were made on the basis of literature data and by our
FTIR studies of model compounds representative of expected
groups. The extinction coefficients of the model compound bands
were estimated from knowledge of the densities and the film
thickness estimated from interference fringes in the infrared.

Dye binding studies were used to obtain estimates of carboxylic acid, aldehydes, and ketones in the weathered films.
For the acid determinations, the samples were exposed to a saturated solution of proflavine (3,6-diaminoacridine) hemisulfate
for 45 to 50 minutes. The samples were extracted several times
with ethanol. The amount of proflavin bound in the film was
determined from the optical density at 450 nm for an extinction
coefficient of 3.89×10^{4}.

There are several literature values for the binding ratio
between poly(acrylic acid) and various acridine derivatives.
Tan and Schneider (7) state that 4 to 8 acid groups in a 3 to 1
copolymer of ethylacrylate-acrylic acid will bind one molecule

of acridine orange. Schwarz and Klose (8) report 2 to 3 groups
poly(acrylic acid) per proflavine molecule. Nishida and Watanabe (9) found a ratio of 20 sodium acrylic acid monomer units
per proflavine hemisulfate molecule. This range in binding
ratio gives a fairly wide range in the estimate of acid groups.

Aldehyde and ketone concentrations were estimated by the
extent of formation of hydrazones (10). The sample films were
exposed to saturated ethanolic solutions of 2,4-dinitrophenyl-
hydrazine for ten minutes. The samples were rinsed extensively
with ethanol. The concentration of bound dye was determined
from the absorbance between 450 and 220 nm using literature
values for the extinction coefficients.

In both dye binding studies, experiments were carried out
with undegraded films to confirm the absence of concurrent physical adsorption of the dyes in the absence of the specific
binding functional groups. Studies were also carried out to
confirm the attainment of apparent equilibrium in terms of dye
exposure time and rinsing time to eliminate any diffusion (sorption time) control of dye binding. Nevertheless, there is still
some possibility that very slow sorption rates would give a
quasiequilibrium condition so that the estimated values of the
functional groups may be too low.

Dilute solution viscosities of the untreated material and
the two irradiated films were measured in a Cannon 50 viscometer
in toluene at 25°C. The crosslink densities were estimated from
the swell ratios and the sol/gel ratios. The treated films were
scraped off the glass slides, weighed, then swollen in toluene.
The solvated gel was separated from solvent by centrifuging at
low speed in a graduated centrifuge tube. After the volume of
the gel was recorded, the mass was determined by pressure fil-
tering the gel through a preweighed teflon filter, then drying
the material and weighing it.

Mechanical relaxation behavior of poly(n-butyl acrylate)
was studied using a freely oscillating torsional pendulum at
frequencies of about 1 cycle/sec in the temperature range of 100
to 300°K. Two damping measurements were made every 5°.

RESULTS AND DISCUSSION

Sol-Gel and Weight Change Studies: Studies of intrinsic viscosity and per cent sol of films irradiated for up to 21 hours
under the medium pressure mercury lamp showed evidence of extensive network formation, see Figure 1. These results indicate
that significant crosslink formation began in the time between
two and seven hours of exposure. By fifteen hours the material
in solution was of lower molecular weight than at seven hours.
At twenty-one hours, gel formation is essentially complete.
This time period corresponds to about 840 hours exposure in the
QUV apparatus. Samples exposed in the QUV showed an increase in
gel content from 81% to 91% for periods ranging from 300 to 2500
hours.

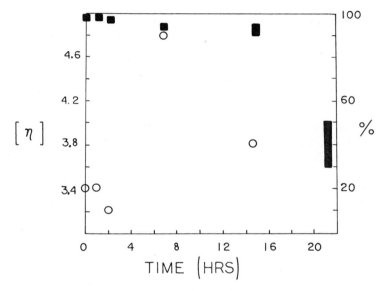

Figure 1)   Changes in gel content (■) and in the intrinsic
            viscosity of the sol fraction (o) with irradiation
            time.

It is reasonable to assume that random scission, unzipping and crosslinking may be occurring simultaneously. We attribute decreases in the intrinsic viscosity to a rise in the population of low molecular weight polymer due to chain scission and an increase in the intrinsic viscosity to a rise in the population of high molecular weight polymer due to crosslinking as a precursor to gel formation.

An estimate of crosslink density was made from swelling ratios using the Flory-Huggins-Rehner equation with an assumed value of 0.3 for the polymer-solvent interaction parameter, χ. This simple calculation ignores any changes in polymer-solvent interactions during photooxidation due to conversion of the n-butyl ester side group to carboxylic acid, alcohol and aldehydes. The logarithm of the mole fraction of crosslinks is a linear function of irradiation time as shown in Figure 2.

The FTIR spectrum of unexposed PnBA is shown in Figure 3 with assignments of the predominant bands. The carbonyl band identified is that of the carbonyl in the ester side group of the chain repeat unit. The COC stretch band is that of the ester side group to the butyl group.

The results of exposure of PnBA in the QUV apparatus were measured by FTIR difference spectra; typical spectra are shown in Figure 4. These spectra were obtained by computer subtraction of the spectrum of unexposed PnBA from the spectrum of PnBA taken after exposure. Peaks below the baseline indicate a decrease in those functional groups while peaks extending above the baseline indicate a gain in those groups.

A qualitative interpretation of the spectra indicates a progressive loss of ester carbonyl and COC groups with an increase in another type of carbonyl, assumed to be ketonic type. This is consistent with a crosslinking mechanism involving the conversion of the ester side group into a ketonic carbonyl crosslink group plus free butanol. Inspection of the spectra does show evidence for hydroxyl formation in the range of 3000 to 3600 wavenumbers. This broad band is generally not suitable for precise quantitative analysis.

Calculation of the areas of these major bands before and after UV treatment indicates that the CH (2900 cm^{-1}), C=O (1736 cm^{-1}), CH (1450 cm^{-1}), COC (1170 cm^{-1}) and n-butyl (1064 cm^{-1}) bands are decreasing; e.g., by an average of nearly 40% for the sample exposed for 1885 hours. However, the magnitude of the area under the fingerprint region has changed very little. This suggests that under the conditions of photooxidation, the major reaction is cleavage of the ester carbonyl group from the main chain, with loss of CO_2 and butene by evaporation.

Spectral changes in the carbonyl region indicate conversion of the ester carbonyl to two or more other types of carbonyl group. The shoulder at 1713 cm^{-1} which is evident in the difference spectra suggests the presence of carboxylic acid and/or

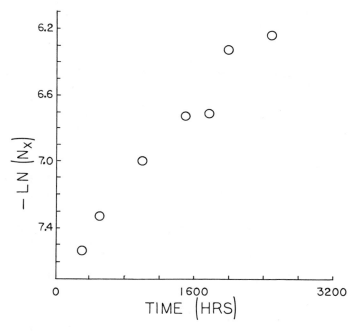

Figure 2) Natural logarithm of the mole fraction of crosslinks (N_x) as a function of irradiation time.

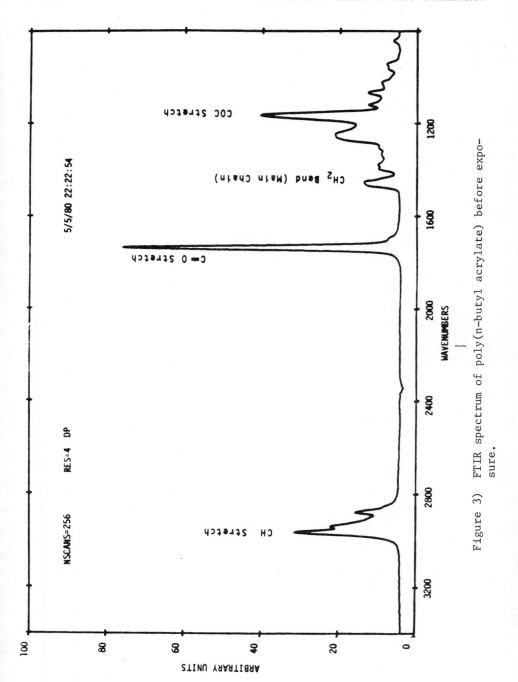

Figure 3) FTIR spectrum of poly(n-butyl acrylate) before exposure.

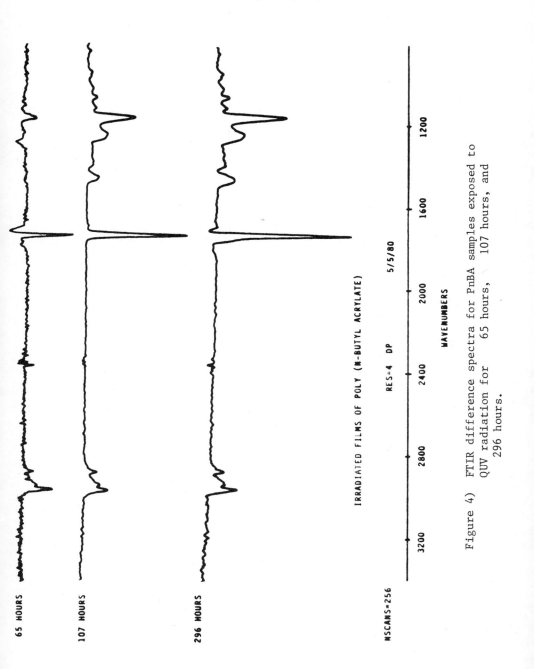

Figure 4)   FTIR difference spectra for PnBA samples exposed to
QUV radiation for   65 hours,   107 hours, and
296 hours.

ketone. The work of McGill (2) indicates that the crosslinker
group in irradiated PnBA is a ketone. The shoulder at 1785 cm^{-1}
may be due to low concentrations of acid anhydride, lactone, or
to an interchain ester. It is interesting to note that the carbonyl stretch band for γ-butyrolactone is located at 1773 cm^{-1},
fairly close to the new carbonyl band at 1785 cm^{-1} in the
weathered PnBA. The carbonyl bands for propionic anhydride at
1819 and 1752 cm^{-1} are both too far from the band at 1785 cm^{-1}
in the weathered PnBA to account for it.

The conversion of ester groups to products follows apparent
first order kinetics. An inspection of Figure 5 shows that the
rate of conversion of ester decreases by a factor of about two
for exposure times greater than 1800 hours in the QUV.

The concentration of hydroxyl groups formed was estimated
by comparing the intensity of the O-H stretching bands centered
at 3250 cm^{-1} to the intensity of the ester carbonyl stretch band
at 1735 cm^{-1} in the untreated film. The molar extinction coefficients for the hydroxyl and carbonyl stretch bands were calculated from the model studies.

This measure of the formation of hydroxyl groups, shown in
Figure 6, indicates an apparent leveling off at times greater
than about 2000 hours exposure. This may be related to the
change in rate of ester loss but the uncertainty of FTIR measurement in the region of 3000 to 3600 cm^{-1} does not permit a
more rigorous interpretation of the behavior.

The relative amount of carboxylic acid in each film was estimated from the concentration of bound proflavine hemisulfate.
Our data indicate that the rate of conversion of ester to acid
is constant following a lag time of the order of 200 hours,
Figure 7. The data indicate that about 0.6% of the original
ester groups in the PnBA film weathered for 2500 hours were able
to bind one molecule of proflavine hemisulfate. Since there are
several literature values for the binding ratio, this estimate
of 0.6% corresponds to a minimum of 1.8% and a maximum of 12% of
carboxylic acid groups in the film. Since the film did not
swell in the experiments and proflavine is a bulky molecule, it
is quite possible that the dye did not penetrate the film completely. Therefore, the actual content of carboxylic acid may
be higher.

The concentration of aldehydes and ketones in the photooxidized films was estimated from the extent of formation of the
yellow 2,4 dinitrophenylhydrazones. The results are presented
in Table I along with the results of other measurements.

The results of measurements of dynamic mechanical behavior
using torsion braid analysis on samples exposed to the medium
pressure mercury lamp for 0 and 160 hours are shown in Figures
8 and 9. In Figure 8, for 0 hours, the log decrement spectrum
shows two transition peaks. The more intense peak at higher
temperature corresponds to the glass transition temperature.
The lower temperature peak is not now assigned, but may be relat-

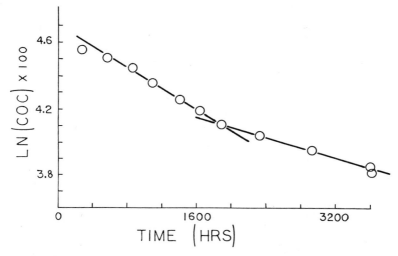

Figure 5) Change in COC concentration in PnBA with time of exposure to QUV radiation as measured by FTIR at 1170 cm^{-1} (symmetric stretch).

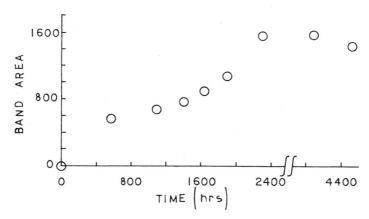

Figure 6) Change in hydroxyl content with time of exposure to QUV radiation as measured by the corrected area of the FTIR band at 3250 cm^{-1}.

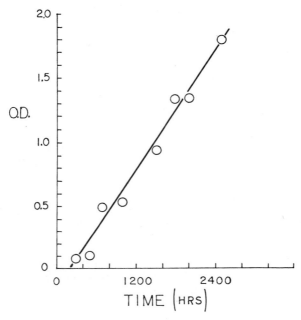

Figure 7)   Change in carboxyl content with time of exposure to
            QUV radiation as measured by the optical density at
            450 nm related to proflavine hemisulfate nominally
            bound to -COOH.

Table I: Products of Photooxidation of PnBA at
           2000 Hours in the QUV

Group	% Change	Method
ester C-O-C	− 30 ± 10	FTIR
COOH	+ 1.4 ± 10	Proflavine
CHO,CO	+ .4 ± .2	2,4-Dinitro-phenylhydrazone
CHO	+ .2 ± .2	
COH	+ 5.3 ± 2	FTIR
X links (CO)	+ .15 ± 0.10	Swell Ratio

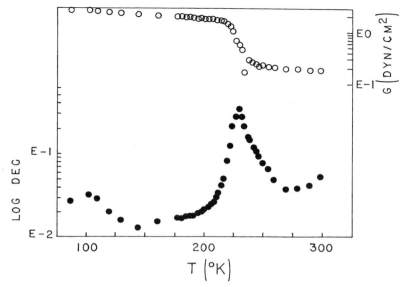

Figure 8)   Dynamic mechanical behavior of unexposed PnBA by
torsion braid analysis: log decrement (•) and
modulus G (o).

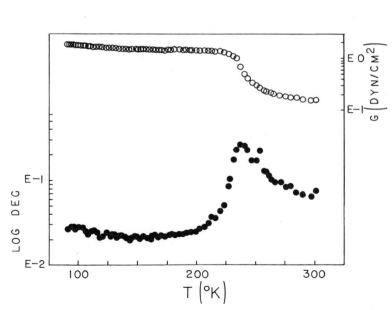

Figure 9)   Dynamic mechanical behavior by torsion braid analysis of PnBA exposed for 160 hours to the medium
pressure mercury lamp: log decrement (•) and modulus G (o).

ed to ester side group rotation. Exposure for 42 hours did
not cause any major changes in transition position or intensity.
Exposure for 160 hours, Figure 9, does cause a shift in the
temperature of peak maximum and an increase in baseline between
the transition peaks so as to obscure the lower temperature
transition.

The torsion braid analysis was used to estimate the glass
transition temperature of the linear and crosslinked polymer for
a calculation of crosslink densities in the irradiated polymer.
As seen in Figure 8, T_g for our sample turns out to be 230K, as
compared with the literature value of 219K. This discrepancy
does not affect our conclusions however, since we are concerned
only with changes induced by the chemical reaction. Crosslink
densities calculated from these data by three different equations (11) were 2.6%, 3.5%, and 2.0% for the sample irradiated
for 160 hours to give an average of 2.7%. This is much higher
than the crosslink densities observed in the samples treated in
the QUV for up to 425 hours.

An interpretation of these results is that there is little
effect of degradation crosslinking or other product formation on
chain or group motions involved in the observed transition and
modulus behavior up to about 160 hours exposure. This is a reflection of the relatively low crosslink density obtained up to
that exposure time. It is anticipated that creep behavior, a
low frequency response, would exhibit greater sensitivity to low
levels of crosslinking.

Analysis and Correlation of Studies: The photooxidation of PnBA
is apparently a first order reaction with an initial rate (up to
2000 hours) that is higher than the subsequent rate. The rate
constant for crosslink formation is greater than that for COC
cleavage, Table II. However, the calculated number of crosslinks per chain as weathering progresses is very much lower than
the number of cleaved COC groups. The rate of —COOH formation
is the same order of magnitude as the rate of crosslink formation.

The quantum yields for the cleavage of the n-butyl ester
from the PnBA during irradiation in the QUV are 0.089 (0 to 284
hrs) and 0.095 (284-564 hours). The range of values is only
slighly lower than that obtained by Gupta et. al. (4,5) who reported an initial quantum yield of 0.1 for the production of
butene during the photooxidation of PnBA.

The photodegradation reaction schemes proposed by Gupta and
Liang indicate that random chain cleavage is a significant process. The swell ratios and the gel fractions should reflect the
extent of random cleavage. If the chain cleavage rate is appreciable but less than twice the crosslinking rate a network
will be produced that is insoluble, infusible, and elastic.

The crosslinks may be formed by either acid anhydrides, ketones, or interchain esters (e.g., 12, 13). During the photo-

Table II: Reaction Rate Constants and
Correlation Coefficients

	k(sec)	r^2
Swell Ratio	$- 1.39 \times 10^{-7}$	.96
Crosslink Density	1.68×10^{-7}	.97
COC (0–1800 Hours)	$- 7.14 \times 10^{-8}$	.93
COC (1800–3600 Hours)	$- 4.34 \times 10^{-8}$	.99
Proflavine Binding (or COOH Formation)	2.23×10^{-7}	.97

degradation of PnBA it is very likely that pendant carboxylic acid groups are formed adjacent to n-butyl ester groups or physically close by on a different chain. However, we must assume that this product would be converted to two carboxylic acid groups in the presence of atmospheric moisture and thus would not account for the observed crosslinking. McGill and Ackerman (2) have suggested that the crosslinks are formed by ketones based on observation of the volatile products and the rate of random chain scission.

The sum of the products of PnBA photooxidation in the chain should equal the side groups lost unless there is extensive monomer formation. The data in Table I indicate that about 30% of the original ester COC bonds were lost at 2000 hours of photooxidation, while the sum of the new groups formed is 7.25%.

This disagreement between the number of ester COC groups lost and the number of photooxidation groups formed may be due to errors in measuring the concentrations of products or to a major product which has been overlooked. Unfortunately, it was necessary to use several different methods to analyze for products; the accuracy of these methods in films has not been well documented.

Since the infrared spectra of PnBA films weathered for 2000 hours show a loss of similar magnitude in the CH_3, CH_2 bands at 2900 cm^{-1} and in the carbonyl band at 1735 cm^{-1} with no concomitant appearance of any new major bands to account for this, it is tempting to suggest that the decrease in these band intensities may be due to chain unzipping with release of monomer. However, the studies of Gupta and Liang suggest that butyl acrylate is not a major product of short term photooxidation.

The gravimetry studies indicate that the films lost no more than 10% of their masses during 2000 hours of irradiation. The loss of such a large proportion of the ester COC, CH_3, CH_2 and carbonyl groups cannot be accounted for by the loss of 10% of the mass even if it is assumed that only butene is released and in-chain carboxylic acid is the major product.

If a significant amount of water has been absorbed by the PnBA films, it should be apparent in the FTIR difference spectrum. The most intense bands for water are centered at about 1650 and 700 cm^{-1}, with approximately identical intensity. Thus, the new absorbance bands observed at 1710 to 1600 cm^{-1} could be assigned to water except that no new absorbance is observed at 700 cm^{-1}.

A reviewer of the original manuscript has suggested that a Norrish type II process (γ-hydrogen abstraction) may be anticipated, as contrasted to a type I process (α-cleavage), since the photochemistry of esters is very similar to that of aliphatic ketones. It has been shown (14) that, at reaction temperatures above T_g (as in this present study), the major photochemical reactions of vinyl ketone copolymers are derived from the type II process, with quantum yields in the polymer similar to those

in solution. The functional groups observed in this study are
not inconsistent with a type II reaction scheme. Additional
studies to account for the products of ester COC cleavage and
to distinguish between a type I and II process (e.g., by deu-
terium labelling) would be necessary for any further interpretation.

SUMMARY AND CONCLUSIONS

Although there is scatter in some data, it is clear that
these materials are undergoing a loss of ester side and CH_3/CH_2
groups and that the period of most rapid loss is between 600
and 1400 hours. The swell ratio and percent gel studies show
that the polymer had about 3 crosslinks per chain after 425
hours and about 5 after 1436 hours. At 1436 hours, approximately 0.12% of the butyl ester groups have been converted to
carboxylic acid, corresponding to about 8.4 acid groups per
chain.

The decrease in the infrared absorbance band areas for the
COC and CH_3 plus CH_2 groups suggests that either the film has
undergone a physical deformation due to flow, a loss of side
group due to scission of the ester bond, or a combination of
the two effects. It must be assumed that both mechanisms operate simultaneously before the onset of gelation. However, most
of the decrease in peak areas occurred after the films were weathered for 425 hours; at this time 65% of the PnBA had become a
gel with 3 crosslinks per chain. Hence, one must conclude that
the changes observed represent chemical effects.

These data can serve as a preliminary basis for evaluation
of PnBA as a pottant material with prediction of structure-
property relationships for extended exposure conditions. On the
basis that one hour exposure in the QUV corresponds to one day
exposure in practice, the present study is predictive of a ten
year period of use.

Support of this research program by a contract from the Jet
Propulsion Laboratory as part of the Low-Cost Solar Array Project (Flat-Plate Solar Array Project) is gratefully acknowledged. We thank Drs. A. Gupta and J. Moacanin for their discussions and cooperation.

Literature Cited

1. K. Morimoto and S. Suzuki, J. Appl. Polym. Sci., 16, 2947
    (1972).
2. W. J. McGill and L. Ackerman, J. Appl. Polym. Sci., 19, 2773
    (1975).
3. W. Burlant, J. Hinsch and C. Taylor, J. Polym. Sci., A2, 57
    (1964).
4. A. Gupta, R. Liang, F. D. Tsay and J. Moacanin. Macromol.,
    13, 1696 (1980).

5. R. Liang, F. D. Tsay, J. Moacanin and A. Gupta, submitted.

6. J. N. Pitts, Jr., G. W. Cowell and D. R. Burley, Environmental Science and Technology, 2(6), 435 (1968).

7. J. S. Tan and R. L. Schneider, J. Phys. Chem., 79, 1380 (1975).

8. G. Schwarz and S. Klose, Eur. J. Biochem., 29, 249 (1972).

9. K. Nishida and H. Watanabe, Colloid and Polymer Sci., 252, 392-395 (1974).

10. J. D. Roberts and Charlotte Green, JACS, 68, 214-216 (1946).

11. Lawrence E. Nielsen, in Reviews in Macromolecular Chemistry, Ed., by Butler and O'Driscoll, Vol. 4, Marcel Dekker, Inc., New York.

12. D. H. Grant and N. Grassie, Polymer, 1, 125-134 (1960).

13. A. Jamieson and C. McNeill, European Polymer Journal, 10, 217-225 (1974).

14. E. Dan and J. E. Guillet, Macromol., 6, 230 (1973).

RECEIVED December 27, 1982

Photochemical Stability of UV-Screening Transparent Acrylic Copolymers of 2-(2-Hydroxy-5-vinylphenyl)-2*H*-benzotriazole

A. GUPTA—California Institute of Technology, Jet Propulsion Laboratory, Materials Research and Biotechnical Section, Pasadena, CA 91109

G. W. SCOTT—University of California, Riverside, Department of Chemistry, Riverside, CA 92521

D. KLIGER—University of California, Santa Cruz, Division of Natural Science, Santa Cruz, CA 95064

O. VOGL—University of Massachusetts, Polymer Science and Engineering Department, Amherst, MA 01003

The mechanism of photodegradation of certain hydroxyphenyl benzotriazole based ultraviolet absorbers has been investigated and a new polymerizable ultraviolet absorber in this group has been synthesized. The photoreactivity is entirely confined at the surface of polymethylmethacrylate films containing the ultraviolet absorbers as pendant groups. A mechanism involving sensitized photooxidation has been proposed to interpret the data.

Polymerizable ultraviolet absorbers are needed whenever a thin film of ultraviolet absorbing layer is required to retain the permanence of its absorption characteristics over a service life of five years or more. Vinyl substituted ultraviolet absorbers, e.g. vinyl derivatives of 2-hydroxybenzophenone were initially synthesized at DuPont[1-2]. The synthesis was modified and the yield was significantly improved by Vogl, et al. by using an improved dehydrobromination procedure[3-5]. More recently, the superior screening capacity of hydroxyphenyl benzotriazoles led us toward the development and testing of copolymers of 2(2-hydroxy-5-vinylphenyl) 2H-benzotriazole (I) with methylmethacrylate and styrene. Synthesis of I and characterization of its polymerization reactivity was recently reported by Vogl, who also demonstrated that I can be grafted onto saturated aliphatic C-H bonds in polypropylene, and ethylene-co-(vinylacetate)[5]. The copolymer of I with MMA was selected for aging tests in order to assess its applicability as a transparent front cover on photovoltaic modules.

Synthesis of the copolymer was carried out as described in Scheme 1. Details of the synthesis have been published and will not be discussed further. The copolymer was purified by extraction with nonsolvents and reprecipitation from methylene chloride. It was solution cast to form thin films ($5-8 \times 10^{-3}$ cm) which were dried in a vacuum oven overnight before irradiation commenced. These copolymer rotating films were cut into strips, mounted on the outer surface of a cylinder and exposed to pyrex filtered radiation from a medium pressure Hg arc lamp. Preliminary results are shown in Figures 1 and 2. The absorption spectra of the films gradually change, as is shown in Figure 1a. However, most of this change can be attributed to changes in the scattering properties of the films. Relative absorbance (S_λ) defined as A_λ/A_{600}, measuring absorbances, assuming that no absorption developed at 600 nm, does not change appreciably as shown in the inset in Figure 1b. A copolymer film maintained in the dark served as a control. Fourier transform IR spectra were recorded on the films as a function of irradiation period, as shown in Figure 2. Figure 2b shows ATR-IR spectra indicating considerable photooxidation after 2100 hours of irradiation. The rate of growth of hydroxyl groups, which are the principal products of photooxidation, is shown in Figure 3. As photooxidation proceeds, the polymer undergoes crosslinking, ultimately resulting in the formation of a gel fraction.

The dependence of the rate of photooxidation on the concentration of the chromophores was sought to be investigated by testing thin films of a blend of PMMA and the copolymer, formed by solution casting a mixture of 85 weight percent PMMA ($Mn \simeq 400,000$) and 15 weight percent of the copolymer ($Mn \simeq 90,000$). Figure 4 shows the FTIR (ATR) absorbance difference measured on these films aged for typical periods. The rate of photooxidation is estimated to be reduced by more than an order of magnitude relative to the pure copolymer. These blends were than examined by ESCA in order to determine the actual benzotriazole group concentration at the surface. Figure 5a shows the N (IS) ESCA peak in the pure copolymer films. While Figure 5b shows that no trace of nitrogen could be found on the surface of the blends. The rate of photooxidation at the surface monolayer was also monitored by the surface energy analysis. These results are given in Table I. The contact angle measurements were carried out using water and polypropylene glycol and work angle, W_a was calculated as follows:

$$W_a = \gamma_L (1 + \cos \theta)$$

where γ_L is the surface tension of the solvent in dyne/cm. Then:

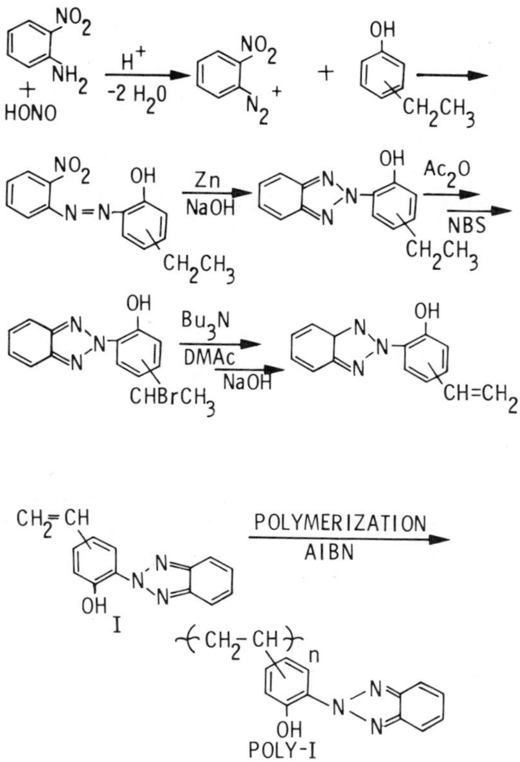
NO2
NH2
HONO
H+
-2 H2O
NO2
N2+
OH
CH2CH3
NO2
OH
N=N
CH2CH3
Zn
NaOH
N
N
N
OH
Ac2O
NBS
CH2CH3
OH
CHBrCH3
Bu3N
DMAc
NaOH
OH
CH=CH2
CH2=CH
OH
I
POLYMERIZATION
AIBN
(CH2-CH)n
OH
POLY-I

Scheme 1.

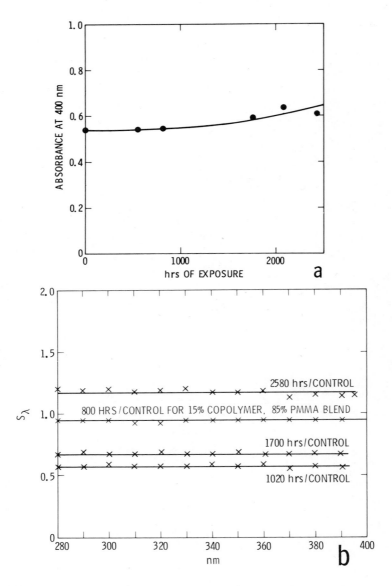

Figure 1.   Absorbance of Copolymer I Films as a Function of
             Irradiation Period.
             a)   Absorbance Data Recorded on Films.
             b)   Absorbance Data Recorded on Solutions
                  of Aged Films; $S\lambda$ = $A\lambda/A_{600}$ (see text).

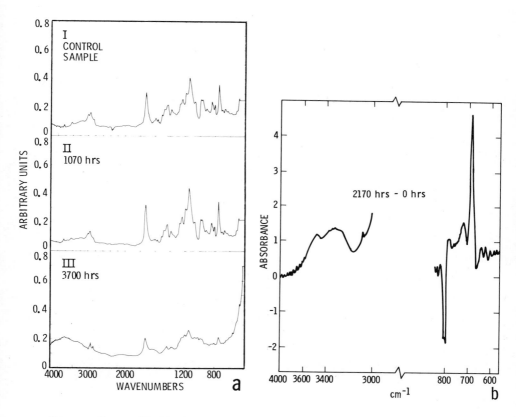

Figure 2.   FT-IR Spectroscopic Analysis of Aged and Control
Films of the Copolymer I.

a) Transmission FT-IR Spectra.
b) ATR FT-IR Difference Spectra.

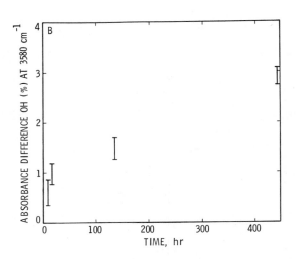

Figure 3.   IR Absorbance Increase at 3580 cm^{-1} as a Function of
              Time on Copolymer Films from ATR FT-IR Spectral
              Data.

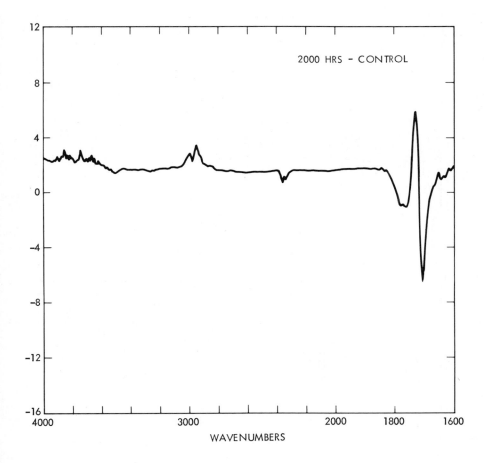

Figure 4.   ATR FT-IR Difference Spectra for Films of Blends of
the Copolymer (I) and PMMA (15:85 by Weight); Control
is a Film of the Blend Maintained in Dark.

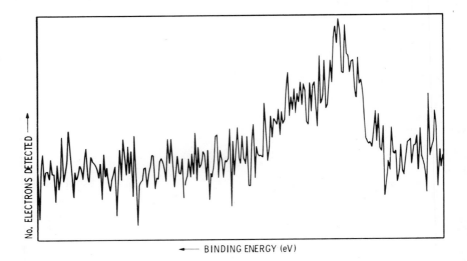

SAMPLE: PURE COPOLYMER (I)

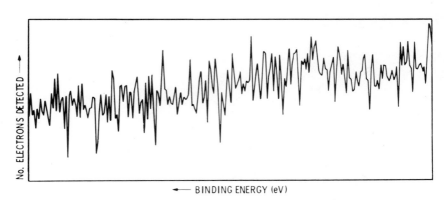

SAMPLE: BLEND OF I AND PMMA (15:85)

Figure 5.   ESCA Data on Films of the Copolymer (I) and the
               Blend.

$$\gamma_s = \gamma_s^d + \gamma_s^p$$

when $\quad \gamma_s^d = \left| \begin{array}{cc} (W_a/2)_i & (\gamma_i^p)^{1/2} \\ \\ (W_a/2)_k & (\gamma_i^p)^{1/2} \end{array} \right|^2 \quad D^2$

where $\quad D = \left| \begin{array}{cc} (\gamma_i^d)_i^{1/2} & (\gamma_i^b)_i^{1/2} \\ \\ (\gamma_i^d)_k^{1/2} & (\gamma_i^b)_k^{1/2} \end{array} \right|$

Here subscript i denotes H_2O and subscript k denotes PPG.

Table I. Calculated Surface Tension Values as a Function of Irradiation Period.

PERIOD OF IRRADIATION (hr)	WATER (γ_L = 72.8 dynes/cm)		PG-E-200 (γ_L = 43.5 dynes/cm)	
	EXPOSED SIDE	DARK SIDE	EXPOSED SIDE	DARK SIDE
0	67.2	68.2	36.0	35.6
88	93.5	69.3	69.5	38.9
278	90.4	66.7	70.3	35.5
419.5	89.2	72.2	69.4	35.0

The observed photooxidative crosslinking process was judged to be a consequence of introduction of tertiary hydrogen atoms on copolymerization of vinyl derivations of ultraviolet absorbing chromophores. Hence, a propenyl derivative of the 2-hydroxyl-phenyl benzotriazole nucleus was synthesized, as shown in Scheme 2. Details of the synthesis of this compound will be re-ported, subsequently. The absorption spectrum of the propenyl de-rivative[*]. Photodegradation rate measurements on this material are in progress.

[*]Copolymer with methyl methacrylate is shown in Figure 6.

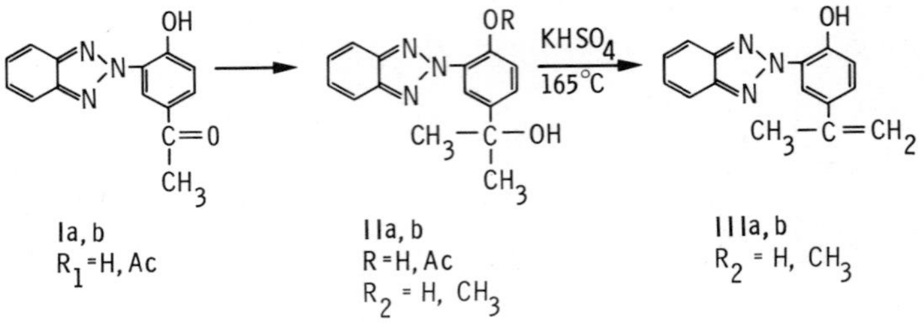

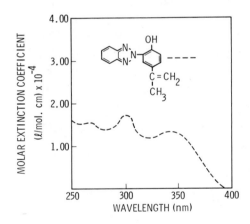

Scheme 2

Figure 6.   Absorption Spectrum of 2[(2-hydroxy 5-propenyl)
phenyl] 2H-benzotriazole in Methylene Chloride at
30°C.

The absorption spectra of a model compound has been reported
in several different solvents at room temperatures and also as a
function of temperature down to 11K. These spectroscopic measurements indicate that there is an equilibrium between two or more
conformers in the ground state. Two conformers absorbing at 302
nm and 340 nm may be stabilized by a combination of intra- and intermolecular hydrogen bonding, as shown in Figure 7a. Preliminary
C-13 nmr spectral data indicate that the degree of aromaticity is
quite solvent dependent. The complex distribution of chromophore
molecules in the ground state make it difficult to propose a
straightforward interpretation of emission and excited state decay
data obtained from ground state absorbance recovery rate and fluorescence decay rate measurements. Some of these measurements will
be reported.

The mechanism of photodegradation of the copolymer is pre-
sumed to involve an electronic energy transfer process from the
benzotriazole chromophore to a photoreactive group on the polymer
backbone, e.g., hydroperoxy groups formed on oxidation of the tertiary hydrogen atoms as shown in Scheme 3. This mechanism is necessarily confined to the surface, since it requires penetration
of oxygen and actinic radiation (300-400 nm). Hydroxyl groups and
simultaneous crosslinking and chain scission are the principal
products of photooxidation. Photooxidation causes a decrease in
surface energy of the filters, a somewhat unexpected result. The
decrease in surface energy should decrease the soiling character
of front covers of photovoltaic modules.

The rate of energy transfer from the benzotriazole chromophore to the hydroperoxy groups is controlled by the lifetime of
the excited state, as long as it is higher than 1.5 ev approximately. Details of decay mechanisms of the excited states will be
published later. Here we will note that the principal feature of
the deactivation mechanism involves an intramolecular proton
transfer process which may occur before vibrational equilibration
of the vertical excited state is completed. The fluorescence has
a blue (λ_{max} = 405 nm) and a red (λ_{max} = 585 nm) component, with
the blue component only being present at room temperature in di-
lute solution, and at low temperatures in polar matrices. The red
component is present in emission at room temperature from polycrystalline powders and at low temperatures in hydrocarbon matrices.
It may be postulated that the blue component arises from a vibra-
tionally excited O-protonated species, while the red component
arises from a proton transferred zwitterionic excited state.
Phosphorescence is detected from the model compound (II) in polar
matrices at 77K. Table II gives some excited state lifetime data
on the copolymer and model systems.

Photooxidation of the copolymer may be inhibited either by
reducing access of oxygen or by reducing the number of tertiary
hydrogen atoms on the main chain. In the blend of the copolymer
with PMMA, the pendant chromophores are excluded from the surface,
as shown by ESCA measurements. Formation of excited states of

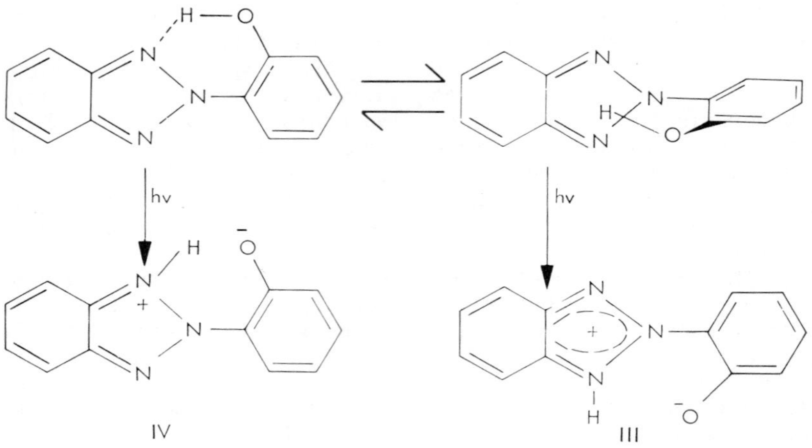

Figure 7.   Proposed Mechanism of Electronic Energy Deactivation
in the Orthohydroxybenzotriazole Nuclei.

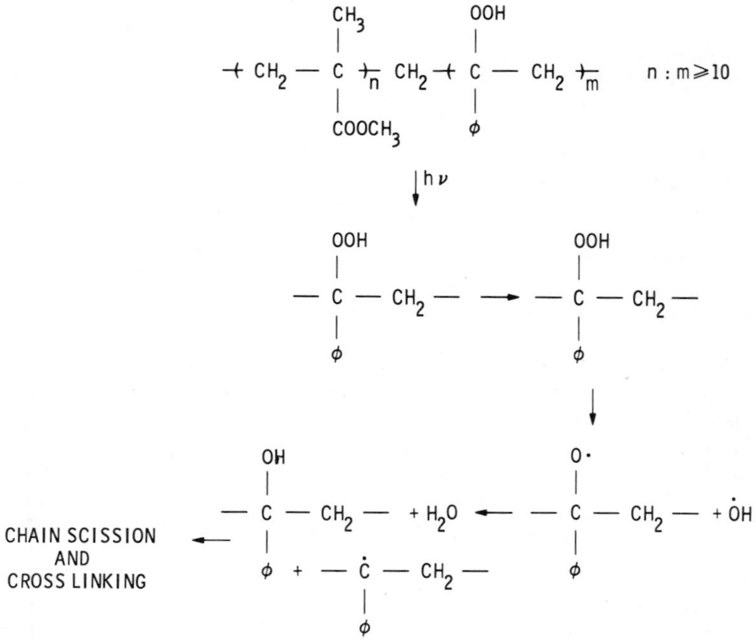

Scheme   3

benzotriazole groups and the electronic energy transfer process therefore takes place inside the bulk of the film. The consequent decrease in photooxidation rate tends to support the energy trans-

Table II.   Fluorescence Lifetimes of the Orthohydroxybenzotriazole Derivatives.

MOLECULE	SOLVENT	TEMPERATURE	WAVELENGTH	LIFETIME
R = CH_3	METHYLCYCLOHEXANE	30^0C	TOTAL FIT	14 ± 3 ps
	ETHANOL	30^0C	TOTAL FIT	52 ± 4 ps
	METHYLENE CHLORIDE	30^0C	TOTAL FIT	19 ± 5 ps
COPOLYMER	- DO -	30^0C	TOTAL FIT	15 ± 4 ps
R = CH_3	EPA	77K	390 nm	2.4 ± 1.2 ns
R = CH_3	2 METHYL PENTANE	77K	420 nm	2.2 ± 1.0 ns
R = CH_3	- DO -	77K	600 nm	1.4 ± 0.7 ns

fer mechanism and rule out direct excitation of hydroperoxy groups as an initiation step.

In conclusion, we have investigated the mechanism of sensitized photooxidation of ultraviolet absorbing clear acrylic films containing pendant ultraviolet absorber groups. The main conclusions of the mechanistic study indicated that propenyl derivatives of ultraviolet chromophores, copolymerization of which would lead to development of methyl groups on the backbone would be more appropriate candidates for outdoor applications requiring long service life. Synthesis of the first such comonomer has been reported here.

Acknowledgments

The research described in this paper was performed by the Jet Propulsion Laboratory, California Institute of Technology and was sponsored by the Flat-Plate Solar Array Project, Department of Energy.

Literature Cited

1.   D. Bailey and O. Vogl, J. Macromol. Sci. Reviews, C14(2), 267 (1976).
2.   S. Tocker, Makromol. Chem. 101, 23 (1967).

3. O. Vogl and S. Yoshida, Rev. Roum. de Chimie, 25(7), 1123
    (1980).
4. S. Yoshida and O. Vogl, Polymer Preprints, ACS Divison of
    Polymer Chemistry, 21(1), 203 (1980).
5. W. Pradellok, O. Vogl and A. Gupta, J. Polym. Sci, Polym.
    Chem. Ed., 19, 3307 (1981).

RECEIVED April 19, 1983.

Effects of Deformation on the Photodegradation of Low-Density Polyethylene Films

DJAFER BENACHOUR and C. E. ROGERS

Case Western Reserve University, Department of Macromolecular Science, Cleveland, OH 44106

The effects of uniaxial or biaxial deformation on a polymeric material's resistance to ultraviolet radiation can
be significant in many practical applications. In this
study it was found that both uniaxial and biaxial elongation of low density polyethylene film enhances the photodegradation rate at 40°C. The enhancement process for
uniaxial deformation has been shown to be closely related
to the mechanism of deformation and the morphological
changes induced upon elongation. The necking development
region, where original material structure is most dis-
rupted, showed the largest enhancement. Highly oriented
material is less sensitive to photodegradation. The
experimental evidence suggests that the increase in
degradation rate may be attributed primarily to strain
effects (morphological changes) with some contribution
from stress per se (stored energy). Biaxial stretching was found to result in greater degradation, probably because of a larger decrease in film thickness
and more constraint applied. A comparison of the nature of uniaxial and biaxial deformations gives some
further insight into the drastic effects of photooxidative degradation on mechanical properties. Cyclic
deformation (fatigue) involves a competition (dependent on deformation frequency, amplitude, and the
number of cycles) between the formation of fatigue
damages (microcracks, etc.) which promote degradation and orientation of structure which reduces the
degradation process.

Photooxidation and deformation of polyethylene (PE) have
been intensively investigated, and mechanisms for each process
have been suggested which are, more or less, well accepted (1–7).
However, how photooxidation can be affected by deformation (type,
extent ...) has not been given much attention until recently (8–
11). In a previous paper, we reported data on the effects of

uniaxial stretching on the photodegradation behavior of low density polyethylene (LDPE) films (11). The present paper is a continuation of the mentioned work, and deals with the effects of cyclic uniaxial elongation and biaxial stretching. The data are explained in terms of the photooxidation mechanism and the resultant degradation products and the type of deformation applied and its effects on the material (orientation, damages, morphological changes ...). We also give a brief account of the dependence of fatigue life (N_F) on the photooxidation extent of LDPE films.

Experimental

Low density polyethylene films were used. The material characteristics are listed in Table I. The samples were exposed

Table I: LDPE film characteristics

Thickness (mils)	1.25
Density (g/cc)	0.914
Molecular Weight $\overline{M}_w$	60.000
Melt Index (g/10 min)	1.6
Crystallinity (%) (estimated by IR and x-ray)	50

in a Q-UV weatherometer (Q-Panel Co.) at 40°C and ambient atmosphere. The oxidation extent was followed by measuring the carbonyl absorbance at 1716 cm^{-1} using Fourier Transform Infrared Spectroscopy (Digilab FTS-14).

Uniaxial deformations were done using a specially designed stretcher (11) which was made to fit in the FTIR sample holder, thus allowing spectra to be taken while films are kept elongated. Cyclic stretching and fatigue tests were performed on an Instron Tensile machine. The strain-control mode was used for fatique. A T. M. Long Company biaxial film stretcher was used in the constant rate of deformation mode for concurrent and sequential biaxial deformations. All stretchings and mechanical testings were carried out at room temperature.

Results and Discussion

Effects of Photodegradation on Fatigue Life: The photooxidation behavior of the material used in this study is illustrated in Fig. 1 where the extent of oxidation is plotted as a function of UV exposure time. Notice the exponential shape of the curve. Similar behavior has been observed by other workers (1,3) for PE,

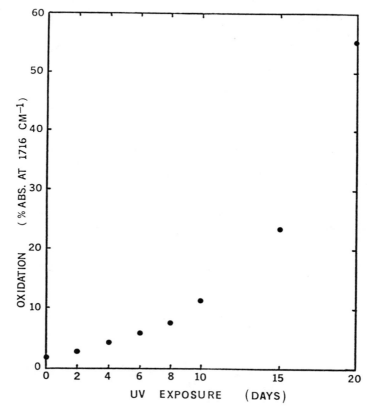

Figure 1.   Photodegradation of LDPE at 40°C as a function of exposure time in the QUV apparatus.

as well as polypropylene, films and is due to the autocatalytic nature of the photooxidation process in such materials.

Despite many studies of the effects of photooxidation on mechanical properties (such as Young's modulus, tensile strength, ultimate elongation, etc. ...), there is very little information about these effects on fatigue life. For that reason, we studied the fatigue life of LDPE films as a function of UV exposure. The results are shown in Fig. 2 where $\log N_F$ (N_F being the number of cycles sustained before failure) is plotted vs. time of UV exposure. In this case the strain amplitude is 8% and the frequency (ω) is 10 cycles/min.

The linear relationship which is observed can be described by an equation such as:

$$\log N_F = a + bt \qquad (1)$$

where

        a = fatigue life on non-oxidized film under the considered fatigue test conditions. a is given by the intercept of the plot.

        b = constant, depending on fatigue test conditions and material characteristics; b is given by the slope of the curve.

        t = time of UV exposure.

Similar behavior was observed for two other frequencies and the values of a and b, as a function of ω, are listed in Table II. All frequencies used were lower than 2 Hz in order to mini-

Table II: Values of a and b as a function of frequency (Equation (1))

Frequency: ω (cycles/min)	a	N_F at t = 0	b
5	5.40	255×10^3	− 0.283
10	5.00	100×10^3	− 0.263
50	4.50	32×10^3	− 0.253

mize any thermal effects. The constant-strain mode was chosen to prevent sample failure by creep. It appears that a decreases as ω increases while b decreases also but less steeply (as a matter of fact, we can consider that, within the error margin of ± 10%, b remains constant.)

The relationship between fatigue life, oxidation extent and UV exposure is illustrated in Fig. 3 where both $\log N_F$ and log [carbonyl] are plotted vs. UV exposure (for ω = 10 cycles.) It can be seen that as the time of UV exposure increases, i.e., as [carbonyl] increases, N_F decreases sharply. Fig. 4 shows that $\log N_F$ depends in a linear fashion on log[carbonyl]. Such a de-

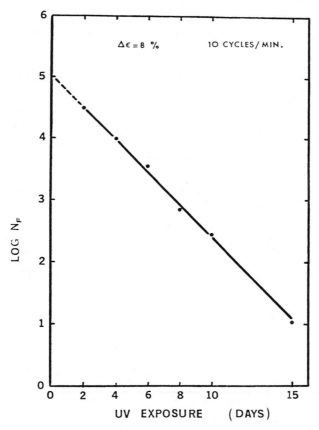

Figure 2.   Dependence of fatigue life (N_f, the number of cycles to failure) on UV exposure time for LDPE. ($\Delta\varepsilon = 8\%$, $\omega = 10$ cycles/min).

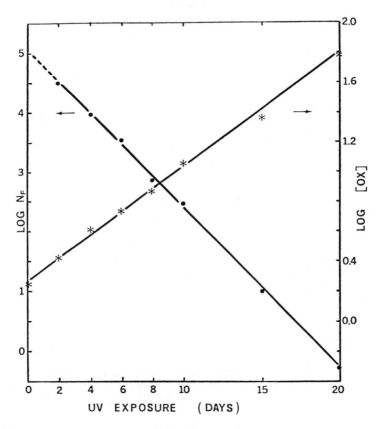

Figure 3.   Fatigue life and carbonyl content of LDPE as a function of
UV exposure ($\Delta\varepsilon$ = 8%, ω = 10 cycles/min.).

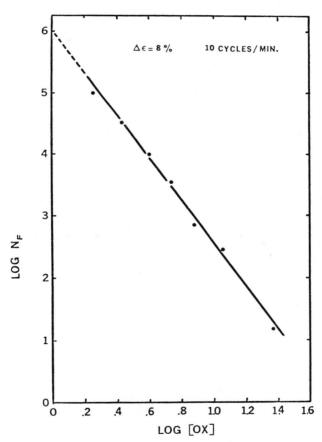

Figure 4.   Fatigue life vs. carbonyl absorbance of photooxidized LDPE.

$$N_F = A[carbonyl]^B \qquad (2)$$

where A and B are constants, depending on material characteristics and UV exposure and fatigue test conditions. A and B were calculated (from the intercepts and slopes, respectively, of plots similar to Fig. 4.) for all three frequencies used; the values are listed in Table III. It appears that while B remains

Table III: Values of A and B as a function of frequency
(Equation (2))

Frequency: ω (cycles/min)	A	B
5	6.25	− 3.52
10	6.00	− 3.50
50	5.50	− 3.47

Note: A will be the fatigue life of LDPE films containing 1% of carbonyl absorbance at 1716 cm^{-1} (most commercial films show such amounts of chromophores, probably resulting from oxidation during processing.)

more or less constant (therefore, B can be assumed not to depend on frequency) A decreases as ω increases. This is attributed mostly to the higher deformation speed with higher frequency (similar to stress-strain experiments where the higher the deformation rate, the sooner the sample failure.)

The relationship between fatigue life and carbonyl content can be explained as follows: according to the photooxidation mechanism of PE, carbonyl groups result mainly from a Norrish type II reaction, i.e., for each carbonyl formation, there is a scission of a segment of a molecule chain. Such scission creates a defect in the structure which can grow and propagate into a microcrack under application of a load. Under cyclic loading, it is understandable that the number of cycles the sample can sustain will be directly related to the number of defects (such as microcracks, microvoids ...), as is clearly described by equation (2).

More work is needed to see if equation (2) holds for different strain amplitudes and different material characteristics (different crystallinities, molecular weights ...).

Effects of Uniaxial Elongation: LDPE films were first stretched to different elongations (deformation speed: 1 inch/min.), then photooxidized while being kept stretched. The extent of oxidation, at a given UV exposure (10 days at 40°C), as a function of draw ratio is illustrated in Fig. 5. (The interpretation of the data has been reported, in more detail, in our previous paper (11), and only a brief summary is given here.) There is enhancement of the degradation process due to the deformation. The enhancement process is related to the deformation mechanism; the larger the disrupture of the film structure (necking region development) the greater the enhancement. The orientation effects, which takes place for $\lambda > 4$, tend to reduce the enhancement.

The reduction of degradation enhancement due to orientation is better seen when samples are stretched and then the time to fail, under UV radiation, is recorded. The results are shown in Fig. 6 where one should notice the break in scale for the reference (non-oxidized) sample. There is a drastic decrease in failure time (F.T.) for low draw ratios $1 < \lambda < 1.7$. This can be attributed to stored elastic energy which makes the chemical bonds more reactive toward UV, even at low stress levels. As λ increases and the polymer structure becomes more and more oriented, F.T. increases steeply before reaching a plateau once the orientation process is more or less completed. If we consider that photooxidation is oxygen diffusion controlled (1-5), the orientation effect is to decrease such diffusion by making the structure much more compact so that the degradation will be reduced.

In oriented samples, oxidation is much more concentrated in the surface layer, thus decreasing the formation of microcracks within the bulk of the sample which increases its ability to resist failure under UV radiation.

Effects of Cyclic Uniaxial Stretching: LDPE films were fatigued ($\Delta\epsilon = 20\%$, $\omega = 10$ cycles/min., for 10^4 cycles) before being put in the weatherometer in "the relaxed state", i.e., with free ends. Their photooxidation behavior is compared to that of non-fatigued (reference) samples in Fig. 7. The fatigued samples show more degradation which is attributed to damages resulting from cycling (fatigue damages: microcracks, microvoids, microcrazes ...). Such damages were clearly observed by optical microscopy (12) and are known to enhance the susceptibility of LDPE films to photodegradation. As N increases, the fatigue damages will increase resulting in more and more degradation as shown in Fig. 8.

An increase in strain amplitude ($\Delta\epsilon$) also increases plastic deformation (non-recoverable elongation) as illustrated in Fig. 9. The number of cycles is given in a logarithmic scale, and the plastic deformation (P.D.) is given in percent: P.D. = $(\ell - \ell_0)/\ell_0$, where ℓ and ℓ_0 are the sample lengths after and before fatigue, respectively.

Such plastic deformation is accompanied by an orientation effect which, as we have seen, tends to lower degradation. Figs. 10 and 11 show that samples fatigued for 10^4 cycles degrade less pendence can be described by the following equation:

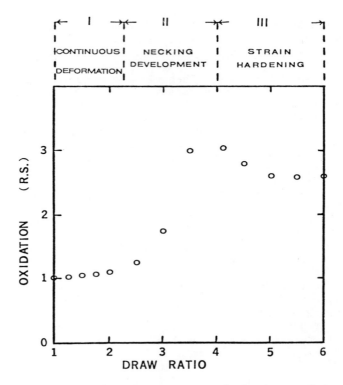

Figure 5.   Dependence of oxidation extent (relative scale) of LDPE after UV exposure for 10 days at 40°C on uniaxial elongation (draw ratio).

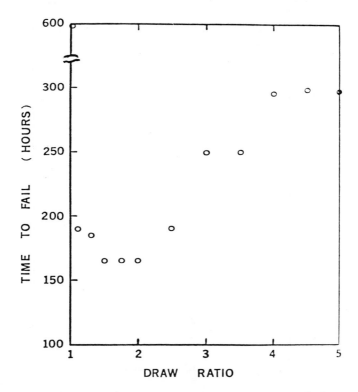

Figure 6.   Failure time under UV exposure vs. uniaxial draw ratio of LDPE.

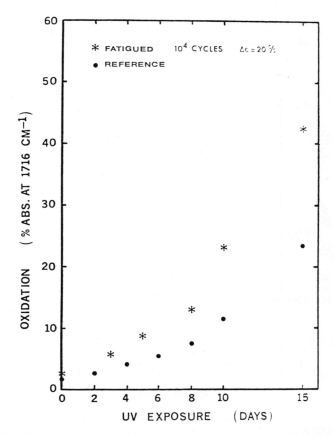

Figure 7.   Oxidation vs. UV exposure for reference and fatigued samples
($\Delta\varepsilon$ = 20%, 10^4 cycles, ω = 10 cycles/min.).

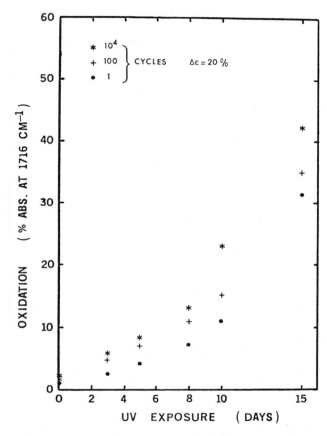

Figure 8. Effect of number of uniaxial deformation ($\Delta\varepsilon$ = 20%) cycle on photooxidation as a function of UV exposure time.

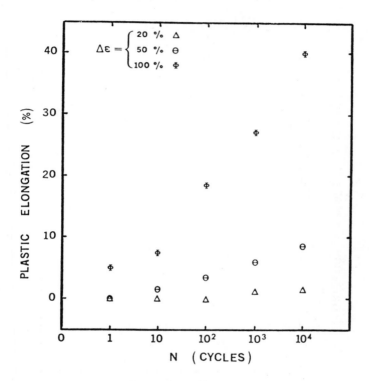

Figure 9.   Dependence of plastic deformation on deformation amplitude
as a function of the number of deformation cycles.

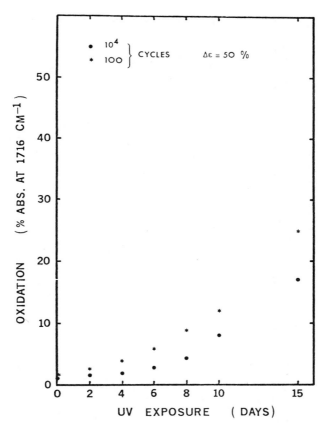

Figure 10.   Effect of number of uniaxial deformation cycles ($\Delta\varepsilon$ = 50%) on photooxidation as a function of UV exposure time.

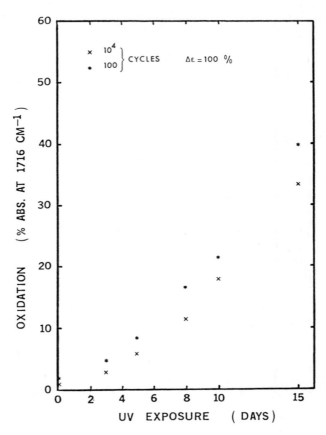

Figure 11.   Effect of number of uniaxial deformation cycles ($\Delta\varepsilon$ = 100%) on photooxidation as a function of UV exposure time.

than those subjected to 100 cycles. We are plotting only 2
curves in each case for the purpose of figure clarity. The
overall trend of all data was that degradation increases as N is
increased from 1 to 100 cycles; after 100 cycles, the oxidation
decreases.

The orientation was estimated by infrared dichroic ratio
measurements for different absorption bands. Dichroic ratios for
the 2016 cm^{-1} band, which contains both amorphous and crystalline contributions (13), are shown in Fig. 12. It is clear that
as $\Delta\varepsilon$ and N increase, more orientation takes place which will
reduce degradation. The orientation tends to "overcome" the
fatigue damages effects with respect to photodegradation. The
two processes—orientation and fatigue damages are competitive
and, depending on $\Delta\varepsilon$ and N, one or the other process has more
impact on the photooxidation susceptibility of the material.

Effects of Biaxial Stretching: LDPE films were biaxially
stretched, in one step, then clamped in frames (to maintain the
elongation) and put in the weatherometer. Their photooxidation
as a function of engineering strain ($\varepsilon = \varepsilon_1 = \varepsilon_2$ = engineering
strain in both directions) is shown in Fig. 13, for a UV exposure
of 10 days at 40°C. The oxidation content is given in a relative
scale, normalized with respect to the unstretched sample, and
corrected for change in thickness using the 1378 cm^{-1} band. We
see that the biaxially stretched samples exhibit more degradation
than the uniaxially ones. This is attributed mainly to two factors: i) further decrease in thickness, and ii) more constraint
applied during the biaxial stretching.

We could not biaxially stretch, in one single step and at
room temperature, films for $\lambda > 1.63$ (probably because of crystallinity). Therefore, to get higher elongations, biaxial
stretching in two steps (sequential) was employed. The oxidation
behavior as a function of draw ratio (λ is the same in both
directions) is given in Fig. 14. The oxidation scale is as described above for Fig. 13.

The points to notice in Fig. 14 are: i) there is enhancement of degradation and, ii) there are three stages in the enhancement process. These points are explained as follows: the
enhancement can be attributed to a combination of different factors, mainly decrease in thickness, applied constraint, and disrupture of structure. The presence of three stages is related to
the stages of deformation achieved during biaxial stretching.
Okajima, Tanaka et al. (14-25), working on biaxial stretching effects in polypropylene films, reported that the deformation mechanism has three distinct stages, each one corresponding to a different type of orientation. The different stages are described
as follows:

region 1: for low λ; mostly only amorphous orientation.
region 2: for $1.7 < \lambda < 3$; in this region, inclination of

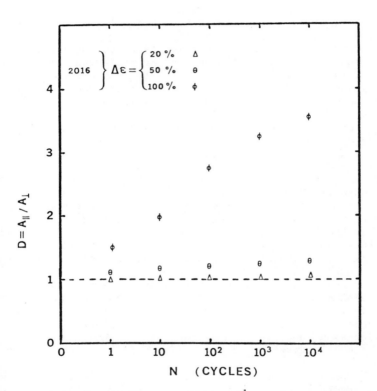

Figure 12.   Dichroic ratio (for the 2016 cm^{-1} band) as a function of number of deformation cycles for different deformation amplitudes.

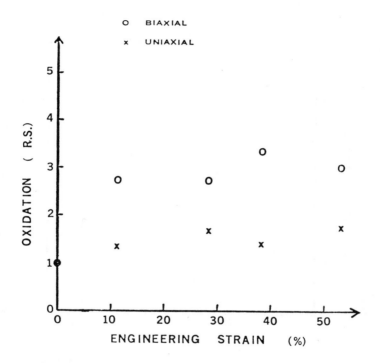

Figure 13.   Comparison of the effects of uniaxial and biaxial stretching (concurrent directions) on photodegradation (10 days at 40°C) as a function of engineering strain.

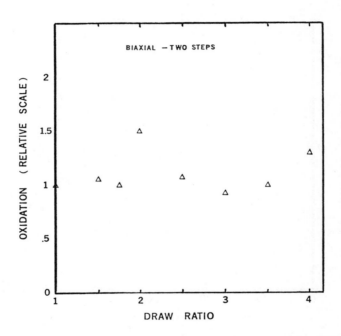

Figure 14. Effects of biaxial stretching (sequential directions) on photooxidation as a function of the draw ratio.

lamellae and tilting of chain axis in lamellae occur in order to "orient" chains parallel to the film surface. At such draw ratios, an "optically balanced state" is reached (19).
      region 3: for λ > 3: mostly unfolding of chain molecules from lamellae.
      A close look at Fig. 14 shows that there is good correspondence between deformation regions and oxidation stages. First, there is an increase in oxidation which can be attributed to stored elastic energy. In the second stage, 1.7 < λ < 3, since there is an "optically balanced state", i.e., the structure of the stretched film is very similar to that of the non-deformed one, we expect the samples to show similar oxidation content. This is seen by the decrease, then levelling off of the carbonyl content to a value close to that of the reference sample. At higher draw ratios, the stress effect (applied constraint) becomes more and more significant and the samples will undergo more degradation as shown by the increase in oxidation for λ > 3.

Conclusions

      All types of deformation—uniaxial, biaxial and cyclic uniaxial stretching—enhance the photooxidation of LDPE films. A close relationship exists between the enhancement process and the deformation mechanism: more disrupture of the structure results in a larger enhancement. Damages such as microcracks, microvoids, microcrazes—resulting from disrupture or fatigue effects—increase the degradation rate while orientation decreases it. The larger degradation extent exhibited by biaxially stretched samples (by comparison to uniaxially elongated films) is attributed to the further decrease in thickness and more constraint applied.

Acknowledgments

      The Fellowship support of SONATRACH (National Oil and Gas Company of Algeria) is gratefully acknowledged.

Literature Cited

1. Ranby, B., Rabek, J. F., "Photodegradation, Photooxidation and Photostabilization of Polymers", John Wiley & Sons, New York, 1975.
2. Kamal, M. R., Ed. "Weatherability of Plastic Materials", Appl. Polym. Symp. 1967, 4, Interscience, New York.
3. Hawkins, W. L., Ed. "Polymer Stabilization", Wiley-Interscience, New York, 1972.
4. McKellar, J. F., Allen, N. S., "Photochemistry of Man-Made Polymers", Applied Science Publishers, London, 1979.

5. Hawkins, W. L., "Oxidative Degradation of High Polymers", in
    Oxidation and Combustion Reviews, Tipper, C.F.H., Ed., Vol.
    I., Elsevier, New York, 1965.
6. Howard, K. W., "The Effects of Weathering on the Engineering
    Behavior of Plastic Films", Ph.D. Thesis, University of Cali-
    fornia, Davis, 1976.
7. Peterlin, A. J. Mat. Sci. 1971, 6, 490.
8. Pabiot, J.; Verdu, J. Polym. Eng. Sci. 1981, 21(1), 32.
9. Akay, G.; Tincer, T. Polym. Eng. Sci. 1981, 21(1), 8.
10. Akay, G.; Tincer, T. Europ. Polym. J. 1980, 16, 597.
11. Benachour, D.; Rogers, C. E., Photodegradation and Photo
    stabilization of Coatings, Winslow, F. H.; Pappas, S. P.,
    Eds., ACS Symp. Ser. 1981, 151, 263.
12. Benachour, D., "Effects of Deformation on the Photodegrada-
    tion of Low Density Polyethylene Films", Ph.D. Thesis, Case
    Western Reserve University, Cleveland, 1982.
13. Read, B. E.; Stein, R. S. Macromolecules 1968, 1(2), 116.
14. Okajima, S.; Kurihara, K.; Homma, K. J. Appl. Polym. Sci.
    1967, 11, 1703.
15. Okajima, S.; Homma, K. J. Appl. Polym. Sci. 1968, 12, 411.
16. Okajima, S.; Homma, K.; Masuko, T.; Tanaka, H. J. Polym. Sci.
    1969, A-1(7), 1997.
17. Okajima, S.; Masuko, T.; Tanaka, H. J. Polym. Sci. 1970,
    A-2(8), 1565.
18. Okajima, S.; Masuko, T.; Tanaka, H. J. Polym. Sci. 1969,
    A-1(7), 3351.
19. Okajima, S.; Mori, K.; Morita, M.; Kurihara, K.; Tanaka, H.
    J. Polym. Sci. 1971, B-9, 729.
20. Okajima, S.; Masuko. T.; Tanaka, H. J. Appl. Polym. Sci.
    1972, 16, 441.
21. Okajima, S.; Iwato, N.; Tanaka, H. J. Polym. Sci. 1971, B-9,
    797.
22. Okajima, S.; Masuko, T.; Tanaka, H. J. Appl. Polym. Sci.
    1973, 17, 1715.
23. Okajima, S.; Iwato, N.; Tanaka, H. J. Appl. Polym. Sci.
    1973, 17, 2533.
24. Okajima, S., Iwato, N.; Tanaka, H. J. Appl. Polym. Sci.
    1975, 19, 303.
25. Okajima, S.; Tanaka, H. J. Polym. Sci. 1977, B-15, 349.

RECEIVED December 27, 1982

PHOTOVOLTAIC AND RELATED APPLICATIONS

Luminescent Solar Concentrators: An Overview

A. H. ZEWAIL and J. S. BATCHELDER [1]

California Institute of Technology, Arthur Amos Noyes Laboratory of Chemical Physics, Pasadena, CA 91125

The Luminescent Solar Concentrator (LSC) offers the possibility of reducing the cost of photovoltaic solar energy conversion through the use of light pipe trapping of luminescence. Three concepts govern the performance of an LSC: light concentration, light pipe trapping of luminescence, and photovoltaic conversion. We present prototype performance data as well as a simple model which predicts the light intensity gain and efficiency from molecular parameters.

The LSC Concept: Collection and Concentration of Solar Energy

Solar energy is an appealing alternative energy source, with the acclaimed advantages of being nonpolluting, renewable, widely distributed, and of delivering peak power at the times of peak loads. The cost of converting solar energy to other forms is the principle barrier to its use, especially in photovoltaic applications. The Luminescent Solar Concentrator (1), or LSC, is being developed as a possible means of reducing the cost of solar photovoltaic conversion. It concentrates the incident sunlight so that a fixed area of solar cells can produce more power. The LSC differs from conventional optical concentrators such as mirrors or lenses in that it does not require even seasonal tracking to achieve reasonably high flux gains.

An LSC is a plate of transparent material, such as glass or plastic, which contains luminescing centers that absorb and then emit light (e.g., phosphors or organic laser dyes). Sunlight enters the upper face of the plate and is partially absorbed by these centers. A fraction of the resulting luminescence is trapped by total internal reflection. Successive reflections transport the luminescence to edge-mounted solar cells, as shown in Figure 1.

There are three concepts (1) that can be thought to govern the performance of an LSC: these are light absorption, light

[1] Current address: 85 Allison Road, Katonah, NY 10536

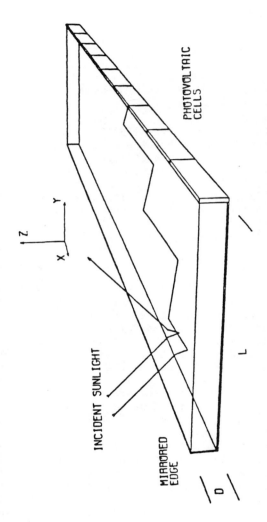

Figure 1. Operation of an LSC. Sunlight enters from above, passes through the plate, through an air gap to a mirror, and back through the plate. Part of this light is absorbed by luminescing material, which then emits into the plate. About 70% of this emission is trapped by total internal reflection. This light propagates (with self-absorption) in the plate until it is absorbed by the edge-mounted cells.

pipe trapping of luminescence (and self-absorption), and photovoltaic conversion. Since solar cells are in general the most expensive part of the converter, one wishes to span the area directed at the sun with a sort of concentrator which will funnel the light into a smaller area of cells. This idea imitates nature's use of chlorophyll as an antenna for absorbing sunlight and transmitting the resulting excitation to a center for chemical reaction. Several schemes (1-8) have been proposed for increasing LSC efficiency using variations of these ideas.

The ability of an LSC to concentrate light can be understood very simply. Suppose that the surface of the LSC plate which faces the sun has an area A_{face}, and that the edge on which the solar cells are mounted has an area A_{edge}. We refer to the ratio of the area of the face to the area of the edge as the geometric gain for the plate: $G_{geom} = A_{face}/A_{edge}$. This geometric gain is analogous to the concentrating power of a mirror or lens. If an LSC was completely efficient, such that all of the useful sunlight entering the face of the plate emerges from the edge, then the light emerging from the edge would be brighter than the sunlight by a factor equal to the geometric gain. Can the output light intensity be made arbitrarily large by just making the plate large enough? Not surprisingly, the answer is no. The limitations on the output intensity and on the overall efficiency are dictated by the details of self-absorption and of light pipe trapping of luminescence.

In this review, we present an overview of the LSC concept--we emphasize the physics of the device, which is governed by light absorption and concentration. Focus will be on prototype systems made of dyes in plastics, but no details will be given concerning other systems, such as inorganic ions in glasses (for a review see Reference 9) or concerning theoretical developments (for details see References 1, 5, and 10). Finally, we present photovoltaic performance data, and conjecture on future developments.

Light Pipe Trapping of Luminescence: Geometrical Effects

Luminescence from the interior of an LSC plate incident on the faces at an angle greater than the critical angle will be totally internally reflected. (The critical angle in this case is given by $\sin^{-1}(1/n)$, where n is the index of refraction of the plate.) The critical cone originates at the point of luminescence and forms a critical angle everywhere it intersects a surface. If the luminescence is isotropic and the LSC is planar, the probability that the luminescence will escape out of the critical cones is (1):

$$P = 1 - \sqrt{1 - 1/n^2} \qquad (1)$$

Equation (1) contains the implicit approximation that the luminescing dye molecules show no angular dependence on their absorption and emission. Actually, to a good approximation, dye molecules typically appear as electric dipole antennas, usually with the absorption dipole nearly collinear with the emission dipole. This decreases the calculated amount of light which is trapped, because the incident sunlight will be mostly absorbed by dipoles which are oriented in the plane of the LSC, and these dipoles have a greater chance of emitting into the critical cones. If θ_s is the angle of incidence of the sunlight with respect to the face of the LSC, we have shown that the probability that the subsequent luminescence will escape out of the critical cone is (1,5):

$$P(\theta_s) = 1 - \sqrt{1 - 1/n^2} \ (1 - \frac{1}{10n^2} + \frac{3 \ sin^2(\theta_s)}{10n^4}) \qquad (2)$$

For example, most prototype devices use polymethyl methacrylate (PMMA, trade name PLEXIGLASS) as a substrate material. PMMA has an index of refraction of about 1.49. This produces an escape probability of 26% for isotropic absorption and emission, and an escape probability of 29% for dipole absorption and emission of perpendicularly incident light.

If the index of refraction of the substrate material is increased, the fraction of emission trapped in the plate is also increased. It has been shown (1,5) that the optimal index for LSCs with no antireflection coating is about 2. Surface reflection losses become important for higher indices unless antireflection coatings are used.

We have shown that additional light can be trapped in the LSC plate if the surface of the plate facing the sun is concave (1, 6). Such a distortion causes more light to escape from the bottom face than from the top. However, this light can be partially returned to the LSC by means of a separate backing mirror.

A variety of substrate materials and luminescing centers have been studied. The first were neodynium laser glass and rhodamine-6g doped PMMA (3-5, 7, 8). Inorganic ions (9) in glasses such as neodynium have an important advantage due to their resistance to photodegradation; however, the plates of these materials have low efficiencies due to the relatively low quantum efficiency. Organic laser dyes have become the standard luminescing material. These are dissolved in the monomer prior to polymerization, and the resulting combination is either cast or applied as a film to another clear substrate material such as glass. We also made LSC's by diffusing the dye into a pre-casted polymer, or by trapping dye solutions between glass plate and mirror (5).

The Spectroscopy of Dyes in LSC's

Figure 2 shows a typical absorption and emission spectrum for the organic laser dyes, which in this case is DCM. The spectrum on the right is the absorption spectrum or extinction coefficient which is measured in units of liters per mole centimeter. The

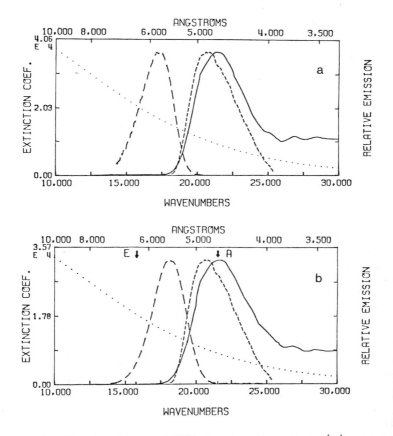

Figure 2. Spectroscopy of DCM dye in chloroform (a) and in PMMA (b). Note the change in Stokes shift. The spectra of many other dyes were recorded similarly.

curve on the left is the normalized luminescence spectrum (fluorescence). We have taken the luminescence spectrum to be a probability distribution for the wavenumber of the emitted light, such
that the area under the curve is equal to one. The emission peak
is shifted to lower energy with respect to the absorption peak by
the Stokes shift. Typical quantum efficiencies for organic laser
dye molecules in LSC's are about 90%. We have characterized the
spectra of over 15 dyes in LSC's. From these studies (see references) we learned about the degree of overlap between absorption
and emission, the degree of energy transfer in multiple dye LSC's,
and the spectral homogeneity of dyes dispersed in the polymer.
From the dependence of the luminescence spectra on excitation
energy and LSC temperature it is clear that there is some spectral
inhomogeneity (Figure 3) even at room temperature.

The Photon Transport Problem and Self-Absorption: LSC Gain and Efficiency

It is apparent from Figure 2 that there is some overlap between the absorption and emission spectra. This is true for many
luminescing materials in general, and leads to the effect of self-
absorption. Luminescence can in principle travel long distances
in the LSC plate on the order of meters. The probability that
this light will be absorbed by some other similar dye molecules is
given (in a simple picture) by the Beer-Lambert law: $1-10^{-CL\varepsilon(\bar{\nu})}$,
where L is the pathlength, C is the concentration, and $\varepsilon(\bar{\nu})$ is the
molar extinction coefficient at $\bar{\nu}$ (wavenumber, λ^{-1}). It is useful
to refer to the emission following the initial absorption of sunlight as the first generation emission. Any of this first generation emission which is self-absorbed and re-emitted is called second generation emission. (Third generation emission results from
self-absorption of second generation emission, and so on.)
Let r be the average probability of luminescence outside of
the critical cones being self-absorbed, and similarly $\bar{r}$ be the
probability that luminescence in the critical cones will be self-
absorbed. The collection probability $Q^{(1)}$ that first generation
luminescence will arrive at the edge of the plate following excitation from sunlight is the quantum efficiency of luminescence (η)
for the dye times the fraction of the luminescence which is emitted
outside of the critical escape cones times the probability that
this trapped luminescence will not be self-absorbed:

$$Q^{(1)} = \eta(1 - P)(1 - r) \tag{3}$$

Similar expressions can be derived for the higher order generations. Summing all of the generations together gives the collection probability, Q, that an absorbed solar photon will arrive at
the edge-mounted solar cell:

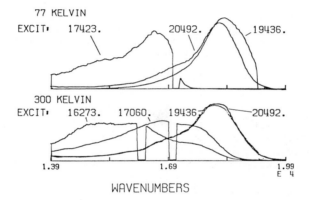

Figure 3.   Emission of rhodamine-575.   The lower three plots
represent emission in cast PMMA at room temperature for three
different excitation energies.   The upper plots are similar
spectra taken at liquid nitrogen temperatures.   All spectra
are normalized to unit area.

$$Q = Q^{(1)} + Q^{(2)} + \ldots = \frac{\eta(1 - P)(1 - r)}{1 - \eta[\bar{r}\, P + r(1 - P)]} \qquad (4)$$

For simplicity, we will assume that the edge mounted solar cell
has the same quantum efficiency for all incident light above its
bandgap energy. Let the air mass one (AM1) efficiency of the cell
facing the sun directly be η_{cell}. If I is the total solar flux of
sunlight absorbed by the LSC, then the efficiency of the LSC-
cell combination (electrical power out/sunlight power incident) is
approximated by:

$$\text{Eff.} = \eta_{cell} \cdot Q \cdot S/I \qquad (5)$$

Similarly the light amplification or flux gain of the plate is the
ideal flux gain, G_{geom}, times the fraction of the useful sunlight
absorbed, S/I, times the probability that this absorbed light
will be transported to the cells, Q:

$$G_{flux} = G_{geom} \cdot Q \cdot S/I \qquad (6)$$

In arriving at Equations (5) and (6), we have ignored scattering
and absorption by the matrix material, reflection at the LSC-cell
interface, and variations in the output of the cells due to the
different spectrum and intensity of the incident light. These
effects are discussed elsewhere (1, 5). Suppose a plastic plate
with a geometric gain of 12 is impregnated with rhodamine-6g and
has 12% AM1 cells mounted on its perimeter. As we shall see later,
typical values for the fraction of the solar spectrum absorbed is
S/I = 30% with a collection efficiency of Q = 50%. This results
in an overall efficiency of 2% with a flux gain of 2. The enhanced value of S/I cannot be obtained in single dye LSC; multiple
dye LSC's serve this purpose quite efficiently as shown by Swartz
et al (3) and Batchelder et al (5).

    That the plate just mentioned should have absorbed 30% of the
useful incident sunlight is calculated easily enough by convolut-
ing the solar spectrum with the absorption spectrum across the
thickness of the plate. To increase the fraction of light absorbed and insure an efficient cascade from one dye to another, several dyes are chosen in a sequence from the highest to lowest energy of their absorption bands, such that the luminescence spectrum of one dye overlaps strongly with the absorption of the dye
following it. If dyes chosen in this manner are mixed together in
an LSC plate, light absorbed by one dye will be emitted and absorbed by the next one, and this process is repeated until the
photon reaches the dye with the lowest energy absorption. (The
energy of excitation lost in cascading from one dye to the next is
dissipated in the plate.) The result is an LSC with a very broad
band absorption of solar energy, and with an emission energy that

function of the product of the concentration times the pathlength. can be made close to the band gap (Figure 4).

Measurements of Self-Absorption and Performance

The principle factor reducing the overall efficiency of present LSC's is self-absorption of luminescence due to the non-zero overlap of the absorption and emission spectra. In the past three years, one of our major research efforts on LSC's has focussed on measuring and understanding these self-absorption probabilities. Four techniques have proven useful in measuring self-absorption: direct measurement of the steady state absorption and emission spectrum, observation of the spectral shift of the emission with increasing sample pathlength, time-resolved measurements from picosecond pulsed excitation, and polarization anisotropy measurements. Before highlighting these different methods, we should mention the approximations used in the theory (1, 5) developed for explaining self-absorption in LSC's. In developing the self-absorption model, which explains the major features of the performance, we assume the presence of a simple reabsorption event, and we develop a reabsorption probability that is averaged over the entire luminescence band. In References 1 and 5, the details of the model and the conditions for its validity are given. The agreement between theory and experiments is satisfactory, and we use the self-absorption probability result only in an average sense for performance evaluation.

Steady-State Spectral Convolution. The steady state absorption and emission spectra of dilute dye samples can be measured using standard spectroscopic techniques. Once the extinction coefficient, $\varepsilon(\bar{\nu})$, and the normalized luminescence spectrum, $f(\bar{\nu})$, are known for a particular dye, the self-absorption probability r over a pathlength L in the sample containing the dye at a concentration C is given by

$$r = \int_0^\infty d\bar{\nu} \, f(\bar{\nu}) \left[1 - 10^{-LC\varepsilon(\bar{\nu})} \right] \qquad (7)$$

An interesting outcome of Equation 7 is that it produces an average pathlength for a sample with a known self-absorption rate. For example, analytic techniques (1) exist for summing the output light for a particular infinite ribbon geometry of LSC, including self-absorption effects. This analytic collection efficiency can in turn be converted to the appropriate self-absorption probabilities using Equation 4, and from there into the average pathlength traversed by the collection radiation. The result of this manipulation is that the average pathlength of sample traversed by collected luminescence in an infinite ribbon LSC is approximately equal to the width of the ribbon, for concentrations and plate thicknesses such that the peak optical density through the thickness of the plate is approximately equal to one. Self-absorption inside the critical cones is usually negligible in such cases.

EXCITATION SPECTRA SHOWING ENERGY TRANSFER
IN MULTIPLE DYE METHANOL SOLUTIONS

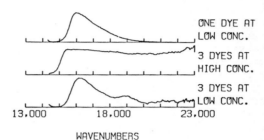

WAVENUMBERS

Figure 4a.   Three excitation spectra of methanol dye solutions with emission detection at 6400 Å. The three spectra
correspond to the same solutions used in Figure 4b.

EMISSION SPECTRA SHOWING ENERGY TRANSFER
IN MULTIPLE DYE METHANOL SOLUTIONS.

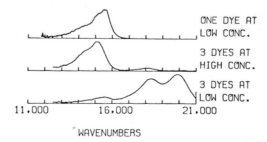

WAVENUMBERS

Figure 4b.   Three emission spectra of methanol dye solutions resulting from 4500 Å excitation. The top spectrum
is from a micromolar oxazine-720 solutions, and the lower
spectra are hundred micromolar and micromolar concentrations,
respectively, of coumarin-540, rhodamine-640, and oxazine-720.

The important point here is that for a given L, knowledge of ε and f produce r.

Pathlength Dependent Spectral Shift. Equation 7 implies that as the luminescence passes through more LSC material it will get shifted (into the red) since most of the overlap of the absorption and emission bands occurs at the high energy end of the emission spectrum. This effect can be measured directly, as is shown in Figure 5. A rod of LSC material was excited with a focussed spot of light from a mercury lamp (or a laser), and spectra were taken of the emission from the end of the rod as a function of the distance of the excitation from the end of the rod. The organic laser dye used was rhodamine-575, and the sample pathlengths for each spectrum in order of the most to least intense were 0.3, 1.0, 3.0, 10.0, and 30.0 cm. The spectra have been corrected only for the system response of the detection (monochromator and the PMT), so that the amplitude of each spectrum corresponds to the intensities emerging from the end of the rod.

Figure 6 shows the results of an analytic calculation of the spectrum shift for the same system measured in Figure 5. The calculation involves an average of the Beer-Lambert law over all possible pathlengths and wavenumbers of the emission (5).

This technique suggests what is probably the simplest way to determine the probability of self-absorption empirically for a particular plate. The peak position of the luminescence spectrum can be calculated for a variety of self-absorption rates using the above formalism. Thereafter, the self-absorption rate for a particular device can be found (crudely) by a moderate resolution measurement of the peak position of the luminescence spectrum.

Time-Resolved Emission. When an LSC is excited with a pulsed light source whose duration is much less than a nanosecond, the light output from the sample will decay exponentially with time. If the concentration of the dye in the sample is increased, this exponential decay takes a longer time. We have shown (5) that the self-absorption rate is related to the ratio between the measured lifetime of the plate, τ, and the lifetime measured in the limit of low concentration, τ_0, in the following way:

$$r = [1 - \tau_0/\tau]/\eta(1 - P) \tag{8}$$

Equation 8 is true in the limit that the transit time for light in the plate is short compared to the lifetime, and that self-absorption in the critical cones is negligible (i.e., r of the critical cone, $\bar{r}$, is zero). Thus Equation 8, which can be easily rewritten to include $\bar{r}$, is not valid for large values of P. Note that when $\tau = \tau_0$, r = 0, and when (1 - P) increases both r and τ will change. The smallest error bars in Figure 7 show the measured self-absorption rates for rhodamine-575 in methanol as a

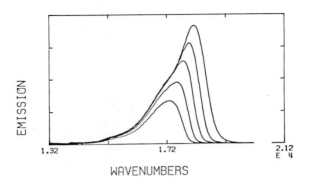

Figure 5. Emission spectra from a 92 micromolar methanol solution of rhodamine-575. The sample pathlengths, in order of most to least intense, were 0.3, 1.0, 3.0, 10.0, and 30.0 cm.

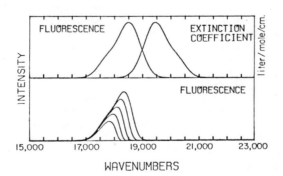

Figure 6. Analytic result for emission spectra as a function of sample pathlength. Experimental absorption and emission spectra were each fitted as the sum of two gaussians. These are used with our model to imitate the experiment shown in Figure 5. The Stokes shift used is 1000 cm^{-1}.

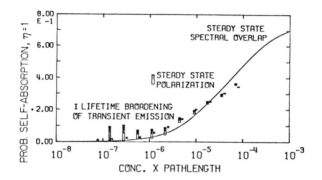

Figure 7.   Self-absorption probabilities for rhodamine-575.
This shows a juxtaposition of predicted self-absorption
probabilities for three measurement methods: spectral overlap convolution (solid), emission depolarization (boxes),
and time-resolved spectra (bars).   The second two techniques
are plotted assuming the quantum efficiency of luminescence
is one.

Due to the high accuracy possible with photon counting measurements of the lifetime, this technique yields the most accurate
measurement of any technique we have studied of the product ηr,
the quantum efficiency of luminescence and the self-absorption
probability, for low rates of self-absorption.

   Polarization Anisotropy. Since the dye molecules which absorb and emit light in the LSC behave as dipole antennas, they
have a tendency to emit light which is polarized in the same direction as a polarized source. For example, suppose that a randomly
oriented sample of dye molecules located at the origin is excited
by light polarized in the $\hat{z}$ direction. It has been shown that
luminescence in the $\hat{x}$ direction will have three times as much intensity in the $\hat{z}$ polarization component as in the $\hat{y}$ component (11).
If this luminescence is subsequently self-absorbed, the resulting
luminescence would be less polarized. The degree of polarization
can be measured by the reduced anisotropy, which is the parallel
intensity ($\hat{z}$ component) minus the perpendicular intensity ($\hat{y}$ component) divided by the sum of the parallel plus twice the perpendicular intensities:

$$RA = \frac{I_{\parallel} - I_{\perp}}{I_{\parallel} + 2I_{\perp}} \equiv A \tag{9}$$

The reduced anisotropy of first generation emission is therefore
$(3-1)/(3+2*1) = 2/5$. We have shown (5) that the reduced anisotropy of the ith generation of emission is $(2/5)^{2i-1}$. For
example, the reduced anisotropy of very high generations is nearly
zero (unpolarized). Due to effects such as rotation of the dye
molecule or the absorption and emission dipoles not being collinear, the first generation reduced anisotropy is usually less than
2/5 even at very low concentrations and pathlengths. Let $RA_0 (A_0)$
be the highest measured reduced anisotropy of a particular dye and
surrounding material, and $RA_e (A_e)$ be the measured reduced anisotropy at the concentration and pathlength of interest. The self-
absorption probability can be shown to be (5)

$$r = [1 - \frac{A_e}{A_0}]/\eta(1 - P) \ (1 - A_e A_0) \tag{10}$$

As before we have assumed that self-absorption in the critical
cones is negligible. Also, we have taken an average of the reduced anisotropy over all generations. Thus, Equation 10 cannot
be applied in the limit of $A_e \to 0$ since Q will be going to zero.
Note that when $A_e = A_0$, r becomes zero as expected. The error
boxes in Figure 7 show the measured self-absorption rates for
rhodamine-575 as computed using this technique. The solid line in
Figure 7 is found by inserting the measured absorption and emission spectra of rhodamine-575 in Equation 7. There is good

agreement among the three techniques, and very good agreement between the polarization and lifetime related measurements. The disparity between the spectral convolution technique and the other two is in the variation of the average pathlength of sample traversed by luminescence as the concentration is varied.

CODEs for Efficiency and Performance Evaluation. It is very useful to give prospective luminescent materials a "figure of merit" which measures the material's self-absorption characteristics (hence Q,..., etc.) in an LSC. We have defined (5) the critical optical density (CODE) to be the product of the optical pathlength L times the dye concentration C times the peak extinction coefficient $\varepsilon(\bar{\nu}_m)$ such that the luminescence from the dye has a 50% chance of being self-absorbed over that optical pathlength:

$$CODE \equiv L \cdot C \cdot \varepsilon(\bar{\nu}_m)$$

(11)

$$1/2 = \int_0^\infty d\bar{\nu}\ f(\bar{\nu})\ 10^{-LC\varepsilon(\bar{\nu})}$$

Typical CODEs for organic laser dyes are between 10 and 40. Most coumarin dyes are about 100, while some dyes with unusually large Stokes shifts, like DCM, have CODEs above 200. Clearly the Stokes shift comes into play directly!

The CODE of a dye can be used directly to determine the maximum flux gain that can be achieved in an LSC using that dye. Most LSCs use PMMA as a matrix material, which has an index of refraction of 1.49. Most of the laser dyes of interest have a quantum efficiency of about 90%. If we ignore self-absorption in the critical cones, we find from Equation 4 that a self-absorption probability of 50% produces a collection efficiency of about 50%. We next observe that if we are trying to minimize self-absorption, we should have the dye concentration in the plate as low as possible. The lowest concentration of the dye which will still produce reasonably good absorption of the incident sunlight will produce a maximum peak optical density across the thickness of the plate of one. If the dye concentration in the LSC is adjusted such that this is the case, then the CODE becomes the number of plate thicknesses through which the luminescence can pass before being 50% self-absorbed. Finally we note that since the average pathlength of collected emission in the sample is on the order of the sample width, and since the geometric gain of a square plate is given by the width divided by the thickness of the plate (if three sides are mirrored) then the CODE is the geometric gain of such an LSC such that it will produce a collection efficiency of about 50%.

If an LSC is designed such that its peak optical density across the thickness of the plate is equal to one and its geometric gain is equal to the CODE of the dye used, then the overall efficiency of the plate (Equation 5) can be approximated as

$$\text{Eff.} = \eta_{cell} \cdot S/2I \qquad\qquad (12)$$

and the light amplification of the plate (Equation 6) similarly becomes

$$G_{flux} = CODE \cdot S/2I \qquad\qquad (13)$$

This technique is especially useful in projecting the performance of high gain LSCs.

Prototype Efficiency Measurements. A variety of LSC devices have been tested by ourselves and others. Table I is intended to be a representative list of typical performance parameters. Again, the geometric gain is the ratio of the area exposed to the sun to the active area of the edge. The flux gain is the factor by which the short circuit current increased when attached to the plate, as opposed to facing the sun directly. The cell efficiency is the measured or assumed AM1 efficiency of the solar cells used (which in all cased were silicon). The collector efficiency is the total electrical power out divided by the total sunlight power incident on the plate.

In Table I devices B and D were built and tested by Owens-Illinois (8), the rest were built at Caltech. The dyes were contained in thin plastic films attached to the surface of a clear substrate in the OI case. Measurements were made under actual insolation, with the plate edges roughened and blackened where cells were not mounted. These plates have achieved the highest efficiencies, but their small geometric gains make them somewhat ineffective as concentrators. For example, cells mounted on LSC device D will have only a 70% increase in output over an equal area of cells facing the sun directly.

Device C was a meter square liquid cell which we built. The flux gain was measured under actual insolation. We calculate that the two dye combination used absorbed 30% of an AMO spectrum in a two pass (an LSC plate followed by a backing mirror) geometry. Device E has the highest flux gain of any LSC reported. It is a single dye PMMA plate containing DCM, a dye with a relatively smaller rate for self-absorption (the CODE for DCM is about 200 but depends on the solvent or matrix used).

Thermodynamics of LSC's. Several years ago, we initiated a program to calculate the ultimate gain expected for an LSC (12). We started by considering the application of the Winston-Rabl (13) idea to LSC-type systems. But, because the energy of the incident solar light is not preserved in an LSC, we used a detailed balance calculation which gives the intensity of the plate output as a function of size, ignoring self-absorption effects.

Table I.   Prototype performances.   Devices B and D were made and tested by Owens Illinois. The rest were made and tested at Caltech. The collector efficiency is the assumed AM1 cell efficiency times the flux gain divided by the geometrical gain, and corresponds to the electrical power output per solar power input.

Device	No.Dyes	Matrix	Geometrical Gain	Flux Gain	Assumed Cell Eff.	Collector Eff.
A	2	PMMA	23	2.1	18%	1.9%
B		thin film	11	1.3	21%	2.5%
C	3	ethylene glycol	36	3.8	18%	1.9%
D	3	thin film	11	1.7	21%	3.2%
E	1	PMMA	68	5.1	18%	1.3%
F	1	PMMA/DMSO	92	4.4	18%	0.9%

In Figure 8, we show the thermodynamic gain for an idealized LSC. We computed the gain for a plan geometry by substituting a black body cavity for the edge mounted cells, and then computing the temperature of the cavity by the detailed balance calculation. The straight diagonal line indicates the operation of concentrators using geometric optics such as mirrors and lenses. We considered an LSC absorbing in the region from 3500Å to 8000Å, emitting at 8300Å, and a geometric gain ranging from 1 to 10^5. Under the conditions of the model there is a considerable buildup of energy in the plate-cavity system. For example, an idealized LSC with no gain at all (G_{geom} = 1) would cause the cavity to rise to a temperature of 1000°K. A similar black body facing the sun directly would rise to about 400°K. It is difficult to explicitly include effects such as self-absorption that are seen in more realistic dye spectra, so that the results in their present form are overly optimistic in the predicted light output. In what follows we use the CODE method to obtain more realistic gains, and compare our results with that obtained using a generalized brightness theorem.

The thermodynamic limit to the performance of an LSC can be obtained by considering the incident sunlight and trapped light in the plate as two systems of photon gases in equilibrium. This system was first studied by Kennard (14) and Ross (15). Recently Yablonovitch (16) has applied these results to the LSC and obtained (from a generalization of the brightness theorem of optics to inelastic processes; optical elements which change the light energy):

$$\overline{G}_{flux} \leqslant \frac{\overline{\nu}_2^2}{\overline{\nu}_1^2} \; EXP \; (\frac{hc(\overline{\nu}_1 - \overline{\nu}_2)}{n \, k \, T}) \tag{14}$$

where $\overline{\nu}$ is the wavenumber of the light, and n is the refractive index. $\nu_1 - \nu_2$ is the Stokes shift which, as described in the text, conditions the self-absorption probability. Phenomenologically, the larger the Stokes shift the bigger the gain, as we expect from the theory of self-absorption outlined before. For a Stokes shift of 0.25 eV (~ 2000cm^{-1}) and at room temperature, G will typically be on the order of 636 according to the above equation. If G is limited to a reasonable value of 100 then the Stokes shift is about 1400cm^{-1}.

To compute gains for dyes in LSC's, Batchelder (12) made several calculations invoking Equation 14 and the CODE (see previous sections) of the dye. It was found that an LSC can achieve higher light concentrations when the Stokes shift was increased. Self-absorption ceases to limit the flux gain for a dye like DCM, whose Stokes shift corresponds to about 0.7 eV. Of course, in real calculations (12) one must include the spectral distribution of absorption and emission and not the average wavenumbers.

What about the optimal efficiency of an LSC? The model we used to compute this optimal efficiency assumes that the dye in an LSC absorbs all of the solar flux from the peak of its absorption

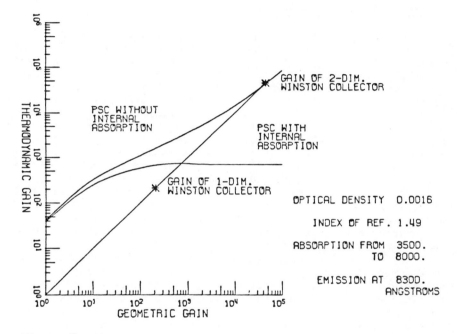

Figure 8.   Thermodynamic gain of an idealized LSC using a detailed balance of energy.

cutoff and the absorption edge of the solar cell (another kind of Stokes shift). Then, we calculate the E_{out} and E_{in} and compute η_{LSC}, the overall efficiency. We found that, for typical dyes, the optimal efficiency is 8.3%. Figure 9 shows the efficiency of LSC plates utilizing silicon or gallium arsenide cells as a function of the difference between the absorption cutoff of the dye and the bandgap of the semiconductor. If we repeat the calculation using a measured solar spectrum, we find the efficiency to be 9.3% for an absorption cutoff of 1.6 eV and an output voltage of 0.5 V.

Problems and Future Objectives

Dye stability is one important problem. We and others (17) have observed that many of the dye molecules tested have approximately a 50% probability of photodegrading over a number of excitations on the order of 10^6. This rate should be increased by about two orders of magnitude or more to obtain acceptable stability levels if an LSC is to have a 20 year lifetime. This can be seen in the following way. In order to minimize self-absorption, we can require that a typical plate have a peak optical density of one. If we also make the reasonable assumption that a typical peak extinction coefficient for the dye is 50,000 liters/mole cm, then there will be about 10^{20} molecules per square meter. (This argument will also pertain to the final dye in a multiple dye plate.) If the dye absorbs 30% of the usable visible solar spectrum, in 20 years the plate will have absorbed 10^{28} photons per square meter. Acceptable performance, therefore, requires that the quantum efficiency of photodegradation be at most 10^{-8} molecules per photon instead of the typical 10^{-6}.

Two areas of research emerge as the next logical step in LSC development. These are to examine our liquid LSC (single and multiple dye systems) in more detail, and to undertake the synthetic chemistry of dye optimization including the efficient energy transfer. We feel that the homogeneous spectra and large Stokes shifts characteristic of dyes in solution are highly desirable. Either the liquid cell design or some matrix-solvent combination should supercede the cast plastic design. The efficiency and gain should improve due to the higher Stokes shifts.

Preliminary work (5, 18) in this laboratory and at JPL indicates that the residual monomer can play an important role in photodegradation. (We can test the presence of residual monomers using the technique of fluorescence depolarization described before.) Thus, it is desirable to find new meterials and new schemes of attaching these dyes in or into the polymer to increase the stability and the Stokes shift (e.g., by using well-known techniques such as dimerization). Finally, an interesting problem which we plan to pursue is related to the physics of efficiency limitations: can one approach (or perhaps even overcome?) the so-called thermodynamic limit?

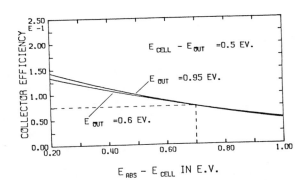

Figure 9. Collector efficiency vs. "Stokes shift." We
assume the sun is 5800 Å black body. The two cases plotted ar
are for Ga As (upper) and Si (lower). The dotted line indicates the operating point using a dye with Stokes shift
similar to that of DCM.

Acknowledgments

This work was supported in part by the Department of Energy Solar Energy Research Institute (Subcontract XF-1-1261-1). Parts of this article are based on results published elsewhere with Dr. T. Cole as co-author. It is our pleasure to acknowledge all the stimulating discussions and the collaboration with Dr. Cole. The efforts of Stuart Vincent and Dr. A. Gupta on the photodegradation problem are also acknowledged. A.H.Z. (to whom correspondence should be addressed, is an Alfred P. Sloan Foundation Fellow and Camille and Henry Dreyfus Foundation Teacher-Scholar. Finally, the thorough and conscientious efforts of the referee are greatly appreciated.

Literature Cited

1. Batchelder, J. S; Zewail, A. H.; Cole, T. Appl. Optics 1979, 18, 3090-3110.
2. Weber, W. H.; Lambe, J. Appl. Optics 1976, 15, 2299-2300.
3. Swartz, B. A.; Cole, T.; Zewail, A. H. Optics Lett. 1977, 1, 73-75.
4. Goetzburger, A.; Greubel, W. Appl. Phys. 1977, 14, 123-139.
5. Batchelder, J.S.; Zewail, A. H.; Cole, T. Appl. Optics 1981, 20, 3733-3754.
6. Batchelder, J.S.; Zewail, A. H. United States Patent 4,227,939.
7. Levitt, J.A.; Weber, W. H. Appl. Optics 1977, 16, 2684-2689.
8. Friedman, P. LSC Contract Report, Owens-Illinois, SERI Contract XS-9-8216-1, 1980.
9. Reisfeld, R.; Jorgensen, C. "Structure and Bonding"; Springer-Verlag 1982; Vol. 49, and references therein.
10. Some Monte-Carlo calculations of self-absorption has been obtained by Olson, R.; Loring, R.; Fayer, M. Appl. Optics 1981, 20, 2934.
11. For theory see, e.g., Gordon, R. G. J. Chem. Phys. 1966, 45, 1643. For measurements of RA of dyes in solutions see Millar, D. P.; Shah, R.; Zewail, A. H. Chem. Phys. Lett. 1979, 66, 435.
12. Batchelder, J. S.; Zewail, A. H.; Cole, T.; unpublished work. Batchelder, J. S.; Ph. D. Thesis, California Institute of Technology.
13. Winston, R. Solar Energy 1974, 16, 89. Rabl, A. Solar Energy 1976, 18, 93.
14. Kennard, E. H. Phys. Rev. 1918, 11, 29.
15. Ross, R. T. J. Chem. Phys. 1961, 46, 4590.
16. Yablonovitch, E. J. Opt. Soc. Am. 1980, 70, 1362.
17. Beer, D.; Weber, J. Opt. Comm. 1972, 5, 307.
18. Vincent, S.; Gupta, A.; Batchelder, S.; Cole, T.; Zewail, A. H.; unpublished work.

RECEIVED November 22, 1982

Polymeric Encapsulation Materials for Low-Cost, Terrestrial, Photovoltaic Modules

E. F. CUDDIHY and C. D. COULBERT—California Institute of Technology, Jet Propulsion Laboratory, Pasadena, CA 91109

P. WILLIS and B. BAUM—Springborn Laboratories, Enfield, CT 06082

A. GARCIA—Spectrolab, Inc., Sylmar, CA 91342

C. MINNING—Hughes Aircraft Co., Culver City, CA 90230

Solar cell modules must undergo substantial reductions in cost in order to become economically attractive as practical devices for the terrestrial production of electricity. Part of the cost reductions must be realized by the encapsulation materials which are used to package, protect, and support the solar cells, electrical interconnects, and other ancillary components. As many of the encapsulation materials are polymeric, cost reductions necessitate the use of low-cost polymers. This article describes the current status of low-cost polymers being developed or identified for encapsulation application, requirements for polymeric encapsulation materials, and evolving theories and test results of antisoiling technology.

The Jet Propulsion Laboratory manages the "Flat-Plate Solar Array (FSA) Project" for the Department of Energy. The project objective is to conduct research on photovoltaic arrays establishing their technical feasibility so that industry could meet a target price for modules of less than 70¢ per Wpk (in 1980 dollars) and with a minimum service lifetime of 20 years. Assuming a module efficiency of 10 per cent, which is essentially 100 W per m^2 at solar meridian, the capital cost of the modules can be alternately quoted as $70.00 per m^2. Out of this cost goal, $14.00 per m^2 is allocated for the encapsulation materials which includes both the cost of a structural panel, and edge seals and gaskets. At project inception, approx. 1975, the accumulative cost of encapsulation materials in popular use, such as RTV silicones, aluminum panels, etc., greatly exceeded $14.00 per m^2. Accordingly, the FSA project established a group called the "Environmental Isolation Task", to identify and/or develop as necessary new materials, and new material technologies in order to achieve the cost and life goal.

This article describes the status of this task group relative to the identification and development of an inventory of low-cost polymeric encapsulation materials (1 - 5), and describes evolving engineering requirements of an encapsulation system which relates to minimum usage of polymeric materials (6).

0097-6156/83/0220-0353$06.00/0

Construction Elements

To perform a survey for candidate materials or material classes which could first meet the FSA project cost goals, it was helpful to examine the design and construction features of commercial terrestrial photovoltaic modules, in order to identify the basic building blocks of encapsulation systems. A basic building block is defined as a construction component for which a distinct material is required. Currently, commercial photovoltaic modules can be classified according to two engineering design options, a substrate system, and a superstrate system. These design classifications refer to the method by which the encapsulated solar cells are mechanically supported. A substrate design means that the encapsulated cells are supported by a structural backside panel, and the superstrate design means that the encapsulated cells are supported by a transparent, sunside structural panel (e.g., glass). For these two design options, up to nine material components of an encapsulation system, called construction elements, can be identified. These construction elements are depicted in Figure 1 along with their designations and encapsulation functions. Note that not all of these construction elements need be incorporated in any given encapsulated module, but all current-day modules have combinations of these elements.

Low-cost candidates for the substrate panels are mild steel and hardboard, and glass is the lowest cost candidate for the structural superstrate panel. On a structural comparison basis, plastic materials used structurally as either a substrate or as a transparent superstrate are considerably higher cost (1,2). The low-cost candidate for the porous spacer is a non-woven E-glass mat (3,4,5). Low cost candidates for all of the other construction elements are polymeric.

Polymeric Encapsulation Materials

Pottant. The central core of an encapsulation system is the pottant, a transparent, polymeric material which is the actual encapsulation media in a module. As there is a significant difference between the thermal-expansion coefficients of polymeric materials and the silicon cells and metallic interconnects; stresses developed from the thousands of daily thermal cycles can result in fractured cells, broken interconnects, or cracks and separations in the pottant material. To avoid these problems, the pottant material must not overstress the cell and interconnects, and must itself be resistant to fracture. From the results of a theoretical analysis (6), experimental efforts (3), and observations of the materials of choice used for pottants in commercial modules, the pottant must be a low-modulus, elastomeric material.

Also, these materials must be transparent, processible, commerically available, and desirably of low cost. In many cases, the commercially available material is not physically or chemically suitable for immediate encapsulation use, and therefore must also be amenable to low-cost modification. The pottant materials must have either inherent weatherability (retention of transparency and mechanical integrity under weather extremes) or the potential for long life that can be provided by cost-effective protection incorporated into the material or the module design.

In a fabricated module, the pottant provides three critical functions for module life and reliability:

(1) Maximum optical transmission in the silicon solar-cell operating wavelength range of 0.4 to 1.1 μm.

(2) Retention of a required level of electrical insulation to protect against electrical breakdown, arcing, etc., with the associated dangers and hazards of electrical fires, and human safety.

(3) The mechanical properties to maintain spatial containment of the solar cells and interconnects, and to resist mechanical creep. The level of mechanical properties also must not exceed values that would impose undue mechanical stresses on the solar cell.

When exposed to outdoor weathering, polymeric materials can undergo degradation that could affect their optical, mechanical, and electrical insulation properties. Outdoors, polymeric materials can degrade from one or more of the following weathering actions:

(1) UV photooxidation.
(2) UV photolysis.
(3) Thermal oxidation.
(4) Hydrolysis.

For expected temperature levels in operating modules, ≈ 60°C in a rack-mounted array and possibly up to 80°C on a rooftop, three generic classes of transparent polymers are generally resistant to the above weathering actions: silicones, fluorocarbons, and PMMA acrylics. Of these three, only silicones, which are expensive, have been available as low-modulus elastomers suitable for pottant application.

Therefore, all other transparent, low-modulus elastomers will in general be sensitive to some degree of weathering degradation. However, less weatherable and lower-cost materials can be considered for pottant application if the module design can provide the necessary degree of environmental protection. For example, a hermetic design, such as a glass superstrate with a metal-foil back cover and appropriate edge sealing, will essentially isolate the interior pottant from exposure to oxygen and water vapor, with the glass itself providing a level of UV shielding.

The situation is different for a substrate module however, which will employ a weatherable plastic-film front cover. Because all plastic films are permeable to oxygen and water vapor (the only difference is permeation rate), the pottant is exposed to oxygen and water vapor, and also to UV if the plastic film is non-UV screening. Because isolation of the pottant from oxygen and water vapor is not practically possible in this design option, it becomes a requirement that the pottant be intrinsically resistant to hydrolysis and thermal oxidation, but sensitivity to UV is

allowed if the weatherable front-cover plastic film can provide UV
shielding.

Therefore surveys (1,4) were done to identify the lowest-
cost, transparent, low-modulus elastomers with expected resistance to hydrolysis and thermal oxidation at temperatures up to
80°C, but these materials were allowed to be sensitive to UV deterioration. It was envisioned that if such a set of pottant candidates were selected on the basis of a less-protecting sub-
strate-module design, they would also be useable in a potentially
more-protecting glass-superstrate design. In addition to the
foregoing requirement for candidate pottant selection, these materials must also be capable of being fabricated into modules by
industrial fabrication methods. This requirment becomes important
as it is desirable to have industrial evaluation of the materials
being developed, and therefore the materials must be readily useable on commerical equipment. The two industrial fabrication
techniques in common use are lamination and casting.

With all of these requirements, four pottant materials have
emerged as most viable and are currently in various stages of development or industrial use. The four pottants are based on ethylene vinyl acetate (EVA), ethylene methyl acrylate (EMA), poly-
n-butyl acrylate (P-n-BA), and aliphatic polyether urethane (PU).
EVA and EMA are dry films designed for vacuum-bag lamination at
temperatures up to 150°C. Above 120°C during the lamination process, EVA and EMA undergo peroxide crosslinking to tough, rubbery
thermosets. P-n-BA and PU are liquid casting systems. P-n-BA, a
polymer/monomer syrup, is being developed jointly by JPL and
Springborn Laboratories. P-n-BA is being formulated to cure within 15 minutes at 60°C. Candidate polyurethane systems are being
supplied for FSA evaluation by various polyurethane manufacturers,
from which one promising PU system has been identified. A brief
description of each of the four pottants follows.

a. Ethylene Vinyl Acetate (EVA). EVA is a copolymer of
ethylene and vinyl acetate typically sold in pellet form by Du
Pont and U.S. Industrial Chemicals, Inc. (USI). The Du Pont name
is Elvax; the USI trade name is Vynathane. The cost of EVA typically ranges between $0.55 and $0.65 per lb. All commercially
available grades of EVA were examined and the list reduced to four
candidates based on maximum transparency: Elvax 150, Elvax 250,
Elvax 4320, and Elvax 4355 (3). Because EVA is thermoplastic,
processing into a module is best accomplished by vacuum-bag lamination with a film of EVA. Therefore, based on film extrudability
and transparency, the best choice became Elvax 150. Elvax 250 was
an extremely close second choice.

Elvax 150 softens to a viscous melt above 70°C, and therefore
is not suitable for temperature service above 70°C when employed
in a fabricated module. A cure system was developed for Elvax 150
that results in a temperature-stable elastomer (3). Elvax 150 was
also compounded with an antioxidant and UV stabilizers, which improved its weather stability and did not affect its transparency.
The formulation of the encapsulation grade ethylene vinyl acetate
is given in Table I. These ingredients are compounded into Elvax

Table I. Formulation of Ethylene Vinyl Acetate (EVA)
Encapsulation Film

Component	Composition (Parts-By-Weight)
EVA (Elvax 150, Du Pont)	100.0
Lupersol 101 (peroxide)	1.5
Naugard-P (antioxidant)	0.2
Tinuvin 770 (UV stabilizer)	0.1
Cyasorb UV-531 (UV stabilizer)	0.3

150 pellets, followed by extrusion at 85°C to form a continuous
film. The thickness of the clear film is nominally 18 mils. The
selective curing system is inactive below 100°C, so that film extruded at 85°C undergoes no curing reaction. The extruded film
retains the basic thermoplasticity of the Elvax 150. Therefore
during vacuum-bag lamination, the material will soften and process
as a conventional laminating resin.

This EVA pottant has undergone extensive industrial evaluation, and manufacturers of photovoltaic (PV) modules have reported
certain advantages of EVA when compared to polyvinyl butyral
(PVB), a laminating film material in common use within the PV
module industry. The reported advantages are:
   (1)   Lower cost.
   (2)   Better appearance.
   (3)   Better clarity.
   (4)   Non-yellowing.
   (5)   Elimination of cold storage.
   (6)   Dimensional stability.
   (7)   No need to use a pressure autoclave.
   (8)   Good flow properties and volumetric fill.
Although this encapsulation-grade EVA has been favorably received by the industry, its status is still considered to be experimental. To advance EVA, several developmental tasks remain to
be completed:
   (1)   Faster processing, primarily in the cure schedule,
         which involves a reduction in cure time and tempera-
         ture; the minimum cure temperature will be dictated
         by the requirement that the curing system must not
         become active during film extrusion.
   (2)   Optimization of the UV-stabilization additives; the
         present additives were selected on the basis of
         literature citations and industrial experience with
         polymers similar to EVA.
   (3)   Identification of the maximum service temperature
         allowed for EVA in a module application, to ensure
         long life.
   (4)   Industrial evaluation of the desirability of having
         a self-priming EVA, recognizing the possibility of

Table II.   Formulation of Ethylene Methyl Acrylate (EMA)
            Encapsulation Film

Component	Composition (Parts-By-Weight)
EMA (TD 938, Gulf Oil Co.)	100.0
Lupersol 231 (perioxde)	3.0
Naugard-P (antioxidant)	0.2
Tinuvin 770 (UV stabilizer)	0.1
Cyasorb UV-531 (UV stabilizer)	0.3

an additional cost component (cost-benefit-perfor-
mance trade-off).

b. Ethylene Methyl Acrylate (EMA). This recently identified
material (3), a copolymer of ethylene and methyl acrylate, has potential as a solar-cell lamination pottant. There are three suppliers of EMA resins; two are domestic, Du Pont and Gulf Oil Chemicals. The Du Pont EMA resin, designated "VAMAC N-123", cannot
be used because of its lack of transparency. The third supplier
is foreign.

Gulf markets three highly transparent EMA resins that are designated 2205, 2255, and TD-938. Grade 2255 is the same base resin as 2205, except that it contains lubricant and antiblocking
additives. Gulf literature for these resins indicate the following features:

(1) Low-extrusion temperatures.
(2) Good heat sealability.
(3) Thermal stability to 315°C (600°F) for short periods
    of time (manufacturer's claim).
(4) Stress-crack resistance.
(5) Low melt viscosities.
(6) Good adhesion to a variety of substrates.

The three Gulf EMA resins were experimentally evaluated and
TD-938 was selected on the basis of film transparency, extrudability, and ease of module fabrication by lamination. The TD-
938-base resin sells for about $0.60/lb (April 1981). A trial
formulation is shown in Table II. Modules have been fabricated
with this EMA by the vacuum-bag lamination process, and have successfully passed module engineering qualification tests. Primer
formulations for bonding EVA and EMA to glass and polyester film
have been developed by Dow Corning and the formulations are given
in Table III.

c. Poly-n-Butyl-Acrylate (PnBA). No commercially available,
all-acrylic liquid-casting and curable-elastomeric system could be
found. Accordingly, the Environmental Isolation Task undertook a
developmental effort. A requirement of encapsulation-grade pottants is retention of elastomeric properties over the temperature
range from −40°C to +90°C. This requirement is met by PnBA, which
has a glass-transition temperature of −54°C (7).

Table III.   Primer Formulations

1)   Primer for Bonding EVA and EMA to Glass

Component	Composition
Z-6030 Silane (Dow Corning)	9.0 wt. %
Benzyl Dimethyl Amine	1.0 wt. %
Lupersol 101	0.1 wt. %
Methanol	89.9 wt. %

2)   Primer for Bonding EVA and EMA to Poly-
      ester Films

Component	Composition
Z-6030 Silane (Dow Corning)	2.5 wt. %
Cymel 303 (Am. Cyanamid)	22.5 wt. %
Methanol	75.0 wt. %

PnBA is not commercially available in a form suitable for use
as an encapsulation pottant, but the n-butyl acrylate monomer is
readily available at a bulk cost of about $0.45/lb. As a result
of the developmental program, a 100%-pure PnBA liquid was developed that could be cast as a conventional liquid-casting resin,
and that subsequently cures to a tough, temperature-stable elastomer. Modules fabricated with the PnBA elastomer have successfully
passed module engineering tests.

In general, the process for producing the prototype liquid
PnBA consists of first polymerizing a batch of n-butyl acrylate to
achieve a high-molecular-weight elastomer, then dissolving the
elastomer in an n-butyl acrylate monomer to obtain a solution of
acceptable viscosity. Following this, a crosslinker, curing
agent, UV stabilizers, and an anti-oxidant are then added. The
current formulation is given in Table IV. This formulation will
cure in 20 minutes at 60°C. The projected high-volume cost for
this material is estimated at about $0.85 to $0.90/lb. compared
with the commercial selling price of $9 to $11/lb for RTV silicones which are used in commercial modules as a casting pottant.

d.   Aliphatic Polyether Urethane. Almost all commercially
available polyurethanes are of the aromatic, polyester type, which
are not favorable because of their tendency toward hydrolysis of
the ester groups, and UV degradation due to UV absorption by the
aromatic structure. Only a few aliphatic, polyether urethanes
have been identified, and one of the more promising for photovoltaic module application is a urethane designated Z-2591, marketed
by Development Associates, North Kingston, Rhode Island. This ma-

Table IV. Formulation of Poly-n-Butyl Acrylate
(P-n-BA) Casting Liquid

Component	Composition (Parts-By-Weight)
n-Butyl Acrylate (monomer)	60.00
Poly-n-Butyl Acrylate (polymer)	35.00
1,6 Hexanediol Diacrylate (crosslinker)	5.00
Alperox-F (curing agent)	0.50
Tinuvin P (UV stabilizer)	0.25
Tinuvin 770 (UV stabilizer and antioxidant)	0.05

terial is currently undergoing extensive evaluation for this application.

UV Screening Plastic Films. The module front cover is in
direct contact with all of the weathering elements: UV, humidity,
dew, rain, oxygen, etc.; therefore, the selected materials must be
weatherable. Only four classes of transparent materials are known
to be weatherable, glass, fluorocarbons, silicones and polymethyl
methacrylate.

In addition to weatherability, the front cover must also
function as a UV screen, to protect underlying pottants that are
sensitive to degradation by UV photooxidation or UV photolysis.
The outer surface of the front cover should also be easily clean-
able and resistant to atmospheric soiling, abrasion-resistant, and
antireflective to increase module light transmission. If some or
all of these outer-surface characteristics are absent in the
front-cover material, additional, surfacing materials may have to
be applied.

Excluding glass, the only commercially available, transparent, UV screening plastic films which have been identified are
fluorocarbon films, Tedlar (Du Pont), and PMMA films, Acrylar (3M
Co.).

a.   Tedlar. Du Pont markets three 1-mil-thick, clear,
Tedlar fluorocarbon UV-screening films. The designation of these
three films are:
(1)   Tedlar 100 AG 30 UT
(2)   Tedlar 100 BG 15 UT
(3)   Tedlar 100 BG 30 UT
An initial difficulty with Tedlar had been poor adhesion to
EVA and EMA. This has been corrected by the use of an all-acrylic
contact adhesive that can be coated directly onto one surface of
Tedlar films. The coated adhesive, a Du Pont product designated
68040 is dry and non-tacky at ambient conditions; thus coated
Tedlar can be readily unwound from supply rolls. Experimental
testing indicates that when the adhesive is heated during the EVA

Table V.   Back Covers (White-Pigmented Plastic Films)

---

1.   Tedlar 150 BL 30 WH, 1.5 mils (Du Pont)
2.   Tedlar 400 BS 20 WH, 4.0 mils (Du Pont)
3.   Scotchpar 10 CP White, 1.0 mils (3M Co.)
4.   Scotchpar 20 CP White, 2.0 mils (3M Co.)
5.   Korad 63000 White, 3.0 mils (Xcel Corp.)

---

and EMA lamination cycle, strong adhesive bonding develops between
the pottants and the Tedlar films. The thickness of the adhesive
coating ranged between 0.3 and 0.4 mil.
     b.   <u>Acrylar</u>. 3M markets UV screening, biaxially oriented
PMMA films under the tradename "Acrylar". The films are available
in two thicknesses, a 2-mil version designated X-22416, and a 3-
mil version designated X-22417. An initial concern with these
films is their tendency for thermal shrinkage when heated above
105°C, the glass transition temperature of PMMA. Although true
for a free standing film, this has not been a problem when uniformly pressed at 150°C in a module assembly by one atmosphere of
lamination pressure.
     Adhesion strength between these films and EVA and EMA after
module fabrication is fair, but not excellent. Chemical coupling
primer systems for these films are being developed.

     <u>Back Covers</u>. Back covers are back surface material layers
which should be weatherable, hard, and mechanically durable and
tough. Engineering analysis indicates that the color of the back
surface material layer should be white, to aid module cooling.
Back covers function to provide necessary back side protection for
substrates, such as for example corrosion protection for low-cost
mild steel panels, or humidity barriers for moisture sensitive
panels. For superstrate designs, the back covers provides a tough
overlay on the back surface of the soft, elastomeric pottant. If
the back cover for a superstrate design is selected to be a metal
foil, an additional insulating dielectric film should be inserted
in the module assembly between the cells and the metal foil, as
shown in Figure 1. Candidate back cover films are listed in Table
V.

     <u>Edge Seals and Gaskets</u>. Trends based on technical and econo-
mical analysis (3) suggest that butyls should be considered for
edge seals, and EPDM elastomers should be considered for gaskets.
Several materials for each application are under investigation.
At this time, one of the more promising edge seal materials is a
butyl edge sealing tape designated "5354" (3M Co.), and one of the
more promising EPDM gasket material is designated "E-633" (Pauling
Rubber Co.).

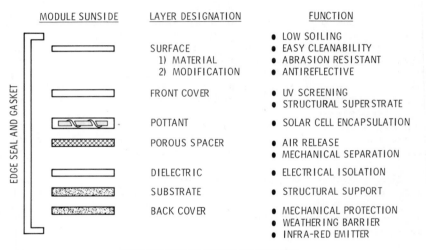

Figure 1.   Construction elements of photovoltaic encapsulation systems.

Encapsulation Engineering

An engineering analysis of encapsulation systems (6) is being
carried out to achieve a reliable and practical engineering
design.   This analysis involves four necessary features of a
module:
1)   Structural adequacy.
2)   Electrical isolation (safety).
3)   Maximum optical transmission.
and 4)   Minimum module temperature.
One of the goals of this analysis is a generation of guidelines
for minimum material usage for each of the construction elements.
The analyses for structural adequacy identified that the
thermal expansion or wind deflection of photovoltaic modules can
result in the development of mechanical stresses in the
encapsulated solar cells sufficient to cause cell breakage.   The
thermal stresses are developed from differences in the thermal
expansion properties of the load carrying panel, and the solar
cells.   However, the analysis interestingly identifed that the
solar cell stresses from either thermal expansion differences or
wind deflection can be reduced by increasing the thickness t of
the pottant, or by using pottants with lower Young´s Modulus E.
In other words, the analysis indicates that the load carrying
panel can be considered to be the generator of stress, and that
the pottant acts to dampen the transmission of the stress to the
cells.   The pottants ability to dampen transmitted stress is
directly related to the ratio of its thickness to modulus, t/E.
For example, the analysis finds for a four foot square glass
superstrate module undergoing a 50 mph wind deflection, that the
pottant t/E ratio should be equal to or greater than 4, where t is
in mils, and E is in units of KSI.   At a ratio of 4, the solar
cell stresses are just at their allowable limit.   If the pottant
were EVA having a Young´s module E of 0.9 KSI, then the minimum
thickness of EVA would be between 4 to 5 mils.   The use of a pottant having a higher Young´s modulus would necessitate that the
thickness of that pottant be correspondingly increased.   It should
be mentioned that the t/E requirement of a glass superstrate module undergoing thermal expansion is only 2.   Thus solar cell
stresses generated by the wind deflection of a glass superstrate
module, rather than thermal expansion effects, dictate the minimum
usage requirements of pottants.
This kind of output from the engineering analysis begins to
enable a cost-comparison basis for candidate materials.   For example, compared to EVA, a higher costing pottant having a higher
Young´s modulus would be much more costly to use both for reasons
of higher materials cost, and the need for more thickness.   On the
other hand, a higher costing pottant having a lower Young´s
modulus may be just as cost-effective due to an allowed thinner
usage.

Low-Soiling Surface Coatings

Evolving soiling theories (8) and physical examinations of
soiled surfaces (5) suggests that soiling accumulates in three
layers. The first layer involves strong chemical attachment, or
strong chemisorption of soil matter on the primary surface. The
second layer is physical consisting of a highly organized arrangement of soil matter effecting a gradation in surface energy, from
a high associated with the energetic first layer, to the lowest
possible state on the outer surface of the second layer. The lowest possible surface energy state is dictated by the chemical and
physical nature of the regional atmospheric soiling materials.
These first two layers are resistant to removal by rain and wind.
After the first two layers are formed, the third layer thereafter
constitutes a settling of loose soil matter, accumulating in dry
periods and being removed during rainy periods. The aerodynamic
lifting action of wind can remove particles greater than about 50
μ from this layer, but is ineffective for smaller particles.
Thus, the particle size of soil matter in the third layer is generally found to be less than 50 μ.

Theories and evidence to date suggests that surfaces which
should be naturally resistant to the formation of the first two
rain-resistant layers should be hard, smooth, hydrophobic, free of
first period elements (for example, sodium), and have the lowest
possible surface energy. These evolving requirements for low
soiling surfaces suggest that surfaces, or surface coatings should
be of fluorocarbon chemistry.

Two fluorocarbon coating materials, a fluoronated silane (L-
1668, 3M Co.), and perfluorodecanoic acid are under test. The
perfluorodecanoic acid is chemically attached to the surfaces with
a Dow Corning chemical primer, E-3820. The coatings on glass, and
on the 3M "Acrylar" film, are being exposed outdoors in Enfield,
Conn., and the loss of optical transmission by natural soil accumulation is being monitored by the performance of standard solar
cells positioned behind the glass and film test specimens. These
test specimens are not washed. Five months of test results to
date are shown in Figure 2 for glass and Acrylar.

After 5-months outdoors, soil accumulation on the uncoated
glass control has resulted in about a 3% loss of cell performance,
whereas the glass coated with L-1668 has realized only about a
0.5% loss. The glass sample coated with perfluorodecanoic acid
has realized about a 1.5% loss. The uncoated Acrylar control has
realized about a 5% loss, whereas the loss on the sample coated
with perfluorodecanoic acid is only about 2.5%, and the loss on
the Acrylar sample coated with L-1668 is about 3.5%. The test
results to date indicate that compared to untreated controls, soil
accumulation is being reduced on those tests samples treated with
the candidate fluorocarbon surface coatings.

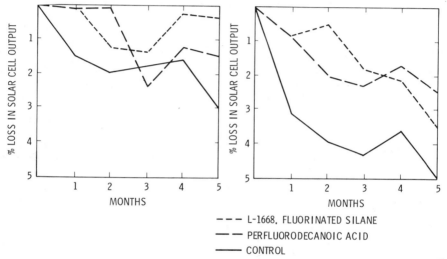

Figure 2.   Experimental evaluation of low-soiling fluorocarbon coating materials.

Acknowledgments

The research described in this paper was carried out by the Jet Propulsion Laboratory, California Institute of Technology, and was sponsored by the U.S. Department of Energy through an agreement with the National Aeronautics and Space Administration.

Literature Cited

1. Cuddihy, E.F., "Encapsulation Material Trends Relative to 1986 Cost Goals", LSA Project Report 5101-61, JPL, Pasadena, California, April 13, 1978.
2. Cuddihy, E.F., Baum, B., and Willis, P., "Low-Cost Encapsulation Materials for Terrestrial Solar Cell Modules", Solar Energy, Vol. 22, p. 389 (1979).
3. Springborn Laboratories, Third, Fourth, and Fifth Annual Reports for JPL's LSA Project, Contract No. 954527, June 1979, 1980 and 1981, respectively.
4. Cuddihy, E.F., "Encapsulation Materials Status to December 1979", LSA Project Report 5101-144, JPL, Pasadena, California, January 15, 1980.
5. Photovoltaic Module Encapsulation Design and Materials Selection, prepared and edited by the FSA Environmental Isolation Task, FSA Project Report 5101-177, JPL, Pasadena, California, August 15, 1981.
6. Spectrolab, Inc., Phase I Technical Report for FSA Contract 955567, November 1981.
7. Brendlay, W.H., Jr., "Fundamentals of Acrylic Polymers", Paint and Varnish Production, Vol. 63, No. 7, pp. 19-27, July 1973.
8. Cuddihy, E.F., "Theoretical Considerations of Soil Retention", Solar Energy Materials, Vol. 3, pp. 21-33, 1980.

RECEIVED November 22, 1982

Encapsulant Material Requirements for Photovoltaic Modules

K. J. LEWIS

ARCO Solar, Inc., Research and Development,
Woodland Hills, CA 91367

Encapsulants are necessary for electrical isolation
of the photovoltaic circuit. They also provide
mechanical protection for the solar cell wafers and
corrosion protection for the metal contacts and
circuit interconnect system over the 20-year design
life of a photovoltaic array. The required
components include the solar cell circuit, the rigid
or structural member, the pottant, and the outer
cover/insulator. Surface modifications may be
needed to develop strong, stable bonds at the
interfaces in the composite. If the module is to be
framed, edge sealants may also be required. The
functions of the individual components and the
performance requirements as they are now known are
described. Costs are compared where possible and
candidate materials identified.

In the next few years, as lower-cost solar cells are
developed, encapsulation materials will become a dominant cost in
a finished photovoltaic (PV) module. Encapsulants are necessary
for electrical isolation, mechanical protection of the cells, and
corrosion protection of the metal contacts and interconnect system
for more than 20 years of outdoor exposure in even the most severe
terrestrial climate. All PV systems, regardless of how inert,
tough, pliable, and weatherable the actual cells themselves are,
need encapsulation. While improvements in encapsulant materials
cannot yield orders of magnitude in cost reduction, judicious
engineering design and materials development can result in
significant cost savings and performance improvements. It is only
with such cost-reduced arrays that the domestic photovoltaic
rooftop market has a high probability of developing.

Figures 1 and 2 compare a commercial module and an advanced
rooftop module design. Each design has a minimum of three
components in addition to the cell circuit. They are the rigid or

0097–6156/83/0220–0367$06.00/0

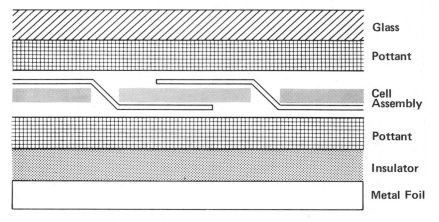

Figure 1.   Commercial Module.

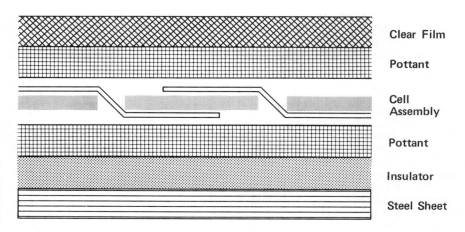

Figure 2.   Rooftop Module.

wind-load-bearing member, the pottant, and the outer
cover/insulator. In addition, if structural elements in the plane
of the circuit are metal or if circuit connections must overlap,
additional non-softening electrically insulating layers are
needed. Adhesives, primers or other surface modifications may
also be needed to develop strong, stable bonds at the various
interfaces in the composite. Edge sealants may be needed if the
module is to be framed.

Rigid Member

The rigid or wind-load-bearing member of the composite
prevents flexure of brittle cells beyond the fracture point. In
addition to requiring a high flexural modulus ($> 10^6$, preferably
$> 10^7$ psi) the rigid entity should be low cost ($< \$1.00/ft^2$,
preferably $< \$0.50/ft^2$) and of minimum weight. It must be weather
resistant (> 20 years) and must have a relatively low thermal
expansion coefficient that is as near as possible to that of the
cell material. This is in order not to put excessive (fatiguing
level) mechanical stress on the cell surface electrical contacts
from the forces of differential thermal expansion over the daily
thermal cycling of a PV array. The thermal expansion coefficient
of crystalline silicon is quite low at 3×10^{-6} $^{o}C^{-1}$.

   Optical, Weight and Permeability Requirements. If the
structural member is the front cover of the module (superstrate
configuration), it must be optically clear ($> 90\%$ transmission)
through the solar spectrum of importance to absorption by the
solar cell (0.4-1.1 microns). It must also be relatively hard
(> 90 shore A durometer), soil repellent, preferably UV absorbing
below 0.36-0.37 microns, and non-permeable to oxygen, water vapor,
and atmospheric pollutants. If the rigid member is the back
cover, it may be opaque.

   Candidates, Cost. Tempered glass is to date the best known
material for a rigid front cover for silicon cells. It costs
~ \$0.75-\$1.25/ft^2 for the low-iron content glass ideal for PV
applications. For a rigid back configuration, surface passivated
steel presently appears to be the optimum choice for use with
silicon cells, considering its thermal expansion coefficient,
weather resistance, and cost for the weight and stiffness
required. It can range from \$0.30/ft^2 for zinc galvanized to ~
\$3.00 for porcelainized. Passivated steel is also non-permeable.
Glass reinforced concrete, sealed hardwood, and aluminum are also
low-cost possibilities, but each has significant disadvantages
compared to steel in the areas of weight or thermal expansion
coefficients.

   Weather and Corrosion Resistance Requirements. There are
many commercial ways to "rust-proof" steel to make it weather
resistant. They range from paint, which is usually the cheapest
but least effective method, to chromeplating, which is one of the
most expensive and effective methods. In between, in both cost

and performance, are porcelain enameling and various low melting metal dip coatings such as tin, zinc, aluminum, or combinations thereof. Porcelainizing is on the expensive side (over $1.00/ft^2, even in large volume) but is one of the oldest, most effective ways to protect the steel. Zinc galvanizing is the most common lower cost method.

Most ordinary "rustproof" steel is not adequate as a substrate for solar cells. The protective layer must stop not only gross rusting, which would cause total delamination, but it must also halt even the slightest progress of corrosion of either the steel or any active coating metals on it once the piece is laminated to the solar cells. This is because even slight corrosion generates small amounts of hydrogen long before the corrosion layer builds up sufficiently to cause adhesive failure. If hydrogen is generated fast enough and if the pottant layer is relatively soft, the gas collects as bubbles behind the impermeable cells or glass front. These bubbles not only become stress points on the cells leading to wafer fracture, but they also reduce the dielectric standoff of the intervening electrically insulating layers and become peel stress points for delamination propagation with thermal cycling. Thus, no corrosion can be tolerated.

Porcelainizing is the best method for eliminating steel corrosion. It can be used only on rigid surfaces, however, since it cracks with very little flexing. Its electrical properties are intrinsically very good but are frequently degraded by cracks, dirt and pinholes. It must also be properly fired since firing temperatures affect its surface characteristics for bonding to solar cell potting polymers. Insufficiently fired films are basic enough to rapidly hydrolyze the vinyl acetate ester groups in ethylene/vinyl acetate. Adhesion is lost very quickly, with rapid generation of acetic acid at the interface.

Paint over zinc galvanizing has been found to blister easily when exposed to high humidity (100% at elevated temperatures up to 100°C). Plastic film laminated to zinc galvanizing does not blister as easily, but outgassing still occurs readily. Blistering starts at the cell or module edges and moves in. Eventually enough "white rust" develops to produce delamination. The quality or type of zinc galvanizing such as degree of passivation (chromate coatings, etc.) or grain size to affect the zinc corrosion rates, which are quite rapid in even the best cases. Zinc/aluminum alloys and aluminum coatings still corrode, but not as much as zinc alone on steel.

<u>Thin Film Cells</u>. Future, lower cost solar cell materials will likely be more flexible than crystalline silicon and therefore may not require a rigid member in the module lay up. They will still need electrical isolation and protection from abrasion and corrosion, however, and will thus still need pottant or probably thinner adhesive layers as well as outer covers/insulators.

Pottant

The pottant is the soft, elastomeric, vibration-damping material that immediately surrounds both sides of fragile solar cell wafers and their electrical contacts and interconnects. It protects the cells from stresses due to thermal expansion differences and external impact. It isolates them electrically and helps protect their metallic contacts and interconnects from corrosion.

Optical Requirements, Cost. The pottant must be highly transparent (> 90% from 0.4 to 1.1 μ), serving as an optical coupling medium to provide maximum light transmission to the cell surface. Because it is used in a fairly thick layer for brittle cells (10-20 mils on each cell side), the pottant must be very inexpensive (≤ $0.30/ft^2, preferably < $0.20/ft^2). At 30 mils total, this translates to between $1.00 and $2.00/lb, including compounding and fabricating into PV factory usable form. No inherently weather resistant material is in this price range, but several moderately stable materials which can be upgraded with stabilizers are in this range.

Mechanical Requirements. The pottant material should have a relatively low modulus (≤ 2000 psi at 25°C). The maximum tolerable modulus depends on the difference in expansion coefficients of the cells and the rigid member and on the thickness of the layer between them.[1] Relatively high modulus rubbers could be used but would require inordinately thick and thus expensive layers to damp out the expansion differences. For example, with an 1/8-in.-thick glass superstrate and silicon cells, which will take 5000 psi maximum linear stress, a pottant of 1000 psi modulus needs to be a minimum of only 1.5 mils each side for a 1:1 safety factor in service; with a 2500 psi modulus the minimum is ~ 3.5 mils, etc.[1] A safety factor higher than 1:1 is highly desirable.

Fabrication technique is also a factor. For example, whether the cell strings are pressed against the pottant while it is still cold and must cushion the cells under pressure, or whether it is squeezed only while it is molten, or not at all, can determine the minimum thickness tolerances for module fabrication. The minimum usable pottant thickness can also be limited by the green strength of the material itself if it is fabricated as a cast sheet. If it is extrusion-coated on the support and cover materials, the pottant can be less tough and therefore thinner, but must be free of pinholes and other flaws to prevent electrical leakage. Experience suggests a minimum thickness of 10-15 mils to achieve the necessary freedom from flaws for sufficient electrical insulation properties as well as ease of handling.

The potting material must have a glass transition temperature below the lowest temperature extreme the PV module might experience, which is ~ -40°C. The material must remain rubbery in order to damp impacts and vibration of the fragile cells.

Similarly, it must exhibit no significant mechanical creep at the upper operating temperature extreme of 90°C in order for the layup, cell positions, etc. to remain intact when tilted at an angle facing the sun.

The pottant must exhibit strong, moisture-resistant adhesion ($\geq$ 10 lb/in. peel strength) to all the surfaces it must bond to, over a 20-year lifetime. The 10 lb/in. peel strength may decrease to 5 lb/in. while the pottant or adhesives are saturated with water in a non-hermetic design as long as it recovers to within 10% of the original value when redried at up to the nominal operating cell temperature of the particular design. Moisture-resistant adhesion is tested mostly by exposure to 100°C/100% RH. Ten months at these conditions would equal 20 years at 70°C, 50% RH, but usually the first week or two will separate the good bonds from the poor ones. Exposure to boiling water for a few hours or overnight is also a good initial screening technique.

In a vacuum lamination process, the steeper the melt viscosity/temperature curve for sheet pottant material, the better. The layers need to be dry and non-tacky during the initial evacuation step so as not to trap air between them. At the same time, the pottant must then melt to as fluid a state as possible in order to effectively penetrate and wet all the irregularities of the cell circuit.

Block or graft thermoplastic elastomers with relatively low molecular weight amorphous segments of a weather resistant saturated backbone have the potential of being superior polymers for potting solar cells. The crosslink forming crystalline segments make relatively soft, low molecular weight, rubbery polymers handle well. They exhibit high cohesive strength or toughness and low surface tack when the crystalline domains are solidified (see Figure 3).

Candidates. Free radical polymerized vinyl or acrylic/ethylene copolymers made in high pressure polyethylene reactors have been shown by E. Cuddihy[1] to be block polymers of pure crystalline homopolyethylene and amorphous high vinyl acetate (~ 70 weight %) or methyl acrylate-co-ethylene segments. When the crystallites are submicron in size as in DuPont's Elvax 150, they do not significantly scatter the incoming light. A number of laboratories have shown, however, that even when some light scattering occurs, it does not necessarily decrease solar cell output.[2] For example, it was found that encapsulating a 3-mil-thick non-woven glass mat in 15 mils of ethylene/vinyl acetate (EVA) drops the transmission by 60%. The same composite when fused to the front of a textured (antireflective treated) silicon solar cell does not drop the output at all unless discoloring from degradation takes place. Aromatic crystalline segments such as polystyrene are undesirable even initially because the degradation products are light absorbing.

A Gulf Oil ethylene/methyl acrylate rubber of 20 weight % EMA which is not nearly as transparent as Elvax 150 EVA is being

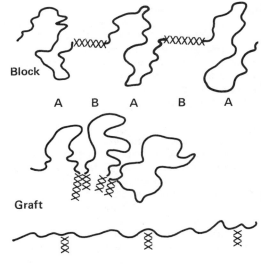

Block

A   B   A   B   A

Graft

Advantages

- Reversible cure by simple melting
- Fast processing, no cure time
- Steep viscosity vs temperature curve
- High ceiling temperature for processing
- No outgassing from decomposing peroxides
- Low (room temp) tack

Unknowns

- Availability of materials of proven weather resistance
- UV sensitivity of aromatic crystalline blocks

Figure 3.   Structures of Thermoplastic Elastomers.

evaluated by a number of laboratories, and appears to offer some
attractive properties. It is more thermally stable and does not
appear to need as much crosslinking to prevent creep.[1] It is
considerably less transparent, however, particularly when uncured.
Lower melt index Elvaxes are also possibilities for a noncuring
thermoplastic elastomer pottant.

Plasticized polyvinyl butyral (PVB) is easy to process
because it is thermoplastic rather than thermosetting, but has the
disadvantage of containing plasticizer (see below).

There are some other advantages of thermoplastic elastomers
as potting materials for solar cells besides the steep melt
viscosity curve. They crosslink by cooling so that long cure
times are unnecessary and the cure is reversible to help the
recovery of flawed panels. They have no ceiling fabrication
temperature above which the cure will set off nd/or outgassing
can occur. Thus thermoplastic elastomers allow more latitude than
is available with a thermosetting material in using temperature to
adjust viscosity for optimum lamination processing.

Plasticized PVB, used extensively in laminated safety glass,
has been .studied. It is expensive and easily degraded when not
hermetically sealed.[3] This can be compensated for by sealing it
between a glass front and metal foil back cover in a superstrate
design module.

Electrical Requirements. The pottant should contain no
plasticizer because plasticizer can reduce the volume resistivity
of a polymer drastically. It reduces the resistivity of PVB by 5
orders of magnitude in some formulations. PVB with 40% diester
plasticizer measures only 10^{11} ohm-cm in laminated form at room
temperature whereas it measures 10^{16} ohm-cm with the plasticizer
driven out. Volume resistivities of 10^{12} ohm-cm or less will
conduct small amounts of current fairly readily, albeit slowly.
(For example, a resolved 5 line pair/mm charge image has been
observed to blur within the first few seconds when placed on the
surface of a film or immersed in a liquid of 10^{11}-10^{12} ohm-cm
resistivity. The same image on or in 10^{14} ohm-cm material will
not blur for several hours. On a 10^{16-17} ohm-cm material an image
will last unblurred from weeks to months.)

Because ionic impurity mobility determines resistivity in a
polymer rather than electron mobility as in metals, higher module
operating temperatures drop rather than raise the resistivity
because of the viscosity drop with temperature. The viscosity
drop in plasticized PVB with temperature is extremely steep.
Indeed, after only a day of dry oven aging at 150°C, an open
plasticized PVB film is brittle from total loss of the plasticizer
and already significantly discolored from oxidative degradation.
The volume resistivity, however, rises to 10^{16} ohm-cm from the
original 10^{11} ohm-cm by the removal of plasticizer.

As an added difficulty, the plasticizer's solvation effect in
PVB appears to enhance the polymer viscosity drop with
temperature. Unfortunately, this raises the leakage current of a

module at 1.5 kV by over an order of magnitude with an operating
temperature rise of only 25°C when an unplasticized, high volume
resistivity barrier layer is not present between the circuit and
any grounded metal surfaces in close proximity (see Figure 4).
25°C is the normal rise for a module between dark and fully
illuminated at a full sun flux of 100 mW/cm^2.

The volume resistivity of an unplasticized pottant material
such as EVA is 10^{14} ohm-cm. The module current leakage at 1.5 kV
with EVA is an order of magnitude lower than with plasticized PVB
at 25-30°C and there appears to be no rise in leakage at ~ 50-
60°C. (See Figure 4 and Table I.) Similarly, the current leakage
of modules containing plasticized PVB can be blocked by the
insertion of an additional high volume resistivity layer such as
polyethylene terephthalate film as discussed below, which is
resistant to the solvation effect of diester plasticizers.

Table I. Volume Resistivity of Pottant Materials
Volume resistivity values (Ω-cm):

Material	Measured		Literature
	Initial	Dry Oven Aged (150°C)	
PVB*	10^{11}	10^{16}	--
EVA	10^{14}	10^{14}	--
PVB (Tedlar)	10^{14}**	5×10^{14}	10^{14}
PET (Mylar)	10^{17}	--	10^{18}

*Plasticized
**Most likely reduced when "effectively" laminated to plasticized
  PVB

The pottant should have a dielectric strength of at least
400-500 volts/mil, which is typical for unoriented amorphous
polymers. Since the pottant is designed to flow, however, it
cannot be relied upon to provide sufficient dielectric standoff by
itself. It will tend to move out of the areas where it is
mechanically stressed (squeezed). Unfortunately, those areas are
usually also the areas of highest electrical stress since field
lines are the densest around irregularities in the geometry of the
circuit metal, e.g., interconnect ribbon or wire kinks, excess
solder beads, etc. A non-softening, high volume resistivity
insulator layer is thus needed to guarantee circuit isolation.

Chemical Requirements. The pottant must be stable; that is,
chemically resistant to oxidation and hydrolysis unless protected
in a hermetic package, to reduction by metals, and to outgassing
of dissolved gases or liquids or decomposition products under
normal operating conditions of -40°C to +90°C for 20 years. The
need for chemical stability is especially stringent when a lower
cost non-hermetic design is used. Even when a hermetic package is

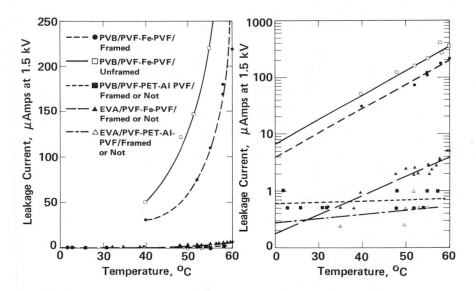

Figure 4.   Module Leakage Current vs. Temperature.

used, however, some stability is required because the small amounts of dissolved oxygen and moisture which will be found in any amorphous, rubbery material can still cause noticeable discoloration of a pottant, particularly when catalyzed by certain metal oxides if present in the metallized cell contacts.[3] Degradation in a hermetic package can also occur from the extreme heat that can develop under "hot spot" conditions which can occur in isolated areas of a module when electrical mismatching of cell output in a series string without diodes has occurred such as may be initiated by shadowing of cells.[4-6] The ideal material would tolerate at least intermittent excursions up to the melting point of solder (~ 190°C) without discoloring, embrittling or reverting, outgassing, or breaking down electrically.

Oxidative breakdown of polymers can follow one or more paths of change in physical and chemical properties. Oxidative attack can be catalyzed by heat, UV light, certain metals or metal oxides, sometimes moisture, etc. Nevertheless, the same functional weaknesses in a given polymer will be attacked by oxygen regardless of the specific mechanisms by which the energy is actually absorbed. The result is usually either embrittlement, chain scission, hydrolysis, or combinations of these. Embrittlement comes from extensive crosslinking. Chain scission or reversion results in extensive molecular weight loss and both chain scission and hydrolysis result in the formation of hydrophilic end groups. When both embrittlement and reversion occur simultaneously, the result is that the overall physical properties, which depend mostly on molecular weight, often remain unchanged for some time until one mechanism dominates. Embrittlement will allow increased mechanical stress to reach the solar cells, and increases the possibilities of interface delamination from stress concentrations. Reversion allows distortion or creeping of the layers, less mechanical protection of cells, and easier bubble formation from outgassing.

The chemical character of a polymer often changes more rapidly with oxidative breakdown than the mechanical properties. Hydrophilic group formation changes the moisture absorption and permeability of the material, a problem when protecting metals from corrosion as in a PV module. Hydrophilic group formation reduces the moisture resistance of adhesive bonds, changing the acid-base characteristics of the bonded interfaces. The development of conjugated carbon-carbon and carbon-oxygen unsaturated color centers changes the optical absorption properties of the material, reducing the amount of light reaching the solar cells and thus reducing their electrical output. Reversion or hydrolysis can also generate volatile species of very low molecular weight, which can evolve from the polymer, leaving voids which degrade voltage standoff, reduce the optical coupling of light to the cells and create stress points for cell fracture.

The exact amount of dissolved volatile species that can be tolerated in an encapsulated PV system depends on the vapor pressure of the particular dissolved species in the particular pottant medium and on the viscosity of that pottant at module operating temperatures. The importance of minimizing such dissolved volatiles was discovered by ARCO Solar in outdoor testing of the first substrate design rooftop module. The encapsulated cells began bulging up and cracking as the array reached summer operating temperatures for the first time. The pottant was a stabilized transparent ethylene/vinyl acetate pottant based on Elvax 150.[7] The outer cover/insulator layer was a flexible acrylic copolymer film.

The bulging cells resulted from a combination of poor adhesion between unprimed surfaces and outgassing of the volatile decomposition products of the large amount of peroxide used to crosslink the EVA. The acrylic copolymer used melts under normal lamination conditions so the edges became buried and a peel was difficult to start. Although other investigators[7,8] have reported that acrylic film and EVA co-crosslink, we found the peel strength to be very weak (~ 1-2 lb/in.) without using primers. A large amount of 2,5-dimethyl-2,5-bis-(t-butyl peroxy) hexane crosslinking agent (Lupersol 101 by Pennwalt) was used by the designers to insure the material would sufficiently crosslink even when heated very slowly. Lupersol 101 forms significant amounts of acetone and t-butanol when it decomposes in addition to methane, ethane, and ethylene (see Figure 5). Acetone and t-butanol are not effectively removed from EVA during most vacuum processing and reliquify upon cooling the module to room temperature. When the modules in the outdoor array began to reach summer operating temperatures of 75-80°C, the vaporizing trapped liquids began to build up pressure behind the cells from the large volume increase of vaporization. When the vapor pressure became greater than the combined adhesive strength of the bonded layers and the flexural strength of the cells, delamination began, with the cells bulging and cracking. The problem was solved by a combination of altering the curing system and raising the adhesion.

The pottant must be chemically inert in that it must not react with the metals or other surfaces it bonds to. Related to this is the need for it to exhibit little or no water absorption to corrode metals bonded to it or to reduce its volume resistivity. The large fraction of homopolyethylene in the EVAs we are studying make them quite inert and low in water absorption.

As explained above, the pottant should contain little or no plasticizer since it can generate electrical problems. The last chemical requirement for the pottant is that its melt equilibrium contact angle with all the surfaces to which it bonds be as low as possible below 90°C. This speeds processing as well as maximizing adhesion and minimizing the collection of water and oxygen at the interfaces to reduce metal corrosion and metal oxide catalyzed polymer changes to form color centers.

Thin Film Systems. As previously mentioned, the same types

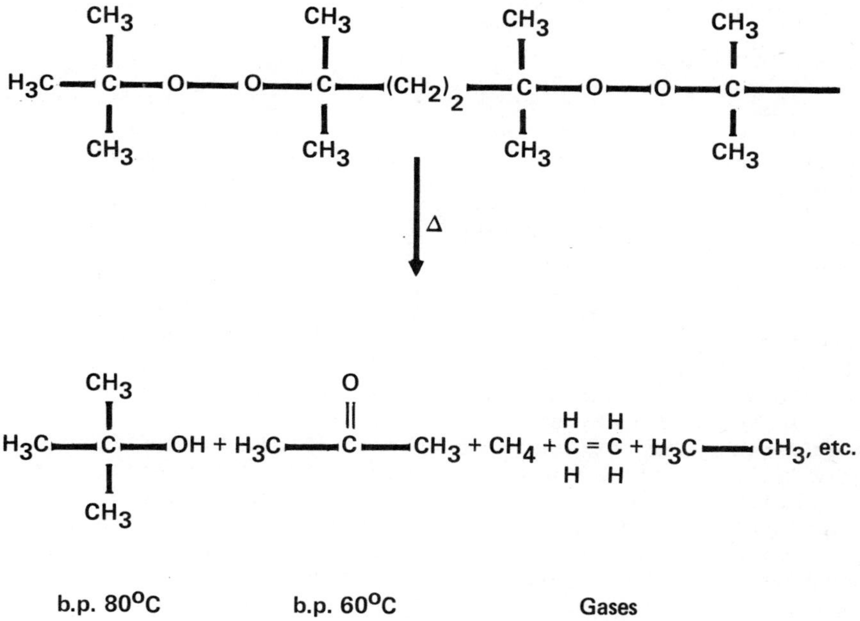

Figure 5.   Decomposition of Lupersol 101.

of materials with the same optical, electrical, chemical, and some of the same mechanical requirements (particularly bond strengths) would be needed for thin film cell encapsulation. All the previously mentioned candidates would qualify but could probably be used in thinner layers. Because less critical mechanical requirements enable the use of thinner layers, the field opens to include more expensive possibilities such as silicones. Liquid systems such as 100% solids casting materials or solution applied coatings (by spray, dip, brush, roller, etc.) also become more practical for a rapid throughput factory. Casting liquids are a possibility even now but are more complicated to use with vacuum processing than are sheets.

Outer Cover/Insulator

The outer cover is the tough, hard, soil-resistant, inherently weather resistant outer layer that protects the softer pottant layers from the effects of abrasion, dirt, and weathering. It augments the pottant in electrically isolating the circuit and preferably acts as a UV screen if used on the front. It is usually a flexible or conforming plastic and may be a film or a solution-applied coating.

Optical, Chemical, and Cost Requirements. If the outer cover/insulator is the front cover on a rigid-back module, it must have the same optical and chemical properties as the pottant; that is, high transmission, good optical coupling, and inherent weather resistance. Being inherently weather resistant means meeting all the criteria previously described under chemical stability requirements for pottants without need of added stabilizers. The only known inherently weather resistant organic materials are acrylics, silicones, and fluorocarbons, ranging from $3-5/lb for acrylics, to $8-10/lb for silicones, to $10-20/lb for fluorocarbons. However, if the cover material is sufficiently tough and flexible, it can be quite thin (1-4 mils) and can be made from a more expensive polymer than the pottant. It can cost up to $10-15/lb and still be economical.

If thin plastic is used as a front cover in a substrate design module, it will not provide a hermetic seal no matter how oriented the microstructure is. The oxygen and water permeabilities may be low but finite. The main question, yet to be fully answered, is whether a non-hermetic package can last for 20 years. Single-crystal silicon solar cells themselves are known to be fairly inert to the effects of heat, light, oxygen, and water. Accelerated corrosion tests are in progress on minicircuits to determine the stability of the cell contacts and interconnect systems (copper ribbons, solder, etc.). The stability of future cells themselves will be the remaining question to be answered for future PV modules.

The back cover, may, of course, be opaque. For stand-alone, glass-front arrays it is preferably white in color. This is because light scattered by that white surface is refracted in the glass and enhances the total module output by about 5%. There is little or no enhancement from a white background, however, with a thin plastic front and architects have objected to the "polka dot" appearance in rooftop applications. Thus, the rooftop module design has a black-colored back plastic layer which matches the cells well. It does not increase the module temperature over a white back film by more than a few degrees.

The back cover must be inherently weather resistant as must the front but its ultraviolet stability can be more effectively enhanced with light absorbing pigments as well as transparent UV stabilizers. Thin plastic alone on the back cannot provide a hermetic seal, but because it can be opaque, a metal foil such as steel or aluminum may be used in the back cover composite to effect a hermetic seal except at the edges, when glass is used as the front cover. A minimum three-layer laminate is required for a back cover composite. The first cover layer serves as the non-softening electrical insulator between the potted circuit and the foil and is protected from oxygen. The metal foil layer is the hermetic barrier and the third layer is the true outer cover where weather resistance is more stringently required.

Mechanical Requirements. The outer cover plastic film must be tough and flexible to resist abrasion and gouging, both in manufacture and in the field. If it is the front cover, it must also be relatively soil resistant. Either front or back covers must form reliable, moisture-resistant bonds to the pottant and to the foil in the case of a back cover. Moisture resistant adhesion is especially important at the edges where moisture can penetrate even a "hermetic" design. Lastly, the cover layers must be dimensionally stable (non-shrinking or yielding) to thermal cycling stresses of manufacture and field operating conditions.

Electrical Requirements. The outer cover must be made of high volume resistivity material that does not soften at lamination temperatures in order to control electrical current leakage through it. Outer covers can be applied as oriented films by lamination, as liquid or powder coatings by electrostatic spray, dip, brush or roller, or by extrusion. Oriented films have by far the best electrical properties in terms of dielectric strength (voltage standoff per unit thickness). Only acrylics could really be considered economical in liquid or powder coatings because of the greater thicknesses required without the orientation of a blown film. Oriented films are also mechanically the toughest films and have the lowest gas permeability of most plastics because of the increased density and induced crystallinity of their structure. A fairly large margin in terms of dielectric standoff is required between the test or use voltages and the parallel plate dielectric strength values of the insulating layer(s) because, as previously mentioned, the irregularities or geometry of the circuit give rise to unpredictable field concentrations for leakage or breakdown.

Candidates. The only commercially available oriented films known at this time which fit the weather resistance requirements are polyvinylidene fluoride (PVF_2), polyvinyl fluoride (Tedlar), polymethyl methacrylate (PMMA), and polybutyl acrylate/methyl methacrylate copolymer (PBA/MMA). PVF_2 is currently expensive. PBA/MMA is inexpensive but in clear form does not appear to be sufficiently oxidatively stable for our purposes. It is also too water sensitive and too easily softened in many laminating processes. PMMA appears to be somewhat more chemically stable than PBA/MMA and is also relatively inexpensive, but has the same dimensional stability problems at 150°C, the normal pottant processing temperature. Both acrylics maintain excellent optical clarity on heat aging, however.

Tedlar is moderate in cost and has known long-term performance out-of-doors. It has excellent toughness, good weather resistance, and moderately good electrical and optical performance. Its thermal stress resistance is marginal but adequate (2-6% shrinkage at 150°C) for most needs. Its cost is higher than optimum, but can be used as a thin film, especially when coupled with less expensive polyethylene terephthalate film for better electrical properties at a lower cost.

Thin Film Systems. An ideal low-cost system could be continuously processed into rolls of arrays. These rolls would consist of a clear, flexible, electrically insulating plastic front cover, a thin layer of pottant or adhesive on either side of the flexible thin film PV circuit, and an opaque, flexible, electrically insulating plastic back cover. These rolls could then simply be unrolled on the roof, nailed into place, and connected to the household circuitry. All components would have to be flexible. Figure 6 illustrates the possible components of an all flexible system -- both hermetic and non-hermetic depending on future cell requirements.

Adhesives, Primers, Surface Modifications

Good adhesion peel strengths at an interface result from a combination of several interacting phenomena. The interfacial forces are a combination of dispersion (van der Waals), polar, and acid-base interaction forces across the interface.[9-13] They will determine long-term reliability. Besides the interfacial forces involved, the rheology of the materials at the interface has a large if not dominant influence in determining the peel strength or delamination tendencies of a particular bond. This is particularly true because of the thermal cycling stresses a PV array is subjected to. The goal is to have cohesive rather than adhesive failure at the interface. If the rheology is such that stress concentrations occur, the cohesive failure can be quite low. This is particularly true if a bubble is trapped in a PV module. Expansion and contraction of trapped or evolving gases during thermal cycling generate highly concentrated stresses at

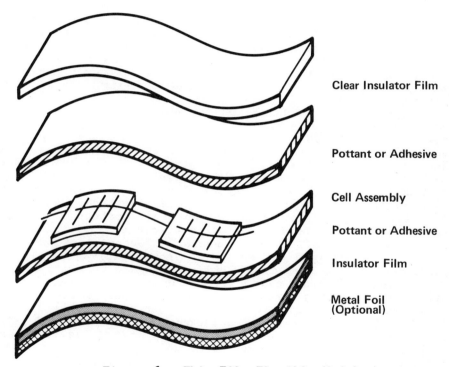

Clear Insulator Film

Pottant or Adhesive

Cell Assembly

Pottant or Adhesive

Insulator Film

Metal Foil
(Optional)

Figure 6.   Thin Film Flexible Module.

the bubble perimeter by nature of the geometry. At least one
component of any module interface should be flexible, as is the
soft pottant, to spread these forces as much as possible.

If properly functional adhesives or coupling agents and/or
other surface modifications are used where possible (such as
changing the acidic or basic character of one or both surfaces at
an interface), the result should always be a stable bond.

<u>Edge Sealants and Frames</u>

An edge sealant is needed if the module is to have an added
frame. It is particularly important if one is making a module
with two impermeable outside layers, e.g., glass and metal foil,
for a "hermetically" sealed package. The seal should be as gas
and moisture tight as possible at the module edges if the pottant
layers being sealed between the impermeable layers are at all
oxidatively or hydrolytically unstable, or if they are hygroscopic
as is PVB. EVA does not really need a moisture-tight seal, just a
mechanical cushion.

<u>Edge Sealants</u>. Mechanically, the edge sealing material for a
module which contains a hygroscopic pottant must have moisture
resistant adhesion of $\geq$ 10 lb/in. or cohesive failure to all the
surfaces it touches: frame, glass, pottant, and foil covers.   It
must be very flexible ($\leq$ 1000 psi modulus). Chemically, the
sealant must be stable to heat and UV light, remain flexible, and
must not flow significantly ($\leq$ 25%) at 90°C under 5 psi pressure
(the weight of a module distributed over its end). When the
material does degrade, it is better if it softens than if it
embrittles. It may be black in color to aid UV resistance and
should be relatively easy to process. Fluoro-elastomers and
silicones are all good but expensive and probably give higher
performance than is needed at PV system operating temperatures.

<u>Frames</u>. The frame adds rigidity for wind-load strength and
provides a coupling structure for mounting the modules at the
proper angle. Mechanically, the frame material should be strong
(T_B > 10,000 psi) and have a high flexural modulus (> 5 x 10^5 psi)
and relatively low thermal expansion coefficient as close as
possible to glass so the channel will not have to be too deep and
can maintain a maximum packing density of exposed cells. The
material must be weather resistant (20 years) and bondable. It
should be relatively low cost (preferably < 0.25/ft^2 of module).
Geometry, thickness, fabrication technique, etc., which strongly
effect the economics, will vary with the materials. The best
candidates to date are extruded or roll-formed metals with
passivated surfaces such as anodized aluminum or galvanized steel.
Highly filled, molded, or extruded, weatherable plastics are also
a possibility but an optimum candidate has not yet been
identified.

Accelerated Testing

Overall, the goal is chemical stability of the pottant and all other organic materials in the array to the extent that they will undergo no more than a 20% change in optical, electrical, or mechanical properties (including bond strengths) over 20 years of outdoor weathering and module operation. This is difficult to determine when most of the materials have not even existed that long, let alone been exposed outdoors in a photovoltaic module. The near-term goal with each design change is to make something better in performance as well as lower in cost than what was being used. Thus, most tests today compare materials relative to each other in accelerated conditions that in some cases are probably excessively severe. For example, the 150°C dry oven and 100°C/100% RH exposure tests are used mostly as screening tools.

Literature Cited

1. Cuddihy, E. "LSA Progress Report 18 and Proceedings of the 18th Project Integration Meeting," Jet Propulsion Laboratory, in press.
2. Investigation of Test Methods, Material Properties, and Processes for Solar Cell Encapsulants, 13th Quarterly Progress Report for May 12, 1978-August 12, 1979, DOE/JPL/954527-12, Springborn Laboratories, Inc., Jan. 1980.
3. Megerle, C.; Lewis, K. Encapsulant Degradation in Photovoltaic Modules, this symposium.
4. Arnett, J. C.; Gonzalez, C. C. "Fifteenth IEEE Photovoltaic Specialists Conference -- 1981," p. 1099.
5. Ross, R. G., Jr. "Fifteenth IEEE Photovoltaic Specialists Conference -- 1981," p. 1157.
6. Gonzalez, C.; Weaver, R. "Fourteenth IEEE Photovoltaic Specialists Conference -- 1980," p. 528.
7. "Investigation of Test Methods, Material Properties, and Processes for Solar Cell Encapsulants, Annual Report," DOE/JPL - #954527-79-10, Springborn Laboratories, Inc., June 1979.
8. Pluddemann
9. Drago, R. S.; Vogel G. C.; Needham, T. E. J. Am. Chem. Soc. 1971, 93, 6014.
10. Drago, R. S.; Parr, L. B.; Chamberlain, C. S. J. Am. Chem. Soc. 1977, 99, 3203.
11. Fowkes, F. M.; "Donor-Acceptor Interactions at Interfaces," J. Adhesion 1972, 4, 155.
12. Fowkes, F. M.; Maruchi, S. Coatings and Plastics Preprints 1977, 37, 605.
13. Fowkes, F. M.; Mostafa, M. A. Ind. Eng. Chem., Prod. R&D 1978, 17, 3.

RECEIVED November 22, 1982

Encapsulant Degradation in Photovoltaic Modules

K. J. LEWIS and C. A. MEGERLE

ARCO Solar, Inc., Research and Development,
Woodland Hills, CA 91367

The aging behavior of several encapsulant candidates
for photovoltaic module designs with a plastic front
surface were studied in the field and via
accelerated aging. Two pottant polymers and two
outer cover/insulator films were tested for
resistance to degradation. Test methods included
dry oven aging, humidity chamber aging, field aging
and accelerated outdoor weathering. Evidence of
degradation included discoloration, embrittling and
other changes in mechanical properties, development
of opacity, changes in electrical resistivity and
the appearance of polymer oxidation products
observed by ESCA and multiple internal reflection IR
spectroscopy.

Encapsulants in a photovoltaic (PV) module provide electrical
insulation and protect the metallized cell contacts and
interconnect system against corrosion over a 20-year lifetime out-
of-doors. The typical environmental stresses and possible
resulting failures in exposed PV modules are listed in Table I.
In the case of brittle cells such as single or polycrystalline
silicon, encapsulants must also provide mechanical protection for
the fragile wafers and interconnect ribbons or wires. Figure 1
shows the typical layup for a plastic front design. The functions
and performance requirements for the various components are
described in detail in the accompanying paper in this symposium
proceedings entitled "Encapsulant Material Requirements for
Photovoltaic Modules" (1).
The layers most vulnerable to degradation are the pottant and
the flexible outer cover/insulator, the latter being particularly
susceptible in the substrate design where it is on the module
front and thus must be transparent. These two layers are the most
easily degraded because the quantities required and the cost

0097–6156/83/0220–0387$06.00/0

Table I. Principal Damaging Environments

Thermal Cycling	Structural Loading
- Interconnect Fatigue	- Cell Interconnect Fatigue
- Encapsulant Delamination	- Structural Fatigue
- Solar Cell Cracking	
	Hail Impact
Humidity	- Optical Cover Breakage
- Cell Metallization	- Cell Cracking
Delamination	
- Encapsulant Delamination	Voltage Stress
	- Insulation Breakdown
Ultraviolet	- Cell Corrosion
	(Ion Migration)
- Optical Material	
Degradation	
- Encapsulant Delamination	Optical Surface Soiling

constraints necessary for low cost solar arrays dictate that they be upgraded, good performance materials.

The pottant is the vibration damping, elastomeric material that immediately surrounds both sides of the fragile solar cell wafers and their electrical contacts and interconnects. It must be soft, transparent, electrically insulating, weather resistant, chemically inert and form strong and stable adhesive bonds to the surfaces it touches. It protects the cells from stresses due to thermal expansion differences and external impact and isolates them electrically. The pottant also helps protect the circuit metallic contacts and interconnects from the corrosive effects of moisture, salt, smog, etc.

The outer cover/insulator must be a tough, soil resistant, weather resistant and electrically insulating layer. It may be a flexible or conforming plastic film or a coating applied from solution. As a front layer, it is desirable that it act as a UV screen for the pottant while it must at the same time be >90% transparent to wavelengths from 0.4 to 1.1 microns. It must form stable bonds to the pottant and to other module materials to which it seals.

Most organic materials contain sites where radicals can form more or less easily, depending upon structure. Only perfluorinated molecules are totally free of such sites since radical formation usually involves abstraction of hydrogen radicals. Both the polymer and abstracted hydrogen radicals become stabilized by the intervention of oxygen. Degradation in the form of crosslinking, chain scission or both follows. The

Substrate Design

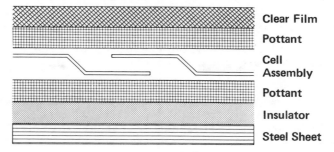

Clear Film

Pottant

Cell
Assembly

Pottant

Insulator

Steel Sheet

Figure 1.   Module Cross Section.

basic mechanism of this process, identified over 35 years ago by
Bolland and Gee ($\underline{2}$), is as follows:

(1)   Initiation $\xrightarrow{\quad r_i \text{ (rate of initiation)} \quad}$ free radicals

(2) $\quad$ R$\cdot$ + O$_2$ $\xrightarrow{\quad k_o \quad}$ RO$_2\cdot$

(3) $\quad$ RO$_2\cdot$ + RH $\xrightarrow{\quad k_p \quad}$ ROOH + R$\cdot$

$\Bigg\}$ propagation

(4) $\quad$ RO$_2\cdot$ + RO$_2\cdot$ $\xrightarrow{\quad 2k_t \quad}$ products − termination at O$_2$ saturation

Fluorocarbon polymers are the most resistant to this type of
degradation because the carbon-fluoride bond is extremely stable,
with an energy on the order of 116 kcal/mole, compared to carbon-
hydrogen bond energies of 91-98 kcal/mole ($\underline{3}$, $\underline{4}$, $\underline{5}$).
Fluorocarbons are, however, extremely expensive because the
monomers are more complicated to synthesize and more dangerous to
handle. The polymers cost on the order of $10-20 per pound
compared with the most widely used hydrocarbon polymers which can
be $1-5 per pound.
Silicone polymers usually have an all O-Si-O-Si-O backbone
and thus do not undergo hydrogen radical abstraction in a position
where it can cause significant crosslinking or chain scission.
The Si-O bond can be hydrolyzed, although not easily. When
silicones do degrade, they become more hydrophilic, allowing more
moisture to reach circuit metals. They can also harden if the
rubbery side chains are lost, thus losing their stress damping
characteristics. They also cost about $10 per pound.
The sites for radical formation on acrylics are either
deactivated by the carbonyl group as in ordinary acrylic esters,
or totally blocked as with the methacrylic esters. This is what
makes them stable compared to other saturated hydrocarbon backbone
materials like polyethylene where there are no electron
withdrawing groups to stabilize against radical formation. The
degree of stability depends on the strength of the electron
withdrawing effect. Most acrylics also exhibit some hydrophilic
character, even when they have very fatty side chains because they
too can hydrolyze, however slightly.
The least stable materials have unsaturation in their
backbones which can be directly and easily attacked by oxygen and
peroxy radicals to degrade according to the mechanisms just
described. The less stable, less expensive materials with
saturated backbones such as polyolefins, and especially those
containing electron withdrawing groups such as esters, halogen
other than fluorine, amides, urethane groups, etc. to stabilize

them, can be dramatically upgraded with antioxidants and photostabilizers to make them potentially acceptable for use in photovoltaics.

There are five classifications of oxidation inhibitors. These are based on differences in the mechanism by which they function to interfere with one or more of the reactions described in the previous equations to prevent or delay catastrophic degradation by oxidation (6). These classifications are:

1. Metal deactivators, which form inactive chelates or insoluble reaction products with transition metals originally present in a form that promotes the decomposition of peroxides to free radicals. Examples are ethylenediaminetetraacetic acid, salicylaldehyde-diamine condensation products or metalalkyl dithiocarbamates such as of nickel or zinc.

2. Light absorbers, which protect from photo-oxidation by absorbing the ultraviolet light energy that would otherwise initiate oxidation, either by decomposing peroxides or by sensitizing the oxidizable material to oxygen attack. The absorbed energy must be disposed of by processes that do not produce activates sites or free radicals. Examples are 2-hydroxybenzophenones, 2-(2'-hydroxyphenyl)benzotriazoles, certain salicylate esters or certain organonickel or chromium compounds.

3. Peroxide decomposers, which promote the conversion of peroxides to non-free radical products, presumably by a polar mechanism. Examples are dialkylarylphosphites, dialkylthiodipropionates or long chain alkylmercaptans.

4. Free radical chain stoppers or "radical traps," which interact with chain-propagating $RO_2\cdot$ radicals to form inactive products. This is usually accomplished by its donation of an $H\cdot$ radical to terminate an active polymer radical, itself forming a more stable one (usually by resonance) which will not rereact with the polymer (e.g., with the help of steric hindrance) and will eventually relax its energy through thermalization, fluorescence or other innocuous means. Examples are sterically hindered phenols or secondary arylamines.

5. Inhibitor regenerators, which react with intermediates or products formed in the chain-stopping (termination) reaction so as to regenerate the original inhibitor or form another product capable of functioning as an antioxidant. Examples are dialkylphosphonates with hindered phenols or diphenoquinones with thiols.

Since the available, inherently weather-resistant materials cannot provide a totally moisture- and oxygen-free environment for the circuit, whether or not they themselves require it for stability, and because stabilizers can so dramatically improve the

performance of much lower cost materials, the lower cost materials
are preferred over silicones and most fluorocarbons for use in low
cost terrestrial solar arrays. It is likely that acrylics in the
$1-5 per pound range will ultimately be the optimum material for
both the pottant and the outer cover/insulator. Their
availability in the current market, however, in forms suitable for
PV application is currently limited and, for the most part,
unproven.

The two pottant materials studied in this report are
plasticized polyvinyl butyral (plPVB) which is easily available
and used in safety glass, and a highly stabilized, peroxide
crosslinked ethylene/vinyl acetate (EVA) copolymer containing
about 33 weight % vinyl acetate (7). The outer cover/insulator
materials studied include polyvinyl fluoride (PVF) and a butyl
acrylate/methyl methacrylate graft copolymer (BAgMMA); both are
blown films.

Aging Tests and Results

Mechanical and Optical. Clear 15-mil-thick films of EVA and
plPVB, 4-mil PVF and 3-mil BAgMMA were exposed in a circulating
air oven at about 150°C for periods of 0 to 26 days. This type of
test is used extensively throughout the polymer industry as a
screening tool for comparing the oxidative stabilities of polymers
and compound formulas.

A summary of the optical transmission changes as a result of
the oven aging can be seen in Figure 2. EVA exhibited very little
yellowing in the 26 days. Mechanically, its tensile strength,
elongation at break and permanent set decreased considerably with
the aging. At the same time, however, its tear strength and
elastic moduli at 10% and 100% elongations, which are within the
regions of concern for PV use, remained relatively constant (Table
II). The plPVB darkened rapidly during the oven exposure,
becoming too brittle to permit tensile testing.

The BAgMMA showed no measurable yellowing throughout the full
26 days of aging, but the large increase in UV transmission during
the first 7 to 10 days suggests that it loses much of its UV
absorber during this period (Figure 3). Further aging then
resulted in a decrease in UV transmission. The BAgMMA film was
becoming noticeably brittle at the 10-day sample and was quite
brittle after 18 days. PVF showed very little yellowing (<10%
loss) and no change in mechanical properties throughout the 26
days of aging.

A small amount of yellowing of the clear pottant or front
cover materials in the 360 to 500 nm region actually does not
significantly affect output of present day single crystal silicon
cells because the number of photons in the high energy region is
only about 10% of the total utilized by the cell (Figure 4). Peak
solar photon flux centers around 700 nm. Thus with as much as a

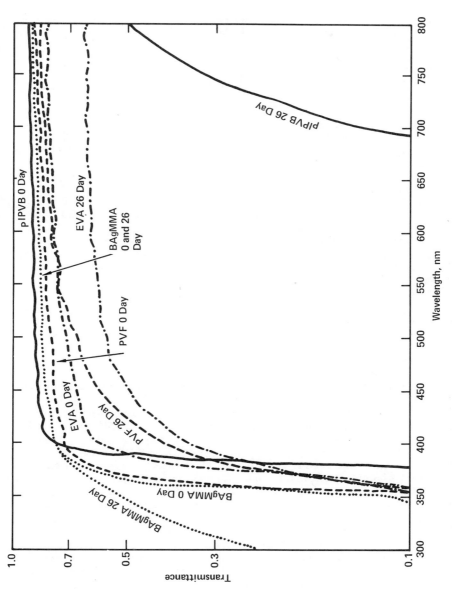

Figure 2. Degradation of Optical Transmission.

Table II. Degradation of Mechanical Properties

	Aging Time		
	0 Day	4 Day	26 Day
Tear Strength, lbs/in			
EVA Unprimed	540	540	490
EVA Primed	630	515	420
PVF 4 mil	3,400	3,400	3,350
Tensile Strength, psi			
EVA Unprimed & Primed	1,300	800	250
PVF 4 mil	19,500	19,000	17,000
Elongation at Break,%			
EVA Unprimed	1,100	930	220
EVA Primed	1,070	730	80
PVF 4 mil	235	230	220
Elastic Modulus, psi			
EVA Unprimed & Primed M100	210	240	220
PVF 4 mil M100	12,700	12,600	12,500

p1PVB & BAgMMA Embrittled

30% transmission loss in the encapsulant films at 450 nm, the circuit output loss would be only ~3%.

Experimental panels underwent accelerated outdoor weathering at DSET Laboratories near Phoenix, Arizona. Mirrors were used to multiply the solar flux and the panels were intermittently sprayed with water. The type of exposure is called EMMAQUA by the DSET Laboratories, whose trade literature provides a more detailed description of their procedures (8). In EMMAQUA exposures of early design, two-cell coupons with plastic front substrate configurations, the support was BAgMMA coated, galvanized steel. The components included both p1PVB and EVA pottants, with both BAgMMA and PVF clear front covers. The p1PVB discolored extensively wherever used; BAgMMA became brittle and cracked in all cases; EVA showed no degradation in any case. PVB showed embrittlement in some cases, which may be correlated with shrinkage-induced stress, but showed very little discoloration. Changes in I_{sc} in the I–V curves of these two-cell PV circuits were proportional to the yellowing of the encapsulants (Table III).

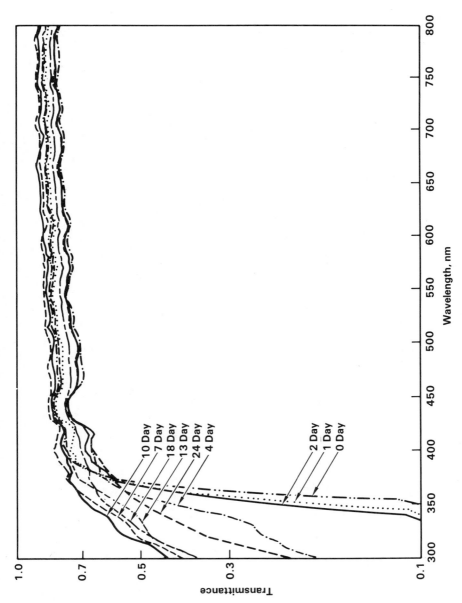

Figure 3. UV-Vis Transmission for Aging BAgMMA.

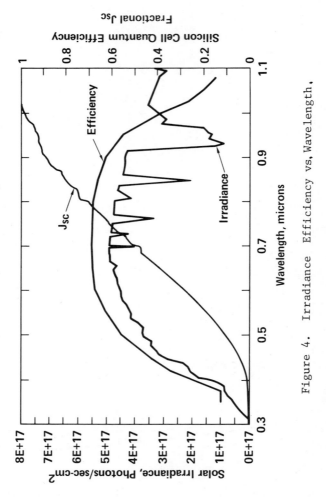

Figure 4.   Irradiance Efficiency vs, Wavelength,

Table III. I_{sc} Changes in DSET Samples

Pottant	Cover	Ratio of Final to Initial I_{sc}
EVA	PVF	1.02 ± 0.03
EVA	BAgMMA	1.01 ± 0.05
PVB	PVF	0.70 ± 0.16
PVB	BAgMMA	0.86 ± 0.12

During humidity testing, EVA films became only slightly cloudy with <20% optical transmission loss due to moisture uptake at temperatures above 60°C, with 100% relative humidity. The transparency of PVF films was unaffected by these conditions. BAgMMA and p1PVB films, on the other hand, were severely clouded as a result of these exposures and both exhibited >50% optical transmission losses.

Electrical and Chemical. Oven aging at 150°C for 26 days had no effect on the electrical volume resistivities of either PVF or EVA films. Both maintained 10^{14} Ω-cm (see Table IV). The p1PVB increased from 10^{11} Ω-cm to 10^{16} Ω-cm in four days due to the evolution of plasticizer and absorbed water. Both plasticizer and absorbed water in p1PVB cause its low initial volume resisitivity. Low volume resistivity material used in a photovoltaic panel can cause electrical current leakage (1). BAgMMA decreased from 10^{17} Ω-cm to 10^{15} Ω-cm in 10 days.

It was observed that field aged p1PVB containing panels, as well as oven aged p1PVB or EVA containing panels exhibited browning over the contact grid line areas when certain metallization materials were used (Figure 5). The grid lines form the solar cell contacts and also bond to the pottant. Since the metallization incorporates transition metal oxide containing glass frit, as well as silver, it was considered likely that the browning was catalyzed by one or more of these oxide components according to the mechanism described in Figure 6 (9, 10, 11). To test this hypothesis, samples of each of the individual oxide constituents of the frit glass, admixed at the 1-10% level in a lead borosilicate glass matrix, were encapsulated in panels with p1PVB and EVA, together with evaporated silver thin film samples. These test modules were then subjected to oven aging and were visually examined for browning over the oxides. Glasses containing vanadium oxide, antimony oxide, or a mixed copper/nickel oxide all showed accelerated browning. Browning over solar cell contacts was shown to be retarded by the use of certain primers prior to encapsulation with EVA (1).

Table IV. Degradation of Electrical Properties

Material	Days Aged 150oC, Oven	Volume Resistivity, $-cm$	Dielectric Strength, DCV/mil
EVA			
Without Primer	0	1.8×10^{14}	700–1000
" "	1	--	590
" "	4	--	720
" "	10	--	960
" "	18	1.4×10^{14}	--
" "	24	0.7×10^{14}	--
With Primer	0	--	720
" "	1	--	590
" "	4	--	710–750
" "	10	--	560–910
p1PVB	0	1.9×10^{11}	450
"	4	3.9×10^{16}	280
PVF (Clear)			
1 mil	0	--	4500
"	7	--	6500
"	13	--	1000
4 mil	0	1.1×10^{14}	2500
"	1	--	2500
"	17	--	1500
"	24	5.0×10^{14}	--
BAgMMA	0	1.6×10^{17}	1700
"	2	--	700
"	10	1.6×10^{16}	500

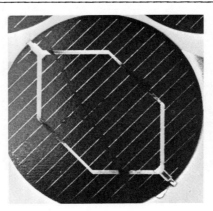

Figure 5. Browning Over Grid Lines.

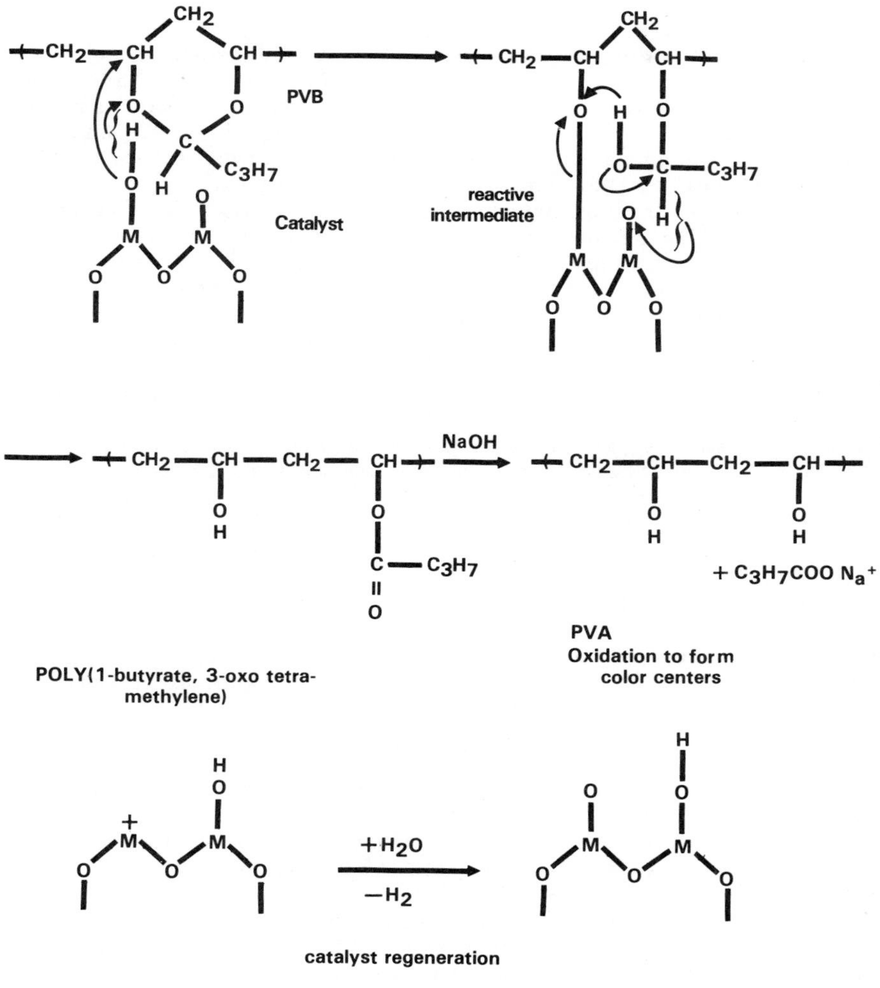

Figure 6.   Possible Mechanism for Catalyzed Degradation.

Analytical Methods. To investigate the chemistry associated
with polymer degradation, ESCA and multiple internal reflection
infrared absorption (MIR) studies were performed on EVA, plPVB,
BAgMMA and PVF films. Fresh films that had been oven aged for 26
days at 150°C were studied. In addition, samples of browned EVA
and plPVB, peeled from cells which had been oven aged or aged in
the field, were also studied.

ESCA analysis of PVF films showed that oven aging results in
defluorination of the polymer surface (Figure 7). Complete loss
of fluorine is observed in the case of extensively aged samples
and these samples exhibit an ESCA spectrum similar to that of
polyethylene. Reappearance of fluorine is observed after a brief
ion bombardment of the sample surface, suggesting that the
thickness of the defluorinated layer is 10-30 nm. The carbon 1s
line from a sample of unaged PVF was deconvoluted into its CHF and
CH_2 components. The observed CHF/CH_2 ratio was significantly less
than one, the value predicted by stoichiometry, indicating that
this unaged surface was also partially defluorinated. The PVF
used was a "bondable" grade which, according to the manufacturer,
has been corona-discharge treated to promote adhesion (_12_).

Both ESCA and MIR results were obtained on oven aged plPVB
film surfaces and plPVB surfaces that exhibited browning because
of contact with cell metallizations in field aged panels. The
ESCA data showed that browning was accompanied by increases in
surface oxygen, primarily as C=O. MIR showed increases in the
ester (~1250 and ~1050 cm^{-1}) and carboxylic acid salt (~1605 cm^{-1})
moieties and decreases in the cyclic ether (~1100 cm^{-1}) moieties
in the near surface region. These results are consistent with the
literature on the oxidation of polymer backbones and formation of
"color centers," which are carbonyl terminated chains possessing
extended conjugation (Figure 8) (_13-17_). When browned plPVB from
field aged panels is peeled from the cell surface, ESCA analyses
of the tear surfaces show that the tear occurs at the
polymer/metallization interface, leaving little residue on the
contact. This catalytic change has no effect on module output
since no light passes through the gridlines and no macroscopic
changes to the mechanical properties occur.

ESCA and MIR analyses of 150°C oven aged EVA films show that
the aging results in an increase in the oxygen content of the
exposed surfaces. This is primarily reflected in an increase in
the C-O moieties in which oxygen is probably present as hydroxyl
groups. When browned EVA, protected by PVF, was peeled from the
metallization of a cell in an oven aged module, the tear surface
on the EVA film showed only a minor amount of oxidation, similar
to that observed on EVA tear surfaces which had not been in
contact with the cell metallization. Analysis of the
corresponding metallization surface, on the other hand, showed
that several hundred nanometers of heavily oxidized polymer
remained behind on the metallization surface after the tear
(Figure 9). This suggests that oxidation of the polymer initiated

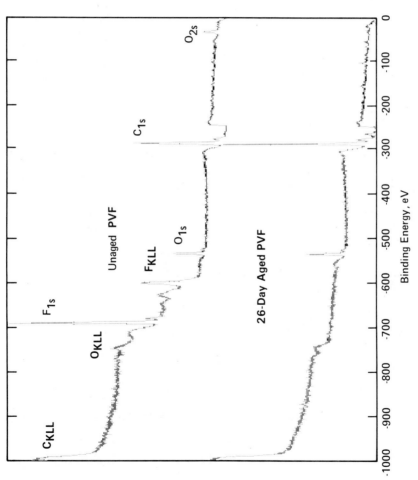

- Surface dehydro-fluorinates upon aging

- Only a thin surface layer is affected.

Figure 7.   ESCA Analysis of PVF.

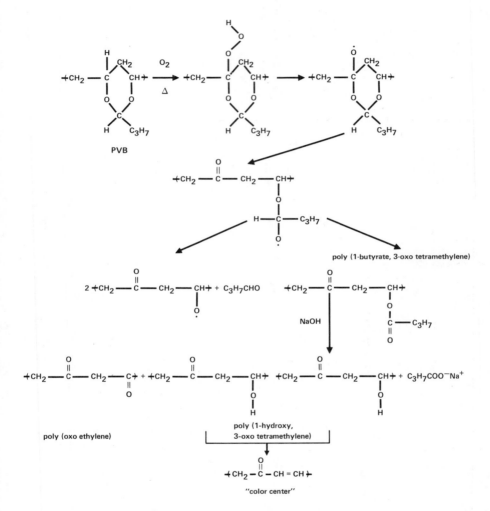

Figure 8.   Possible PVB Browning Mechanism.

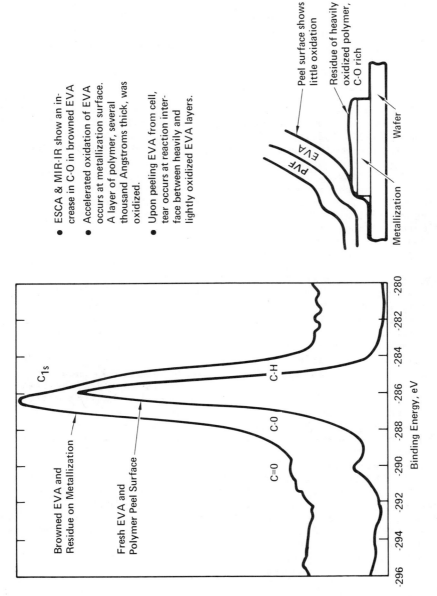

Figure 9.   ESCA Analysis of EVA.

at the interface between the metallization and the polymer, with a
reaction front then proceeding through the polymer as catalyzed
oxidation progressed. When the polymer was peeled from the cell,
the tear occurred along this reaction front.
Oven aged BAgMMA films were also studied by ESCA. Aging resulted
in a 35% decrease in the amount of oxygen present on the surface
of the film, with the largest decrease seen in carbon as C=O, with
a smaller decrease in C-O.

Discussion and Conclusions

It is estimated that the acceleration factor for polymer
aging under the 150°C oven aging conditions previously specified
is about 500X assuming the kinetic "rule of thumb" that reaction
rates approximately double for every 10°C rise in temperature and
assuming average operating conditions for a PV array of 70°C for
12 hours per day. Thus the temperature rise gives a factor of
~250X and the continuous exposure another 2X for a total of 500X.
This implies that every 15 days of oven aging is equivalent to 20
years of in-the-field aging under average operating conditions of
70°C for 12 hours per day. The 70°C/12 hours per day condition
probably represents a worst case (e.g., equatorial desert) for
anywhere in the world. World average estimates are more like
50°C/5 hours per day which would mean the acceleration factor of
150°C/continuous could be as much as 4000X.
Since the oven aging tests were performed in the dark, the
effects of UV light upon polymer degradation were not a part of
those studies. Thus, the oven aging tests do not provide a
complete picture of polymer aging under the conditions experienced
by a solar panel. Nevertheless, the relative oxidative
stabilities of materials can still be compared fairly accurately
because the sites subject to oxidative attack on a polymer
backbone are the same whether that attack is catalyzed by heat or
by UV. The mechanism of energy transfer and thus the method of
stabilization may differ so stabilizer screening is not as
accurately tested but unstabilized materials can especially be
accurately compared. In addition, all the materials being
considered do require some thermo-oxidative stability as well as
photo-oxidative stability.
The EMMAQUA test, which includes the effects of UV, heat and
moisture, is more representative of solar panel aging than are the
oven aging tests. It has a much lower acceleration factor,
however, on the order of about 20X which includes approximately 6-
8X in UV-visible solar flux and about 3-4X in temperature above
which these particular solar panels would ordinarily run. It is
therefore a much slower, more elaborate process that should follow
oven aging as a screening tool with which to eliminate from
further consideration those polymers that have insufficient
resistance to long term oxidation.

Our observations suggest that of the two pottants studied, EVA is more resistant to degradation than plPVB. In addition, of the two outer cover/insulator materials studied, PVF was more resistant to degradation than BAgMMA. In order to more closely simulate field exposure conditions, future aging tests will incorporate both UV exposure and a thermal gradient across the polymer film in the manner described by Laue and Gupta (18). We believe that by determining the mechanisms of degradation in field aged polymer and by determining the reaction rates in accelerated aging tests that can be shown to follow the same degradation mechanisms, it will be possible to estimate service lifetimes for polymeric PV encapsulation materials.

Literature Cited

1. Lewis, K. J. Encapsulant Material Requirements for Photovoltaic Modules. Proceedings of the Symposium on Polymers for Solar Energy Utilization, 1982, Las Vegas, NV.
2. Bolland, J. L.; Gee, G. Trans. Faraday Soc. 1946, 42, 236, 244.
3. Pauling, L. "The Nature of the Chemical Bond"; 3rd. ed., 1960, Cornell University Press, Ithaca, NY.
4. Patrick, C. R. Adv. Fluorine Chem. 1961, 2, 1.
5. Banks, R. E., ed. "Organofluorine Chemicals and Their Industrial Applications"; 1979, Ellis Horwood Ltf., Chichester, Gr. Brit.
6. Shelton, J. R. Offic. Dig. Federation Soc. Paint Technol. 1962, 34, 590.
7. "Investigation of Test Methods, Material Properties, and Processes for Solar Cell Encapsulants, Annual Report," DOE/JPL - #954527-79-10, Springborn Laboratories, Inc., June 1979.
8. Zerlaut, G. A. Accelerated Outdoor Weathering Employing Natural Sunshine. Proceedings of the 21st Annual Meeting of the Institute of Environmental Sciences, 1975, 1, 153-159, Anaheim, CA.
9. Pines, H.; Manassen J. Adv. Catal. 1966, 16, 49.
10. Hensel, J.; Pines, H. J. Catal. 1972, 24, 197.
11. Krylov, O.; Fokina, E. 4th Inter. Congr. Catal. 1968, Moscow.
12. PVF trade literature, E.I. duPont de Nemours & Co., Inc., for Tedlar, which is a registered trademark.
13. Shuvalova, E. V. Polymer Sci. USSR. 1961, 2, 193.
14. Beachell, H.C.; Fotis, P.; Hucks, J. J. Polymer Sci. 1951, 7, 353.
15. Noma, K. Chem. High Polymers (Japan). 1948, 5, 190.
16. Futama, H.; Tanaka, H. J. Phys. Soc. (Japan). 1957, 12, 433.
17. Gilbert, J. B.; Kipling, J. J. Fuel. 1962, 12, 249.

18. Laue, E.; Gupta, A. "Reactor for Simulation and Acceleration
     of Solar Ultraviolet Damage"; Jet Propulsion Laboratory
     Publication 79-92, DOE/JPL Report 1012-31, 1979.

RECEIVED November 22, 1982

Vacuum Lamination of Photovoltaic Modules

DALE R. BURGER and EDWARD F. CUDDIHY

California Institute of Technology, Jet Propulsion Laboratory,
Pasadena, CA 91109

Vacuum lamination of terrestrial photovoltaic
modules is a new production process requiring special
equipment and a significant material development
effort. Equipment design studies resulted in improved
control and lower costs when using a double-chamber
vacuum laminator. Application testing of new materials and primers showed the feasibility of two different back sheet materials, one encapsulant, two new
primers for polymers and one primer for metals.

A few years ago, most terrestrial photovoltaic (PV) modules
were assembled by casting the cells in a transparent silicone
substance, using a metal substrate for support. When this
approach was reviewed by the Flat-Plate Solar Array Project (FSA),
development was begun on new materials that would reduce the
cost and quantity of material required for encapsulation.
During development of these materials, the PV industry improved module design by eliminating metal substrates and incor-
incorporating glass superstrates to provide a hard, easily
cleaned top surface. Use of glass superstrates created new
material problems. Bonding some encapsulating materials proved
to be a difficult problem. Elimination of visual defects, such
as voids and bubbles, became necessary because of increased
market sophistication. Long-term corrosion concerns became important now that thin, water-permeable polymer materials were
being used to reduce cost. Discussion of these problems will
start with equipment development, because improved equipment was
necessary to the subsequent materials development effort. While
the development was being pursued, some module design and processing problems became evident. Because these problems have a
bearing upon the reliability of the lamination process, they
will also be discussed.

0097–6156/83/0220–0407$06.00/0

Equipment Development

When modules were being assembled by casting with silicones, material and labor costs were high and equipment costs were low. Material-cost reduction was sought in the development of new materials. Labor-cost reduction, however, is dependent upon the development of better processes and the introduction of equipment specifically designed to use the new materials and processes. The process that seemed most suitable for the new materials was a lamination process. Initial FSA involvement with lamination occurred late in 1978 at RCA Laboratories.

Early Designs. RCA double-glass lamination experiments showed that an autoclave was expensive and often introduced air into module edges during cooldown (1). Late 1978 was also the time of contract initiation with ARCO Solar, Inc., to develop a laminator that would reduce the near-term costs of PV modules. By July 1979 a double-chamber vacuum laminator was developed, tested and put into production (2). This laminator used a proven thermoplastic, polyvinyl butyral (PVB), which requires humidity- and temperature-controlled storage and handling. Other problems with PVB were high cost and high viscosity at the 150°C process temperature.

An answer to these problems was sought in a material that would cure at process temperatures. One possible solution was to develop a thermosetting polymer, but these polymers place more stringent temperature requirements on a laminator. The original ARCO Solar laminator used long, slender tungsten-filament lamps as a heat source and shiny aluminum strips to adjust for uniformity. This approach works well with a thermoplastic such as PVB, but was considered to be too variable and limited for research into thermosetting materials and substrate module designs.

A resistive strip heater was considered as a possible inexpensive, improved heat source. Wire-wound strip heaters are commonly made with surface-temperature variations of less than ± 2% (when measured on top of a 0.125-in.-thick glass plate). Another well-controlled heat source would be a heated oil system. This system also would have the capability of cooling down the product before laminator is opened.

Small Research Laminator. The Process Development Area (PDA) of FSA received two ARCO Solar-developed laminators at the end of the contract. One of these laminators was modified for use with a strip heater. This configuration, with a boost-and-buck autotransformer power supply, was used successfully for some process verification and materials survey experiments. Assessment of future research needs and present equipment capabilities led to additional modifications of the laminator.

Experience with the laminator showed that the heavy aluminum base plate, a large thermal mass, caused control-response problems. The 0.25-in. Transite plate also caused some control problems, because it was close to the controller thermocouple. Thermal mass keeps the chamber temperature high during the un-load-load cycle, which can start the cure cycle earlier than desired. The controller temperature must be manually adjusted to prevent the temperature of the laminant adhesive-encapsulant from overshooting. An inexpensive, mechanically stiff thermal insulation system that would not outgas during exposure to processing temperatures as high as 175°C was required. Figure 1 shows the laminator modifications required to achieve the desired thermal isolation and improved controllability. An unusual material application was the use of glass marbles as insulation. There was an improvement in warm-up time and in tracking. This design presents a nearly balanced thermal load above and below the strip heater that should allow good tracking, regardless of the desired time-temperature cycle.

Additional laminator changes that were implemented included:

(1) Applying a vacuum to the top chamber of a double-chamber vacuum laminator just before raising the upper chamber (lid) to remove a completed module. This allows the silicone-rubber diaphragm to be attached to the upper chamber, which reduces handling and keeps the diaphragm weight off the laminant stack during eva-cuation of the lower chamber.

(2) Attaching the controller thermocouple to the bottom of the 0.125-in. aluminum platen by laminating. This was done, and the improved thermal coupling enhanced the performance of the laminator.

One measure of the utility of the present laminator is its acceptance by other researchers. Test programs now scheduled on the laminator include:

(1) New encapsulating materials evaluation.
(2) Substrate encapsulation studies.
(3) Preparation of water permeation study samples.
(4) Evaluation of module design developed under JPL contract.
(5) Encapsulant-to-metal primer-compound research.
(6) Preparation of electrostatic test samples.
(7) Preparation of ultraviolet test samples.

Large New Laminator. Because of the equipment development success, a new, larger laminator is being designed to explore problems inherent in fabricating the larger PV modules envisioned for the mid-1980s. A 4-ft-square laminator area is expected; it would be compatible with the 1.2-m square designs or any

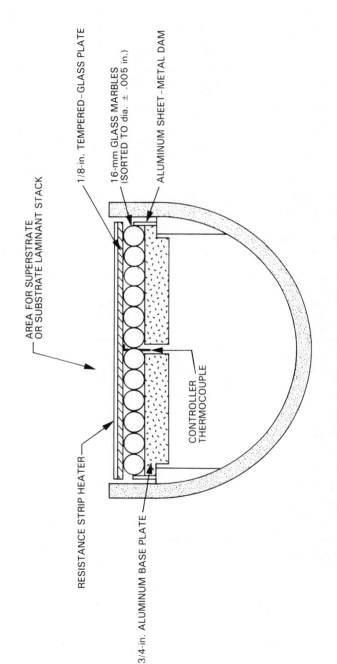

Figure 1.   Laminator With Improved Thermal Isolation.

smaller configurations. The most expensive part of a large
laminator is the chamber that must withstand the atmospheric
pressure load. At present the lowest-cost vacuum chamber that
has been considered uses standard hemispheric pressure vessel-end
caps costing about $600 each (including a 2-in.-wide flange). A
lead alloy counterweight will permit the operator to rapidly and
safely load and unload the laminator.

Support and insulation of the platen was also a problem,
and the use of marbles seems to be an inexpensive choice because
about $700 worth of marbles would provide support and thermal
isolation, and has the added advantages of easy transport and
modification of the chamber. Marbles also reduce the volume of
the vacuum chamber and thereby reduce pump-down time and energy.

Control of the laminator cycle will be very flexible due to
use of four independently adjustable timers and a 24 step industrial controller with 10 individual switch modules.

Materials Research

Design of a PV module that will withstand 20 years of exposure to a variety of terrestrial environments creates many
problems. An FSA cost allocation of $14/m^2 for encapsulation,
superstrate or substrate and edge-seal/gasket places an additional
burden on the encapsulation materials, because the glass superstrate alone has a projected cost of about $10/m^2.

Details and background on early materials research efforts
have been published (3-7). A detailed discussion of present
encapsulation materials will be published soon (8). This report
covers the application testing of developed materials and other
requirements for successful vacuum lamination.

Ethylene Vinyl Acetate System. The first new lamination
material developed by FSA was compounded by Springborn Laboratories Inc. (SLI) from an ethylene vinyl acetate (EVA) feedstock
available from Du Pont. Compared with PVB, EVA costs about
one-third as much, has much lower viscosity at process temperature, and has no humidity-control requirement during processing.

Early laminator experience uncovered problems with curing
and adhesion of EVA. The original material from SLI also would
block (adhere to itself). Subsequent material delivered from
SLI and Du Pont did not block. The Du Pont material had one
waffled surface, which enhanced air removal during vacuum pumpdown.

There is more than one correct cure cycle for EVA. Like
most polymers with peroxide promoters, it is good practice to
raise the bond-line temperature rapidly to avoid peroxide decomposition before an adequate cure has been obtained. One cycle
that has been proven uses two steps, one at 100°C for evacuation
and adhesion, the other at 150°C for long-term oven cure. This
cycle provides a high throughput with only one laminator.

A description of the cure cycle used for materials testing at our laboratory is: Evacuation for 5 min, then 25 min of cure with the top chamber bled to atmosphere. During the 25-min cure, the first 8-10 min is required to raise the encapsulant temperature to 160°C, where it is maintained to the end of the cycle. Modules are then removed without being cooled. Modules fabricated with this cure cycle show even and complete curing and no bubbles.

Adhesion is a more difficult problem. There are many chemically different interfaces in a laminant stack: glass-EVA, EVA-solar cell surface (oxidized silicon or some antireflection (AR) coating), cell back surface metallization-EVA, EVA-back sheet, and EVA-bus bars (copper or tinned copper). Each of these interfaces is important, because mechanically good adhesive bonds will often fail by delamination after exposure to humidity in the field. Water vapor will permeate through polymers and, if there is a non-chemically bonded surface, water may collect and cause failure by displacement. Tables I, II, and III provide details of some of the research efforts in adhesion.

A material survey was made using EVA and ethylene methyl acrylate (EMA) encapsulants with Korad 63000, Scotchpar 20CP and Tedlar 200BS 30WH as back sheets. Table I shows the detailed results of ths survey. Primed and unprimed surfaces and a new Du Pont adhesive, 68040, were investigated. This survey showed good glass bonding with SLI Primer A11861-1, (Dow Corning silane Z-6030, 9 parts; N, N-dimethyl-benzylamine, 1 part; Lupersol 101, 0.1 part, and methanol 89.9 parts). The only back sheet that adhered to EVA was Tedlar with Adhesive 68040. Earlier tests showed good mechanical bonding to untreated Tedlar but poor humidity performance.

Because the glass-to-EVA interface bonding problem seemed to be solved when all samples exhibited adherent and persistent bonds, the focus of the effort was shifted to back-sheet adhesion. An additional series of test (see Table II) confirmed the good results of Tedlar with the Du Pont adhesive 68040. Korad 63000 may be a useful material, but cure temperatures during lamination caused some degradation. Additional tests on this acrylic sheet may be run. A polyester film, Scotchpar 20CP, was interesting, because it would be less expensive than a polyvinyl fluoride film, such as Tedlar. This test series showed that a new primer or adhesive was needed for the Scotchpar film.

Fortunately, E. P. Plueddemann of Dow Corning Corp. had already developed a primer for polyester films. The primer consists of American Cyanamid Cymel 303, 90 parts; Dow Corning silane Z-6040, 10 parts; and methanol, 300 parts. Peel tests of Scotchpar bonded to EVA using this primer were excellent. Unfortunately, this system did not perform well in the 7-day cold-water soak test as shown in Table III.

Table I. Material Survey

Sample No.	Encapsulant	Back Sheet	Glass Primer	Back-Sheet Ahesive	Encapsulant-Glass Peel, g	Back Sheet Encasulant Peel, g
1	EVA	Korad 63000	None	None	Not tested	185
2	EVA	Korad 63000	A11861-1	A11861-1 Primer	2,900	370
3	EMA	Korad 63000	None	None	Not tested	0
4	EMA	Korad 63000	A11861-1	A11861-1 Primer	4,300	Adherent, but brittle
5	EVA	Scotchpar 20CP	None	None	Not tested	55
6	EVA	Scotchpar 20CP	A11861-1	A11861-1 Primer	7,000	590-680
7	EMA	Scotchpar 20CP	None	None	Not tested	110
8	EMA	Scotchpar 20CP	A11861-1	A11861-1 Primer	10,000	3,200; broke
9	EVA	Tedlar 200BS 30WH	None	68040	Not tested	Adherent; couldn't peel
10	EVA	Tedlar 200BS 30WH	A11861-1	68040 and A11861-1 Primer	5,100	Adherent; couldn't peel
11	EMA	Tedlar 200BS 30WH	None	68040	Not tested	Adherent; couldn't peel
12	EMA	Tedlar 200BS 30WH	A11861-1	68040 and A11861-1 Primer	7,000	4,400; broke

Table II. Back-Sheet Adhesion

Sample No.	Encapsulant	Back Sheet	Glass Primer	Back-Sheet Adhesive	Back-Sheet Encapsulant Peel
A-1	EVA	Scotchpar 20CP	A11861-1	68040	0
A-2	EMA	Scotchpar 20CP	A11861-1	68040	0
A-3	EVA	Korad 63000	A11861-1	68040	Adherent, but brittle
A-4	EMA	Korad 63000	A11861-1	68040	Adherent, but brittle
A-5	EVA	Tedlar 200BS 30WH	A11861-1	68040	Adherent; couldn't peel
A-6	EMA	Tedlar 200BS 30WH	A11861-1	68040	Adherent; couldn't peel
A-7	EVA	Gel Test	None	----	----
A-8	EMA	Gel Test	None	----	----
A-9	EVA	Tedlar/68040/ EVA Combination	A11861-1	(68040)	Adherent; couldn't peel
A-10	EMA	Tedlar/68040/ EVA Combination	A11861-1	(68040)	Adherent; couldn't peel
A-11	EVA	Gel test	None	----	----
A-12	EVA	Scotchpar 20CP	A11861-1	1359	0
A-13	EVA	Korad 63000	A11861-1	1359	Adherent, but brittle

Table III. Results of 7-Day Water Soak Test

Coupon No.	Encapsulant	Primer[a]	Back Sheet	Results[b]
C-1	EVA	Cymel	Scotchpar 20CP	Peels
C-2	EVA	Cymel	Scotchpar 20CP	Sample given away
C-3	EVA	Cymel	Korad 63000	Peeled after cure
C-4	EVA	Cymel	Korad 63000	Peeled after cure
C-5	EVA	68040	Tedlar 200BS	Adherent
C-6	EVA	68040	Tedlar 200BS	Adherent
C-7	EVA	68040	Tedlar 200BS	Adherent
C-8	EVA	68040	Tedlar 200BS	Adherent
C-9	EVA	68040	Tedlar 200BS	Adherent
C-10	EVA	68040	Tedlar 200BS	Adherent
C-11	EVA	68040	Tedlar 200BS	Adherent
C-12	EVA	68040	Tedlar 200BS	Adherent
C-13	EVA	68040	Tedlar 200BS	Adherent
E-1	EMA	Cymel	Scotchpar 20CP	Peels
E-2	EVA	Cymel	Acrylar	Brittle
E-3	EVA	A-11861-1	Acrylar	Peels
E-4	EMA	Cymel	Acrylar	Peels
E-5	EMA	A-11861-1	Acrylar	Peels
E-7	EMA	68040	Tedlar 100BG 30UT	Peels after cure
E-8	None	68040	Tedlar 200BS	Peels
E-9	None	68040	Tedlar 200BS	Peels

[a] primer consists of Cymel 303 (American Cyanamid), 90 parts; Du Pont Z-6040, 10 parts; methanol; 300 parts.
[b] After 7-day soak unless otherwise noted.

Another primer system was suggested. This primer was made from Monsanto Resimene 740, 23.5 parts; Dow-Corning Silane Z-6040, 1.25 parts; and anhydrous isopropanol, 75 parts. The initial peel tests were excellent and resistance to 7-day cold water soak was fair to good. Further discussion with Plueddemann led to the addition of 1.25 parts of Dow-Corning Silane Z-6030 to provide double bond for improved adhesion to the EVA. Again, initial peel tests were excellent. Water soak resistance was good to very good on those samples where the primer had been diluted 10 to 1 with anhydrous isopropanol. Additional tests are planned with the possible modification of a small amount of Lupersol 101 to improve the EVA bond again.

EVA does not bond well to all metals. Copper is particularly difficult, which poses a problem: copper is the best candidate for low-cost photovoltaic cell metallization. Some primer tests were run using two primers. The first was zinc chromate powder, 10 parts; Dow-Corning Silane Z-6030, 9.9 parts; N, N-dimethylbenzylamine, 0.1 part; and methanol, 30 parts. Because the zinc chromate is opaque and difficult to keep in suspension, the second primer omitted it.

The primer without the zinc chromate gave excellent initial adhesion if thoroughly wiped after application. Additional tests were made after 10 to 1 dilution with methanol. Again wiping was required and initial peel tests were excellent. Adhesion after water soak was excellent.

Tests were also made on copper which had been fusion solder plated, zinc plated, or nickel plated. None of these showed good initial adhesion.

Ethylene Methyl Acrylate System. EMA, new encapsulant adhesive, is under development. Preliminary work showed excellent adhesion of EMA to glass when the glass is primed with A11861-1. Long-term soaking in cold water reduced the adhesion. Additional work and samples are needed.

Tests with Tedlar and Adhesive 68040 showed adhesion to EMA. However, adhesion after cold-water soak was poor. This problem is being investigated.

The Cymel primer that was used to bond EVA to a polyester was also tried in bonding EMA to a polyester. This system also degraded after long-term soaking in cold water. Korad 63000 has adhered to EMA, but the resulting back sheet was brittle.

EMA adhesion to copper which has been primed as above was excellent both initially and after the 7-day water soak test.

Testing Methods

A gel test, recommended by SLI, was made on EVA coupon samples produced when the original four modules were made. Because uncrosslinked EVA is soluble in toluene, weighed samples

were placed in 60°C toluene for 2 h, and the resultant solution
and sample was poured through weighed filter paper. After fil-
tration, the samples were dried in a 90°C circulating-air oven
for 5 h. The percentage of EVA remaining is a measure of the
degree of gelation or crosslinking during cure. SLI specifies
a nominal 80% gel with 65% as the lower limit. The gel test on
the samples produced above showed better than 95% gelation.

Peel-test samples were prepared by cutting through the layer
to be tested using a 0.25-in.-wide template. The desired layer
was then peeled back by cutting when necessary. Peel strength
was measured using a Unitek Micropull I, Model 6-092. Several
adherent samples had a cohesive strength above the 5-lb limit of
the test equipment, so these samples were tested using a calibrated spring scale. In cases where the adhesion is excellent
it was noted that a peel-test sample could not always be prepared.
These situations are noted in the attached tables.

Performing a peel test after lamination should be considered
as only a good screening test; it is not sufficient for material
selection. Plueddemann recommends a 7-day soak in room-temperature water as an additional test, with final peel tests demon-
strating cohesive failure rather than adhesive failure (Reference
9). All of the laminants made at JPL have been subjected to the
7-day room-temperature water-soak test. Table III summarizes
the results of 7-day room-temperature water-soak tests.

Other Lamination-Related Efforts

The first lamination efforts were mechanically successful
but visually unsuccessful. Many bubbles and voids were found
that were related to solder joints. Another visual problem was
cell misalignment and poor placement. Both of these problem
areas were not caused by the lamination materials but would have
a profound effect on the marketability and field service life
of the final laminated product.

Solder-Flux Removal. Removal of soldering flux residues is
an established process in the printed-circuit-board and elec-
tronic-assembly industries. The quality of the lamination process is dependent upon chemical bonding of all surfaces within
the laminant, so very clean cell-string assemblies are required.
Proper removal of flux residues requires solvents that can remove
both polar and non-polar soluble contaminants, so use of proprietary flux-removal solvents was indicated. Because cell interconnects provide flux traps (expecially the Motorola Inc. and
ARCO Solar combination bus-bar-interconnect designs), it was
decided to try ultrasonic cleaning followed by vapor degreasing.
Six cell-string assemblies for minimodules and four assemblies
for 1 x 4-ft modules were first cleaned in Kester 5345 Rosin
Residue Remover using a Sonix IV Model SS-104 Ultrasonic Cleaner.

Subsequently, these same cell-string assemblies were cleaned in
Kester 5120 vapor degreasing solvent using an Electrovert, Inc.,
Degrestil Model LCD-18 vapor degreaser. The assemblies were
first introduced to the vapor zone, then were dipped in the cold-
solvent tank and finally were removed slowly through the vapor
zone. These cell strings showed no delaminated areas or bubbles
after being laminated with EVA. Cell strings that were only
swab cleaned for flux removal showed both bubbles and delamination when laminated using identical process parameters.

Because flux is such a concern, one contractor is exploring
ultrasonic bonding (10) using prepunched aluminum interconnects
that are attached to electroplated copper cell metallization
with a seam welder. Others are examining fluxless bonding concepts, such as vapor-phase solder reflow.

Conclusions

The following conclusions can be drawn from efforts to date:

(1) Vacuum lamination is an acceptable process for manufacturing void-free PV modules, if matched with correct
materials and used with a qualified cure cycle.

(2) Conceptual design of a large (4 x 4-ft) vacuum laminator indicates the potential for an inexpensive piece
of capital equipment.

(3) Material research by the Encapsulation Task of FSA
has been applied to actual laminated systems with good
results. One laminant system has been developed that
shows excellent adhesion and resistance to delamination
after being soaked for 7 days in cold water. Another
system is nearing final acceptance.

(4) Gel tests are useful in determining proper cure cycles.

(5) Peel tests as a measure of laminate adhesion are only
partially useful. Most laminant systems exhibit either
very low or very high adhesion after a 7-day soak in
cold water. The soak test may not be a sufficient
predictor for 20-yr service life; however, it may be
considered as a screening test for systems that should
receive additional effort.

(6) Complete removal of solder flux is considered necessary
to ensure long-term laminant adhesion. A process
change to avoid solder flux is therefore encouraged.

Acknowledgments

The effort reported in this paper is based upon many concepts and materials developed by the Encapsulation Task of FSA.
Paul Willis of Springborn Laboratories, Inc., contributed
background information on the processing and testing of the
materials developed at Springborn.

E. P. Plueddemann of Dow Corning Corp. conceived the coupling agents and chemical-bonding philosophy so vital to this effort. Many other industry technologists also contributed.

The research described in this publication was carried out by the Jet Propulsion Laboratory, California Institute of Technology, and was sponsored by the United States Department of Energy through an agreement with the National Aeronautics and Space Administration.

Literature Cited

1. D'Aiello, R. V., Quarterly Report No. 5, RCA Laboratories, DOE/JPL-954868-79/2, March 1979.
2. Somberg, H., Quarterly Report No. 2, ARCO Solar, Inc., DOE/JPL-955278-79/2, July 8, 1978.
*3. Cuddihy, E., Encapsulation Material Trends Relative to 1986 Cost Goals, JPL Internal Document No. 5101-61, Pasadena, California, April 13, 1978.
*4. Maxwell, H., Encapsulation Candidate Materials for 1982 Cost Goals, JPL Internal Document No. 5101-72, Pasadena, California, June 15, 1978.
*5. Cuddihy, E. (JPL), Baum, B., and Willis, P. (Springborn Laboratories, Low-Cost Encapsulation Materials for Terrestrial Solar Cell Modules, JPL Internal Document No. 5101-78, Pasadena, California, September 1978.
*6. Cuddihy, E., Encapsulation Materials Status to December 1979, JPL Internal Document No. 5101-144, Pasadena, California, January 15, 1980.
7. Bouquet, F., Glass for Low-Cost Photovoltaic Solar Arrays, JPL Document No. 5101-147, Pasadena, California, February 1, 1980. (JPL Publication 80-12, DOE/JPL 1012-40).
8. Photovoltaic Module Encapsulation Design and Material Selection, JPL Document NO. 5101-177 (in press). (JPL Publication 81-10, DOE/JPL 1012-60).
*9. Plueddemann, E. P., Dow Corning Corp., Chemical Bonding Technology for Terrestrial Solar Cell Modules, JPL Internal Document No. 5101-132, Pasadena, California, September 1, 1980.
10. Rose, C. M., Quarterly Report No. 1, Westinghouse Electric Corp., AESD, DOE/JPL-955909-81/1.

RECEIVED November 22, 1982

Evaluation of Polyacrylonitrile as a Potential Organic Polymer Photovoltaic Material

PHILIP D. METZ, HENRY TEOH[1], DAVID L. VANDERHART[2], and WILLIAM G. WILHELM

Brookhaven National Laboratory, Solar and Renewables Division, Upton, NY 11973

Thin film organic polymer semiconductors are suggested as an attractive option for cost-effective photovoltaic devices. This report first discusses the potential of organic polymer semiconductors to meet the electronic, physical and economic constraints imposed by the photovoltaic application. Then, recent results on one candidate material, polyacrylonitrile (PAN), are presented. PAN pyrolyzed above about 200°C displays the structural, electrical conductivity and optical properties expected of a one-dimensional semiconductor, including an optical absorption edge at ~1.0–2.0 eV. After pyrolysis at temperatures above about 350–400°C a sharp transition to more metallic behavior is observed, with high conductivity (~10⁰ (ohm-cm)$^{-1}$), low activation energy (~0.1 eV), and broad optical absorption. Areas for further investigation are indicated.

The future of photovoltaic solar energy conversion as an alternative energy resource depends on the development of efficient low-cost large-area photoactive materials. While no suitable

[1] Current address: State University of New York—College at Old Westbury, Department of Chemistry and Physics, Old Westbury, NY 11568
[2] Current address: National Bureau of Standards, Structure and Properties Group, Polymer Science and Standards Division, Washington, DC 20234

0097–6156/83/0220–0421$06.00/0

cost-effective material exists today, recent research indicates that it may be possible to develop organic polymers with the semiconductive and physical properties necessary for photovoltaic applications. Unlike inorganic semiconductors where the electrical and optical properties are fixed, the characteristics of polymers may be adjustable by "property engineering" to enhance conversion efficiency. In addition, organic polymer thin films are uniquely suitable for mass production manufacturing, a crucial consideration in the production of photovoltaic devices.

Recently, there has been a great deal of interest in semiconducting organic polymers, particularly polyacetylene $((CH)_x)$, as electronic materials for applications where low cost and large area are important. This report first discusses the potential of organic polymer semiconductors to meet the electronic, physical and economic constraints imposed by the photovoltaic application. Then, recent results on the structural, electrical, and optical properties of one candidate material, polyacrylonitrile (PAN), are presented. Areas for further investigation are indicated.

Feasibility of an Organic Polymer-Based Photovoltaic Device - Engineering Considerations

Economics. The U.S. Department of Energy (DOE) has estimated that in order to be cost-effective, the installed system price for residential photovoltaic systems in 1986 must be $1.60 to $2.20 per peak watt, in 1980 dollars. Of this, $0.80 per peak watt is applied to the photovoltaic collector itself. Typical costs for current photovoltaic systems are $20.00 per peak watt, of which $10.50 per peak watt is allocated to the collector.(1) Although strides are being made in the development of single-crystal silicon photovoltaic devices, the potential for their low-cost manufacture remains an open question. The need to search for other materials which may result in cost-effective devices is evident. The economic attraction of an organic polymer-based photovoltaic device is its use of small amounts of inexpensive material and its suitability for mass production.

Physical Properties. A photovoltaic device must withstand exposure to the environmental conditions encountered in the collection of solar energy, including exposure to ultraviolet and visible radiation and operating temperatures up to 80°C. No organic polymer material which can meet these conditions - and which has suitable photovoltaic properties - has yet been identified. Polyacrylonitrile has emerged as an interesting candidate material because it can withstand these environmental conditions, and may have the semiconducting properties required for a photovoltaic device.

Suitability for Mass Production.   Thin film organic polymers have unique potential for the low-cost large-area mass production of photovoltaic devices.   Small amounts of inexpensive material are required and mass production fabrication processes similar to those already used in the polymer converting industry, including film manufacture, lamination, metal coating, and printing, may be applicable.   This is one of the major attractions of a thin film organic polymer photovoltaic device.

Feasibility of an Organic Polymer-Based Photovoltaic Device – Photovoltaic Properties

For a single gap photovoltaic device it is desirable that light above the ideal gap energy of about 1.5 eV be strongly absorbed.   Polyacetylene ($(CH)_x$) has a direct absorption edge at about 1.4 eV with a peak absorption coefficient of about 3×10^5 cm^{-1} at about 1.9 eV.([2])   The direct edge is a significant advantage in photovoltaic devices over the indirect edge of semiconductors such as crystalline silicon, promising strong absorption of photons with energies above the band gap.   The optical absorption of pyrolyzed PAN is described below.

There are two major problems concerning the feasibility of noncrystalline photovoltaic devices.   The first is that low carrier mobilities may prevent most carriers from reaching the junction region.   The second is that due to the high density of states in the band gap of such materials, most carriers will not even enter the conduction band.   The situation is further complicated by the possibility that the optical and electrical properties of polyacetylene are caused by solitons (bond alternation domain walls).([3], [4])   The first problem may be overcome by noting that the high absorption coefficient of amorphous materials means that (unlike crystalline silicon) very thin "amorphous cells can be fabricated so that most of the charge carriers are photo-excited within the electric field region and no internal diffusion is necessary."([5])   Although it is not possible to guarantee that a high density of states will not occur in the gap of organic polymer semiconductors, work on other disordered systems such as amorphous silicon indicates that the high density of states within the gap is not an intrinsic property.   Thus, by property engineering research, it may be possible to overcome this problem.   An $Al:(CH)_x$ Schottky device with a quantum efficiency approaching unity for photon energies above 2.8 eV and a conversion efficiency of 0.3% at low light levels (0.21 mW/cm^2) has already been reported.([6])   The current saturated at higher light levels, reducing the efficiency, possibly because of a space-charge buildup in the depletion region.([6])   This effect may also have contributed to the low conversion efficiency.   Research may yield better materials and improved photovoltaic devices.

Properties of Pyrolyzed PAN

Structure. The structural changes usually attributed to the pyrolysis (heat treatment in inert atmosphere) of PAN (7-11) are shown in Figure 1. Above about 200°C, the unpyrolyzed polymer (Figure 1a) is converted to a singly conjugated ladder (Figure 1b), and then at temperatures between 300-400°C to a doubly conjugated ladder (Figure 1c). While this picture is an oversimplification (7, 12), there is evidence that it is for the most part consistent with experimental results up to about 350-400°C. In addition, at temperatures above about 350-400°C NMR data described below suggest that crosslinking occurs between the protonated carbons of the chains shown in Figure 1c due to hydrogen loss.

The experimental results which lead to these conclusions are summarized in Table I. Elemental analyses, conducted by a commercial laboratory, are presented in the form of a ratio of the relative numbers of atoms of C, H, N, and O normalized to 3 carbon atoms per monomer unit. IR spectra were obtained in-house using a Perkin Elmer 298 IR spectrophotometer. Solid probe magic angle spinning C^{13} NMR experiments were performed at the National Bureau of Standards. All of these analyses were performed on bulk (i.e. pyrolyzed in batches of ~1g) samples of 485,000 average molecular weight ultrapure PAN prepared in a tube furnace evacuated by a mechanical vacuum pump, and are reported on in detail elsewhere.(13) Table I also summarizes electrical conductivity and optical absorption experimental results obtained using thin films of PAN solution cast in dimethylformamide.

As shown in Table I, elemental, IR, and NMR analyses are all consistent with structure 1a for unpyrolyzed PAN. The NMR spectra suggest a low crystallinity possibly resulting from a lack of stereoregularity in the starting material.

For the 220°C sample, the elemental analysis confirms that no loss of nitrogen has occurred. Increased IR absorption between $1600-1200 cm^{-1}$ indicates conjugation (the absorption is too broad to be more specific), while a decrease in intensity of the C≡N absorption at $2240 cm^{-1}$ suggests that conjugation is obtained by cyclizing the polymer rather than by desaturating the top chain of Figure 1a. The high hydrogen content (2.83 atoms per monomer unit) confirms this view. The electrical conductivity of thin film samples pyrolyzed at 220°C is low, but the conductivity is enhanced by doping (13, 14) as would be expected for a 1-dimensional conjugated chain. Thus, a structure intermediate between 1a and 1b is indicated. Longer pyrolysis time would be expected to cause a more complete transformation of 1a to 1b.

Bulk samples of PAN pyrolyzed above about 250°C exhibit a very sharp exotherm associated with ring closure (i.e. thermal polymerization of the nitrile groups) (13, 15) and considerable mass loss. This behavior is not observed in thin film samples.

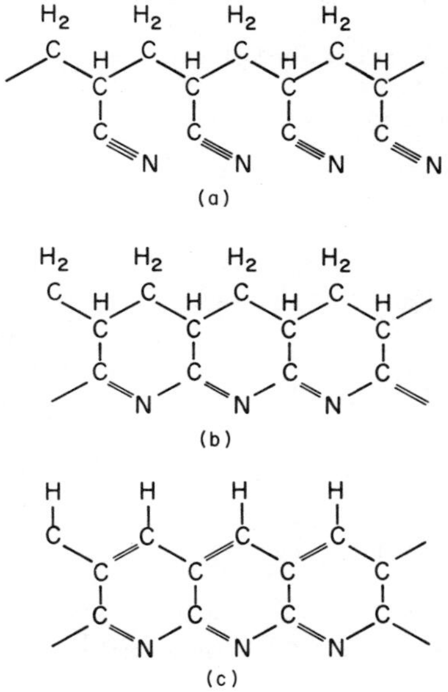

Figure 1.
Structure of Polyacrylonitrile (a) before pyrolysis; (b) singly conjugated; and (c) doubly conjugated ladder.

Table I   Experimentally Observed Properties

Pyrolysis Temperature (°C) Pyrolysis Time (hr)	Unpyrolyzed
Bulk Sample (~1g) Experimental Results	
- Most Likely Structure(s) (see Figure 1)	1a
- Elemental Analysis (Commercial Lab.) Ratio of Atoms C:H:N:O Normalized to 3 Carbons	3.00:3.12:1.02:0.02
- IR Spectra	$C \equiv N$ peak at 2240 cm^{-1}
- Solid Probe C^{13} NMR	Consistent with 1a but low crystallinity (possibly low stereo-regularity)
Thin Film Experimental Results	
- Room Temperature Electrical Conductivity (σ) (ohm-cm)$^{-1}$ (see Figure 2)	$<10^{-9}$
- Activation Energy (eV) (Slope of ln σ vs 1/kT) (see Figure 2)	-
- Optical Absorption Edge (eV) (see Figure 3)	Transparent

Note: Pyrolysis of 280 ° thin film samples is not as advanced as 280 ° bulk sample (see text).

*The IR spectra of thin film samples pyrolyzed above 200 ° indicate the presence of N-H bonds (absorption peak at 3400 cm^{-1}) which are masked in the spectra of bulk samples pyrolyzed at 280 ° by broad IR absorption. These have been associated with the initial stage of cyclization (25).

of Pyrolyzed Polyacrylonitrile (PAN)

220	280	400
3	3	2

1a–1b	1b–1c	1c–Cross-linked
3.00:2.83:1.00:0.04	3.00:1.92:0.91:0.18	3.00:0.74:0.79:0.06
Absorption between 1600–1200 cm^{-1} evidence of conjugation, decrease in peak at 2240 cm^{-1}	No C≡N peak at 2240 cm^{-1}, increased absorption between 1600–1200 cm^{-1}*	Same as 280° plus N–H peak at 3400 cm^{-1} below 440°C
	Fewer Aliphatic carbons, few C≡N bonds, but other C–N bonds exist, slightly >1H/monomer unit, consistent with significant population of 1c	(440°C for 5 hr) conjugated, not uniform, lacks protons, (<1H/ monomer unit), cross-linked at the protonated carbons of 1c
~10^{-8} Increased by I$_2$ doping	~10^{-7} Increased by I$_2$ doping	10^{-5}–10^{0}
1–2	1–2	0.1–0.4
2.0 eV	1.5 eV	<0.5 eV

Thus, a distinction must be drawn between the properties of thin film samples heated to 280°C and bulk samples, parts of which can "self heat" to temperatures 50-100°C higher because of this exotherm.

Elemental analysis of the 280°C bulk sample indicates an anomalously large oxygen content of about 0.2 atoms/monomer unit. Oven drying and reweighing verify that absorption of atmospheric water could account for a significant fraction of the oxygen content. It is not yet clear if all of the oxygen present can be explained in this way. The elemental and NMR analyses both indicate between 1 and 2 hydrogen atoms/monomer unit. The IR and NMR analyses both indicate that few C≡N bonds remain. However, the continued presence of nitrogen in the sample is shown by the elemental analysis, while the IR data indicates further conjugation (and some N-H bonds in thin film samples) and the NMR data strongly support the inclusion of nitrogen in the conjugated system. In short, all results, except for the 9% loss of nitrogen (plus the N-H bonds and oxygen gain) shown by the elemental analysis are consistent with a mixture of structures of 1b and 1c. It is not yet known whether the nitrogen loss is related to the exotherm described above, what structure could account for this loss, or whether it is correlated to the oxygen uptake observed.

The importance of the exotherm above 250°C is shown by comparing the properties of the 280°C bulk and thin film samples. As shown in Table I and as discussed in more detail below, the 280°C thin film sample has electrical and optical properties similar to those of the 220°C thin film sample. In contrast, the 220°C bulk sample is consistent with conversion of 1a to 1b while the post-exotherm 280°C bulk sample is already completely converted to structure 1b and well on the way to structure 1c.

The 400°C sample (440°C for the NMR data) is shown by both the elemental analysis and NMR data to have less than 1 hydrogen atom per monomer unit, implying some cross-linking of the chains from structure 1c. Furthermore, the NMR data indicate (by carryover of certain resonances from the 280°C sample) that much of the crosslinking occurs at the protonated carbons of 1c. The IR spectra indicate N-H bonds at 400°C. In addition, the nitrogen deficit of 0.21 in this sampled coupled with resonance shifts in the NMR spectra argue for the formation of some aromatic structures which do not contain nitrogen. At 440°C, all evidence for N-H bonds has vanished. So, the structure of the sample pyrolyzed at 440°C probably includes crosslinked ladder (1c) structures, some aromatic hydrocarbon moieties, and virtually no tetrahedrally bonded carbons. In any case, the building blocks for very extended conjugated structures certainly exist.

Electrical Conductivity. Electrical conductivity (σ) measurements, reported on in greater detail elsewhere (16), were made

on thin films of ultrapure PAN of average molecular weight
150,000 and 485,000 solution cast in dimethylformamide (DMF).
Measurements were made in vacuo ($\sim 10^{-5}$ Torr) using a two-probe
technique.

The apparatus used (16) permitted the measurement of dc conductivity in the plane of the film during pyrolysis. Samples
were heated in vacuo at about 280°C for about three hours to
drive off the DMF, and then coled to about 200°C. They were then
reheated to about 400°C and maintained at that temperature for
various lengths of time, cooled and reheated. Sample temperature
and conductivity were monitored continuously throughout.

Figure 2 illustrates the conductivity vs temperature data
obtained. Initially (Figure 2d), conductivities are low ($\sim 10^{-9}$-
10^{-7} (ohm-cm)$^{-1}$), rising monotonically for temperatures between
approximately 200–300°C. Samples may be heated to at least 300°C
without altering the σ vs $1/T$ curve obtained during subsequent
cooling or heating below this temperature. However, a qualitative change in the temperature dependence of the conductivity
occurs once the pyrolysis temperature is raised above a critical
value of approximately 390°C at which the conductivity is seen to
rise dramatically from 10^{-6} to 10^{0} (ohm-cm)$^{-1}$. Previous reports
on the conductivity of PAN did not note this behavior.(7, 17-22)

Typical σ vs $1/T$ curves of samples heated above the critical
temperature are shown in Figures 2a, b, and c. Generally, conductivity is much higher than for the unpyrolyzed samples (Figure
2d) and decreases with decreasing temperature. The curves are
slightly concave up and plotting $\ln\sigma$ vs $T^{-1}/4$ yields a better
linear fit,(13) indicative of hopping conduction. Activation energy (slope of $\ln\sigma$ vs $1/kT$) decreases and $\sigma(T)$ increases as the
conductivity at which cooling was begun increases. The $\sigma(T)$
curves are repeatable during subsequent heating or cooling below
the pyrolysis temperature, particularly for the highest σ vs $1/T$
curve (Figure 2a), down to at least -100°C.(13)

Optical Absorption.   The optical absorption spectra of
485,000 MW PAN thin film samples pyrolyzed for 3 hours at temperatures between 220 and 370°C are shown in Figure 3. All samples
have a maximum absorption of $\sim 10^{5}$ cm^{-1}. The optical absorption
edge shifts from ~ 2.0 eV for the 220°C sample to ~ 1.5 eV for the
280°C sample, and to lower energies at higher pyrolysis temperatures. Secondary absorption maxima are observed at energies below the main edge in samples pyrolyzed at 340°C or lower.

For samples pyrolyzed above about 370°C the optical absorption spectra are qualitatively different as shown in Figure 4.
Absorption is strong ($>10^{4}$ cm^{-1}) and comparatively flat with no
maxima evident below 3.5 eV and no absorption edge above 0.7 eV.
The shift of the optical absorption edge to lower energy with increasing pyrolysis temperature between 220 and 440°C is consistent with the trend of the photoconductivity and optical absorption data of Hirai and Nakada (17) and of Ohigashi (19).

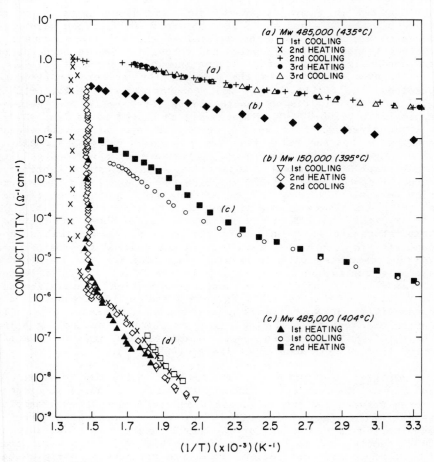

Figure 2.
Electrical Conductivity vs. Temperature of Pyrolyzed PAN.

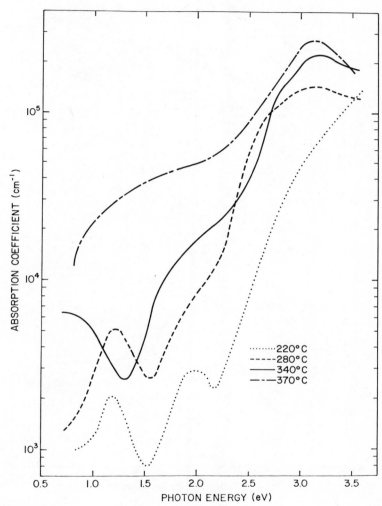

Figure 3.
Optical Absorption (α) of 485,000 Average MW PAN Pyrolyzed
For 3 Hours.

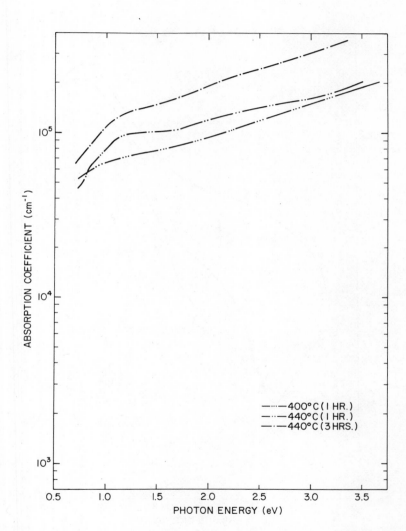

Figure 4.   Optical Abosrption (α) of 485,000 Average MW PAN Pyrolyzed
as Indicated.

However, they did not observe the qualitative change in the spectra seen here for samples pyrolyzed above 370°C.

Discussion of Structural, Electrical Conductivity, and Optical Absorption Results. The electrical conductivity and optical absorption data presented here both exhibit a sharp transition at a pyrolysis temperature of 370-400°C. These transitions have not been previously reported.([7], [17-22]) The picture is far from clear, particularly at pyrolysis temperatures above 300°C but for the most part the structural evidence presented indicates that for samples pyrolyzed between 200-300°C, structure 1b predominates. (Note that the bulk sample pyrolyzed at 280°C is closer to structure 1c due to "self-heating" as discussed above.) This would imply semiconductive behavior as a result of the conjugated lower one-dimensional (1-D) chain (see Figure 1b) – if the upper chain permits dimerization. This behavior is consistent with the location of the observed optical absorption edges between 1.5-2.5 eV shown in Figure 3 and with the low conductivity which is enhanced by doping with I_2 vapor.([13], [14]) The optical absorption edge would then be an interband transition in the 1-D band structure, analogous to polyacetylene.([23]) The secondary maxima seen, which are observed in polyacetylene only when doped, may be due to "self doping" caused by the nitrogen atoms, the saturated chain in 1b, or by impurities in the material.

For samples pyrolyzed above 350-400°C, structure 1c is indicated as previously reported.([7-11]) However, the present analyses give evidence for the onset of cross-linking at the protonated carbon sites in 1c. In addition, some aromatic rings are formed which do not contain nitrogen. As the polymer chains cease to be one-dimensional, the Peierls instability would no longer be expected to apply and the conjugated material would take on a more metallic character. This behavior is consistent with the observed sharp rise in electrical conductivity, drop in activation energy and broad optical absorption for samples pyrolyzed above about 400°C.

Suitability of Pyrolyzed PAN as a Photovoltaic Material. The current state of knowledge of the relevant properties of PAN for photovoltaic applications is as follows. As shown in Figure 3 PAN pyrolyzed between 220-370°C displays strong optical absorption ($\alpha > 10^5$ cm^{-1}) and an absorption edge in the range of roughly 1-3 eV. Hall mobility has been reported for samples pyrolyzed above 600°C ([18]), but drift mobility has not been reported (see [18] for an estimate). Weak photoconductivity has been observed. ([17], [19]) The charge transport mechanism in PAN is poorly understood. Some data suggest a correlation between twice the activation energy of conduction (slope of $\ln\sigma$ vs $1/kT$) and the optical absorption edge ([13], [16]), while there is also evidence that $\sigma \sim T^{-1/4}$ ([13], [16]), suggesting hopping via localized states near the

Fermi energy. An aluminum–PAN Schottky junction has been reported.(24) Enhanced conductivity via doping with halogen electron
acceptors has been shown.(13, 14) Doping with electron donors
has not been reported.

In short, there are some encouraging results, particularly
the adjustable optical absorption edge and some discouraging results such as the low photoconductivity. The charge transport
mechanism is still poorly understood. An evaluation of the
suitability of PAN as a photovoltaic material requires better
data on charge transport, doping and junction formation.

General Discussion and Conclusion

Thin film organic polymer semiconductors are an attractive
option for the development of cost-effective photovoltaic devices. They are uniquely suited to low-cost mass production and
provide potential flexibility via property engineering. However,
materials with better physical and photovoltaic properties than
those currently available are needed. This will require a far
better understanding of the structure and transport properties of
these materials.

Polyacrylonitrile is an interesting material for this application with evidence of semiconductive behavior and stability at
elevated temperatures. The optical absorption data indicate that
the optical absorption edge in PAN may be adjustable, suggesting
the possibility of fabricating multigap cells via this or a similar material. However, a far better understanding of the structure, microscopic transport, and optical properties of PAN is required. This knowledge will provide a better understanding of
the connection between polymer structure and semiconductive behavior, permitting the enlightened synthesis of polymers optimized for photovoltaics.

Acknowledgments

The authors thank S. Aronson for valuable discussions on
data analysis and interpretation, and J. Andrews and F. Salzano
for valuable discussions and advice.

Work performed under the auspices of the U.S. Department of
Energy under Contract No. DE-ACO2-76CH00016.

Literature Cited

1.   Photovoltaic System Definition and Development, Project
     Integration Meeting, Albuquerque, NM, October 21-23, 1980,
     SAND 80-2374, pp 12-15.
2.   Fincher Jr., C. R.; Ozaki, M.; Tanaka, M.; Peebles, D;
     Lauchlan, L.; Heeger, A. J.; MacDiarmid, A. G. Phys. Rev.
     1979, B20, 1589-1601.

3. Su, W. P.; Schrieffer, J. R.; Heeger, A. J. Phys. Rev. Lett. 1979, 42, 1698.
4. Rice, M. J. Phys Rev. Lett. 1979, 71A, 152.
5. Adler, D. Sunworld 1980, 4, 18.
6. Weinberger, B. R.; Gau, S. C.; Kiss, Z. Appl. Phys. Lett. 1981, 38, 555.
7. Brennan, W. D.; Brophy, J. J.; Schonhorn, H. "Organic Semiconductors," Proc. Inter-Industry Conf., J.J. Brophy and J.W. Buttrey, Eds., 1962, p 159.
8. Burlant, W. J.; Parsons, J. L. J. Polym. Sci. 1956, 22, 249.
9. Topchiev, A. V.; Geyderikh, M. A.; Davydov, B. E.; Kargin, V. A.; Krentsel, B. A.; Kustanovich, I. M.; Polak, L. S. Dok. Akad. Nauk. SSSR 1959, 128, 312.
10. Becher, M.; Mark, H. F. Angew. Chem. 1961, 73, 641.
11. Topchiev, A. V. J. Polym. Sci. A: 1963, 1, 591.
12. Monahan, A. R. J. Polym. Sci. A: 1966, 4, 2391.
13. Metz, P. D.; Teoh, H.; VanderHart, D. Structural, Electrical and Optical Properties of Pyrolyzed Polyacrylonitrile, (in preparation).
14. Brokman, A.; Weger, M.; Marom, G. Polym. 1980, 21, 1114 (note: These authors used PAN fibers "stabilized" in air at 220°C.).
15. Grassie, N.; McGuchan, R. Eur. Poly. J. 1970, 6, 1277.
16. Teoh, H.; Metz, P. D.; Wilhelm, W. G. Mol. Cryst. Liq. Cryst. 1982, 83, 297.
17. Hirai, T.; Nakada, O. Jpn. J. Appl. Phys. 1968, 7, 112.
18. Suzuki, M.; Takahashi, K.; Mitani, S. Jpn. J. Appl. Phys. 1975, 14, 741.
19. Ohigashi, H. Rep. Prog. Polym. Phys. Jpn. 1963, 6, 245.
20. Helberg, H. W.; Wartenberg, B. Phys. Status Solidi A: 1970, 3, 401.
21. Jacquemin, J. L.; Ardalan, A.; Bordure, G. J. Non-Cryst. Solids 1978, 28, 249.
22. Airapetyants, A. V.; Voitenko, R. M.; Davydov B. E.; Krentsel, B. A. Dok. Akad. Nauk. SSSR 1963, 148, 605.
23. Etemad, S.; Heeger, A. J.; Lauchlan, L.; Chung, T. C.; MacDiarmid, A. G. Mol. Cryst. Liq. Cryst. 1981, 77, 43.
24. Chutia, J.; Barua, K. Phys. Stat. Sol., 1979, (a) 55, K13.
25. Davydov, B. E.; Krenstel, B. A. Adv. Polymer Sci. 1977, 25, 1.

RECEIVED November 22, 1982

Photovoltaic Properties of Organic Photoactive Particle Dispersions

Polymeric Phthalocyanines

R. BRANSTON, J. DUFF, C. K. HSIAO, and R. O. LOUTFY
Xerox Research Centre of Canada, 2480 Dunwin Drive,
Mississauga, Ontario, L5L 1J9 Canada

Photovoltaic measurements were carried out on a series of silicon and germanium phthalocyanine particle dispersion in a Schottky barrier device. The optical, electrical and photoelectrical properties of these materials were found to depend strongly on the molecular stacking and separation distance of the phthalocyanine rings within the crystals. The electrical conductivity increased with a decrease in the separation distance between the phthalocyanine rings. Strong photoactivity is associated with a molecular arrangement in the solid state corresponding to a staggered parallel plane dimers. Both dihydroxy germanium and dihydroxy silicon phthalocyanine solids exhibited strong absorption peaks in the near infrared (870 nm) which indicates considerable intermolecular interaction is present in the solid state.

Considerable effort has been made to improve the power conversion efficiency of solid-state organic photovoltaic cells in an attempt to find an inexpensive, efficient and stable solar cell for large scale terrestial use. (1) It is believed that this goal can be achieved only via a technological breakthrough in material design and crystal lattice architecture aiming at lowering the bulk resistance of organic semiconductors. A number of conducting metallophthalocyanine compounds have recently been reported. (2-6) It was shown that the electrical conductivity of these materials can be controlled by molecularly stacking the metallomacrocycle rings in a "face-to-face" orientation via covalent linkage. (3) Among the important precursors used in the polymerization of phthalocyanines are dihydroxy silicon, germanium and tin, which upon polymerization produce phthalocyaninato polysiloxanes, polygermyloxanes and polystannyloxanes. Polymeric phthalocyanines were noted for their intense colors, high thermal stability and a dramatic change in electric conductivity upon halogen oxidation or doping with o-chloranil. In spite of the extensive studies of the physical, structural and electrical properties of polymeric phthalocyanines, (2-8) however, there is a lack of photoconductivity data. We therefore, became interested in the synthesis and evaluation of these materials for solar cell applications.

0097–6156/83/0220–0437$06.00/0

In this paper, we report the results of a preliminary investigation of the optical, electrical and photoelectrical properties of a series of silicon and germanium phthalocyanines and a number of their polymers.

EXPERIMENTAL

Materials. Using the methodology developed by Kenney (2) dichlorosilicon and dichlorogermanium phthalocyanines, $PcMCl_2$, were prepared and then hydrolyzed to the dihydroxides, $PcM(OH)_2$ as shown in Scheme I. Polymerization in the presence of diols produced the corresponding phthalocyaninato polysiloxanes and polygermyloxanes in high yield and purity (Scheme II). All materials were characterized by elemental analysis, x-ray, infrared and electronic absorption. Polyvinylacetate was obtained from Polyscience Inc. The conductive tin-oxide coated glass was from Pittsburg Glass Company.

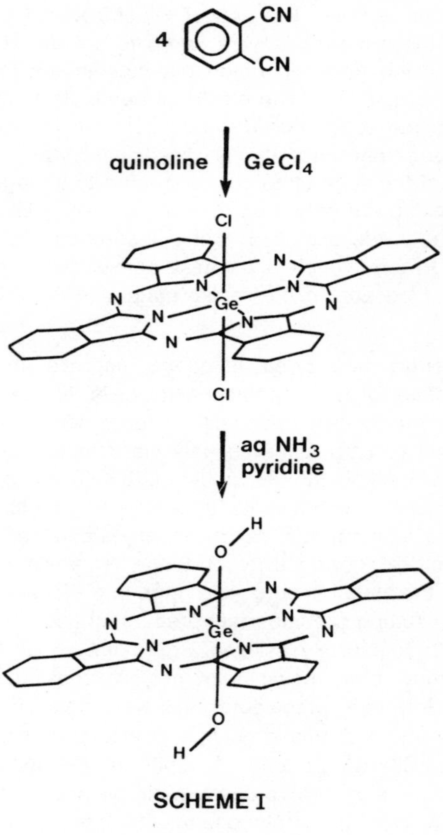

SCHEME I

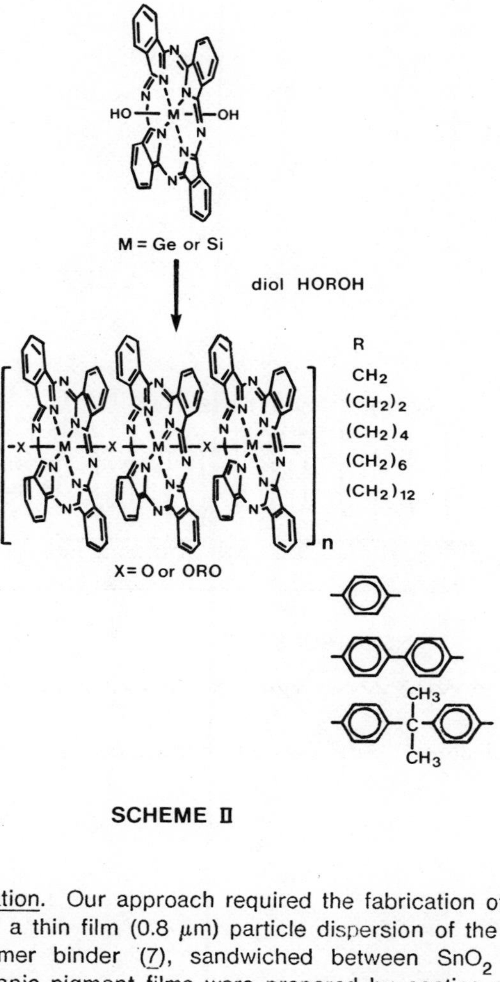

M = Ge or Si

diol HOROH

R
CH$_2$
(CH$_2$)$_2$
(CH$_2$)$_4$
(CH$_2$)$_6$
(CH$_2$)$_{12}$

X = O or ORO

SCHEME II

Device Fabrication. Our approach required the fabrication of a photovoltaic device consisting of a thin film (0.8 μm) particle dispersion of the phthalocyanine pigment in a polymer binder (7), sandwiched between SnO$_2$ and a barrier electrode. The organic pigment films were prepared by coating a suspension of 0.18 gm pigment in 8 ml of a solution of 1.5% polyvinylacetate in methylene chloride, onto a precleaned heavily doped SnO$_2$ substrate with a 2 mil draw bar. The resutling organic films contained 60% pigment by weight. The thickness of these films, measured by a Talysurf apparatus averaged 0.8 μm. A thin, semitransparent (1.2% transmission) indium electrode (90 nm) was vacuum deposited on the top of the organic film to complete the solid state junction as shown in Figure 1.

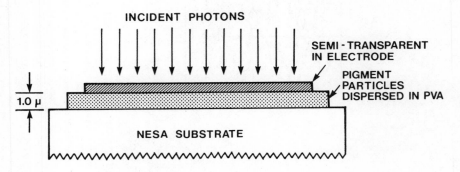

Figure 1   Side view of the photovoltaic device.

MEASUREMENTS

Solubility. The solubilities of the polymers were tested using the solvents: dichlorobenzene,   trichlorobenzene,   methanol,   dichloromethane,   and   1-chloronapthalene. $(PcGeO(CH_2)_6O)_n$ and $(PcGeO(CH_2)_{12}O)_n$ were sparingly soluble only in 1-chloronapthalene. These and other polymers were insoluble in the other solvents listed above.

Melting Points.   All polymers had melting points of greater than 300°C. Polymers were examined under a microscope for changes in their crystal stucture during heating, but no changes were evident.

IR Spectra. The IR spectra of the pigments dispersed in Nujol were taken on Beckman 5220 spectrometer.   Figure 2 shows the IR spectra of dihydroxy germanium phthalocyanine and that of a typical polymer, $(PcGeOC_6H_4 C(CH_3)_2C_6H_4O)$.   There was no hydroxyl peak detectable in the spectrum of the polymer, while an intense OH peak at 3500 cm^{-1} was clearly observable for $PcGe(OH)_2$. The absence of OH absorption in the polymers is indicative of at least moderate chain length.

X-Ray Powder Diffraction. X-Ray Powder Diffraction was used primarily as a fingerprinting technique.   The presence of sharp peaks in all of the spectra indicated that every polymer was crystalline, and that the crystalline structure of each polymer was different.   The unit-cell dimensions do not appear to change continously with the increasing length of the alkane chain. All polymers differ from the starting materials $PcGeCl_2$ and $PcGe(OH)_2$.   The high degree of crystallinity indicates that the polymer chains are rigid or highly organized, presumably as a result of the large size and high symmetry of the phthalocyanine rings. It is not possible to determine the chain length of the polymers exactly using X-Ray Powder Diffraction; some examples are shown in Figure 3. It should be noted that the X-ray powder diffraction lines observed in polymeric sample do not necessarily measure the degree of Pc-Pc interactions.

Bulk Optical Absorption of Phthalocyanine Particles. The visible and near infrared optical absorption of the phthalocyanine pigments was measured on a Cary 17 spectrophotometer using the technique described in Reference 8. A thin film of phthalocyanine dispersed in the polymer was coated onto 3 mil Mylar substrate using wirewound rods. The dry films were index matched by overcoating with a thick film of polyvinyl alcohol. The differential absorption between two films with different thicknesses were recorded to minimize reflection and light scattering losses. The solution absorption spectra of some pigments were also measured in α-chloronapthalene.

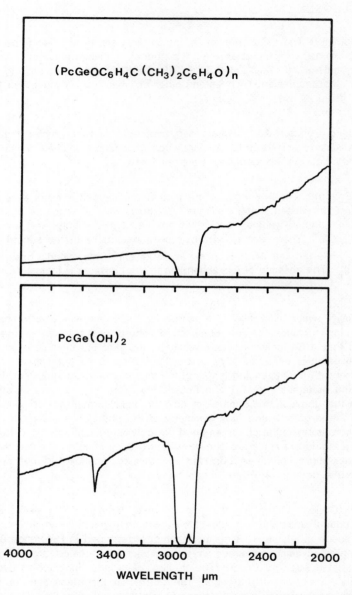

Figure 2   Infrared spectra of $PcGe(OH)_2$ and $(PcGeOC_6H_4C(CH_3)_2C_6H_4O)_n$ polymer taken in Nujol.

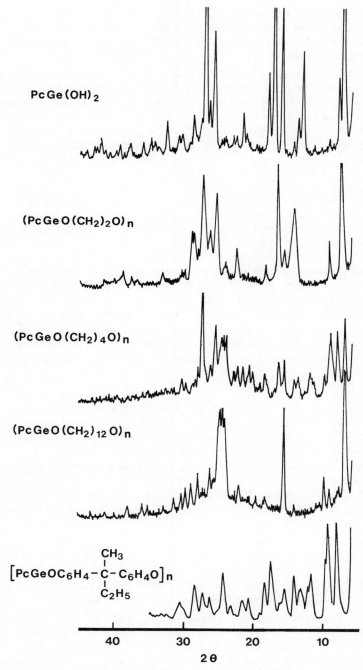

Figure 3   X-ray powder diffraction of $PcGe(OH)_2$.

Photovoltaic Measurements.   The dark and light current-voltage (J-V) characteristics of the SnO_2/phthalocyanine/In cells were determined using an EG/G Polarographic Analyzer, Model 174A and recorded using Hewlett Packard 9845B desktop computer.

The light source was a 1 KW Oriel Solar Simulator equipped with AMO filters. Light intensities were measured by a Karl Lambrecht calibrated light probe.   The unattenuated power flux delivered by the System was 208 mW/cm^2. Incident light intensity was varied using neutral density filters.   The light source for action spectra measurements consisted of a 150 W Xe lamp and a ¼ meter Jarrel Ash monochromator.   The action spectra of the cell was measured as described previously (7).

RESULTS AND DISCUSSION

Many organic pigments have been used as the photogenerator layer in electrophotographic or photovoltaic devices.   In these applications, the most important material parameters are (i) the photogeneration efficiency, φ, which is the average number of mobile electrons and holes generated per photon absorbed, (ii) the spectral response and (iii) the electrical conductivity.   Photovoltaic techniques have recently (4) been shown to be useful for measuring photoconductivity, electrical conductivity and spectral response of organic materials.

In our measurements the electrical and photoelectrical properties of a series of phthalocyanine polymers were determined by using these materials in a Schottky junction devices.

Electrical Conductivity.   Figure 4 shows a typical current-voltage curve of an SnO_2/PcGe(OH)$_2$, PVA/In cell in the dark.   These devices exhibit strong rectifying behaviour, with a forward bias corresponding to a negative voltage at the indium electrode with respect to the SnO_2.   The forward biased current appears to be due to hole injection from SnO_2.   At sufficiently high forward biased current the diode junction resistance approaches a constant value.   This value is the series resistance R_s of the device which is dominated by the bulk resistance, R_b of the pigment. (10) The specific conductivity, σ, of the pigmented film can be calculated from:

$$\sigma = \frac{L}{R_b \cdot A} \qquad (1)$$

where L is the thickness of the pigmented film and A is the area of the electrode. The specific conductivity of PcGe(OH)$_2$ was found to be 3.5 x 10^{-9} $\Omega^{-1}cm^{-1}$ which is an order of magnitude lower than that of PcSi(OH)$_2$ (2.3 x 10^{-8} $\Omega^{-1}cm^{-1}$). These results are in general agreement with those of Meyer and Wohrle (6).   The electrical conductivity of PcGe(OH)$_2$ was found to increase by two order of magnitude upon polymerization to (PcGeO)$_n$, see Table 1.   This effect can be attributed to a decrease in the intermolecular spacing between the macrocyclic

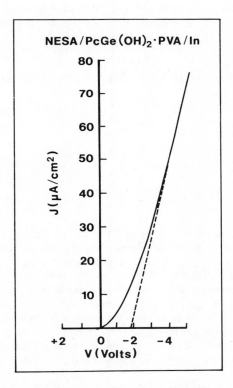

Figure 4   Dark current-voltage curve of SnO_2/PcGe(OH)$_2$, PVA/In photovoltaic
cell.

(phthalocyanine) ring.   Introduction of bulky substituent such as tertiary butyl groups at the benzene rings of $PcGe(OH)_2$, causes a two order of magnitude decrease in the electrical conductivity.   This effect is due to the increase in intermolecular spacing between the macrocyclic rings due to steric factors.

The introduction of hydrocarbon chains separating the phthalocyanine molecules with increasing length, results in a decrease of the electrical conductivity (see Table II). However, the decrease in σ was not catastrophic particularly for the long chain hydrocarbon, implying that the hydrocarbon chains are not all in trans-configuration, permitting some intermolecular interaction between Pc rings to occur supporting Meyer and Wohrle recent results on polysilicon phthalocyanines (6).

TABLE I

ELECTRICAL CONDUCTIVITY OF TETRAVALENT PHTHALOCYANINES

	$\sigma\ (\Omega^{-1}\ cm^{-1})$
$(PcGeO)_n$	1.5×10^{-7}
$PcGe(OH)_2$	3.5×10^{-9}
$(t\text{-}Bu)_4PcGe(OH)_2$	7.4×10^{-11}
$PcSiCl_2$	6.0×10^{-12}
$PcSi(OH)_2$	2.3×10^{-8}

As shown in Tables I and II, σ varied from 10^{-7} to $10^{-12}\ \Omega^{-1}cm^{-1}$ for the series of material studied depending on the chemical nature and molecular packing of the molecules.   The most electrically conductive silicon and germanium phthalocyanines were $(PcGeO)_n$, $PcSi(OH)_2$ and $PcGe(OH)_2$.   However, the resistivities of these materials were still several order of magnitude higher than that necessary for solar cell application.   It has been shown previously that the conductivity of this class of materials can be lowered by many order of magnitude upon halogen oxidation (3) or doping with o-chloranil (6). Although doping might be useful in lowering the resistivity of these materials to the desired levels, the impact of doping on the photoconductivity of the samples is not known.

TABLE II

ELECTRICAL CONDUCTIVITY OF POLYMERIC PHTHALOCYANINES

	$\sigma\ (\Omega^{-1}\ cm^{-1})$
$(PcGeO)_n$	1.5×10^{-7}
$(PcGeO(CH_2)_4O)_n$	2.5×10^{-8}
$(PcGeO(CH_2)_{12}O)_n$	1.2×10^{-8}
$(PcGeOC_6H_4C(CH_3)_2C_6H_4O)_n$	1.2×10^{-9}

Solid State Absorption.   The solid state electronic absorption of organic pigments reveal important information on the molecular arrangement and extent of intermolecular interaction between molecules in the crystal.   The solid state absorption properties of polycrystalline films of some silicon and germanium phthalocyanines and some polymers are shown in Figures 5-8.   Figure 4 shows the solid state absorption spectra of a bis-silicon phthalocyanine linked by butanediol and a series of polygermanium phthalocyanines where the macrocyclic rings were separated by phenyl, biphenyl and bisphenol A.   In these cases the absorption resembled the molecular solution absorption with the exception that it was slightly broader.   These results clearly indicate that a very weak intermolecular interaction between molecules in the solid exists.   The same is true for $PcSiCl_2$ and $PcGeCl_2$. Figure 5 shows the solid state absorption spectra of $PcGe(OH)_2$ and $PcSi(OH)_2$ which is very different from the solution molecular absorption.   The spectra consists of a broad absorption covering the wavelength region from 400-1000 nm, with two maxima one to the red and the other to the blue of the molecular absorption.   The intense near infrared absorption peak around 870-890 nm found for $PcGe(OH)_2$ and $PcSi(OH)_2$ is quite similar to that found for x-H_2Pc, VOPc and most recently for AlClPc.   In the case of x-H_2Pc a parallel plane dimeric structure of two neighbouring Pc molecules in a staggered configuration had been proposed by Sharp and Lardon (11).   This stacking arrangement is very different from that of both α and β polymorphs of phthalocyanines.   In the later forms molecules are projected directly along the stacking axis with equal spacing and in an eclipsed configuration.   It is tempting to speculate that the crystal structure of $PcGe(OH)_2$ and $PcSi(OH)_2$ responsible for the near infrared absorption consists of parallel plane dimers (aggregate) in which the molecules are staggered.   The solid state absorption of all other polymeric germanium phthalocyanine examined showed very weak or undetectable near infrared peaks.   These results clearly indicate a strong intermolecular interaction between molecules in the solid state of $PcGe(OH)_2$ and $PcSi(OH)_2$, while all other pigments showed much weaker interactions.   It is interesting to speculate that the exciton splitting is related to enhanced photogeneration and that a strong intermolecular interaction is essential for high quantum yields of photogeneration, since x-H_2Pc, VOPc and ClAlPc all exhibit high photoactivity.

Photoconductivity.   Figure 9 shows a typical current-voltage curve of an In/$PcGe(OH)_2$PVA/SnO_2 cell under (85 mW/cm^2) AMO illumination.   From the photocurrent-voltage curves under solar and monochromatic illumination, several fundamental parameters, such as the open circuit voltage (V_{oc}) the short circuit current (J_{sc}), the fill factor, as well as the dependence of J_{sc} and V_{oc} on light intensity can be extracted.   The power conversion efficiency of the device can be estimated from:

$$\eta = J_{sc} \cdot V_{oc} \cdot ff/I \tag{2}$$

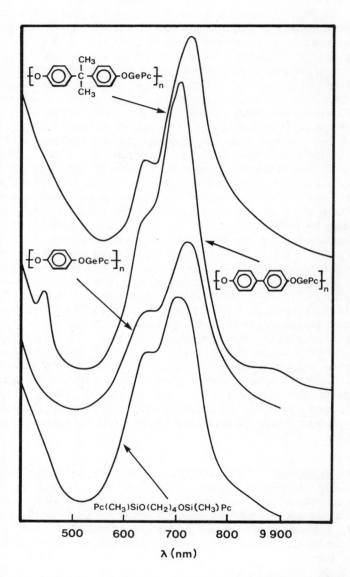

Figure 5 Solid-state absorption spectra of silicon and germanium phthalocyanine
          polymers.

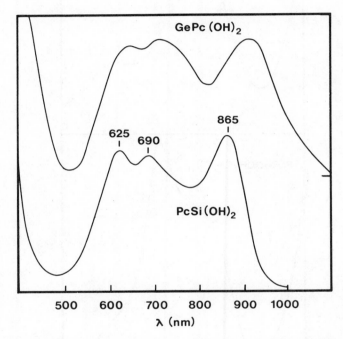

Figure 6 Solid-state absorption spectra of PcGe(OH)$_2$ and PcSi(OH)$_2$ particles.

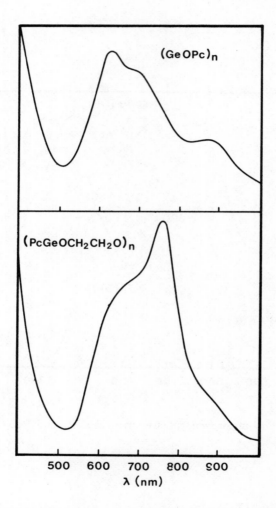

Figure 7   Solid state absorption spectra of $(PcGeO)_n$ and $(PcGeO(CH_2)_2O)_n$.

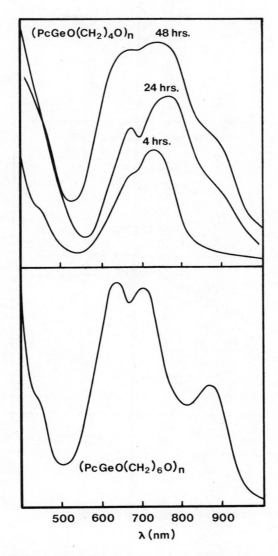

Figure 8   Solid-state absorption spectra of PcGeO(CH$_2$)$_4$O)$_n$ as a function of milling time in CH$_2$Cl$_2$ and of (PcGeO(CH$_2$)$_6$O)$_n$.

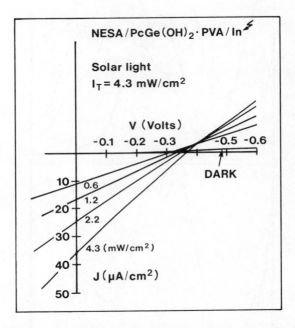

Figure 9 Current-voltage curves for $SnO_2/PcGe(OH)_2$, PVA/In cell in the dark and under solar illumination.

where I is the incident light intensity. The output of In/Pc, PVA/SnO$_2$ photovoltaic cells under AMO solar illumination is given in Table III. The power conversion efficiency is very low (η ~10^{-3} %) typical of a high resistance photoconductor. Again PcGe(OH)$_2$, (PcGeO)$_n$ and PcSi(OH)$_2$ appears to be the most photoactive material in this series. However, these materials are a factor of 20 less sensitive than x-H$_2$PC.

The results in Table III clearly demonstrate the dependence of the photoelectrical properties of the material on chemical and morphological properties. Materials which exhibit a strong intermolecular interaction between the molecules in the crystal, as indicated by the shift of the long wavelength absorption peak to the red, show strong photoactivity.

TABLE III
SUMMARY OF THE PHOTOVOLTAIC DATA* OF Ge and Si PHTHALOCYANINES

Material	J_{sc} (μA/cm^2)	V_{oc} (Volts)	η/10^{-3} %	Long Wavelength absorption peak(nm)
PcGe(OH)$_2$	43	0.38	4.6	910
(PcGeO)$_n$	17	0.10	0.5	875
PcSi(OH)$_2$	15	0.21	.9	865
(PcGeO(CH$_2$)$_{12}$O)$_n$	5	0.21	0.3	850
(PcGeO(CH$_2$)$_4$O)$_n$	1	0.2	0.05	770
(PcGeO(CH$_2$)$_2$O)$_n$	0.16	0.4	0.02	760
(t-Bu)$_4$PcGe(OH)$_2$	0.033	0.13	0.001	730
PcSiCl$_2$	.048	0.55		750
PcSi(OH)CH$_3$	.006	0.17		——
x-H$_2$Pc	147.0	0.45	22.0	800

* At 88 mW/cm^2 solar sumilated AMO light, engineering efficiency

The spectral response of the films was determined from the photovoltaic action spectra, which involves measurements of the photovoltaic short circuit current as a function of excitation wavelength. In all cases studied the action spectrum resembled closely the absorption spectrum of the pigment particles. The action spectrum of PcSi(OH)$_2$ is shown in Figure 10 along with the pigment absorption spectrum for comparison. These results imply constant photocarrier generation efficiency from 400 - 1000 nm illumination range. Transient photoconductivity action spectra for thin films of β-H$_2$Pc also showed wavelength-independent photocarrier generation efficiency in the 400 - 700 nm range. ([12]) These results indicate that the first excited singlet states is the precursor of free carriers.

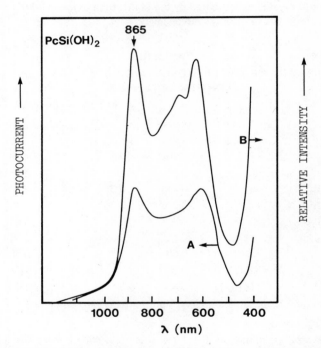

Figure 10 Photovoltaic action spectrum (A) and solid state absorption (B) of
PcSi(OH)$_2$.

CONCLUSIONS

The photovoltaic technique can be used to evaluate the photoactivity of novel pigments such as germanium and silicon phthalocyanines. All the germanium and silicon phthalocyanine samples showed some photoactivity which increases with the increase in intermolecular interaction in the crystal. Polymerization of $PcGe(OH)_2$ to $(PcGeO)_n$ increased the electrical conductivity by two orders of magnitude. Both dihydroxy silicon and germanium phthalocyanines exhibited near infrared absorption and relatively good photosensitivity. Further work to establish the relationship between structure parameters and photoactivity is in progress.

Acknowledgments

Thanks are due to Mr. T. Smith (Co-op student, University of Waterloo) and T. L. Bluhm for their technical assistance and to Dr. J. H. Sharp for his support and encouragement.

Literature Cited

1. Loutfy, R. O., Shing, Y. H. and Murti, D. K., Solar Cells, in press.
2. Esposito, J. N., Sutton, L. E. and Kenney, M. E., Inorg. Chem., 6, 1116, 1967.
3. Marks, T. J., Schjoch, K. F. Jr., and Kundalkar, B. R., Synthetic Metals, 1, 337, 1979/80.
4. Schneider, O., and Hanack, M., Angew. Chem. Int. Ed. Engl., 19, 392, 1980.
5. Yamamoto, Y., Yoshino, K., Inuishi, Y. and Ichimura, S., Phys. Stat. Sol. (a), 59, 305, 1980.
6. Meyer, G. and Wohrle, D., Materials Science, 7, 265 (1981).
7. Loutfy, R. O., and Sharp, J. H., J. Chem. Phys., 71, 1211, 1979.
8. Loutfy, R.O., Can. J. Chem., 59, 549, 1981.
9. Melz, P. J., Champ, R. B., Chang, L. S., Chion, C., Keller, G. S., Liclican, L. C., Neiman, R. R., Shattuck, M. D. and Weiche, W. J., Photogr. Sci. Eng., 21, 73, 1977.
10. Loutfy, R. O., Sharp, J. H., Hsiao, C. K., and Ho, R., J. Appl. Phys., 52, 5218, 1981.
11. Sharp, J. H. and Lardon, M., J. Phys. Chem., 72, 3230, 1968.
12. Sharp, J. H. and Popovic, Z. D., J. Chem. Phys., 66, 5076, 1977.

RECEIVED November 22, 1982

Photophysics of Films of Poly(2-vinylnaphthalene) Doped with Pyrene and 1,2,4,5-Tetracyanobenzene

NAKJOONG KIM and STEPHEN E. WEBBER

University of Texas, Department of Chemistry and Center for Polymer Research, Austin, TX 78712

Energy migration in polymer films via triplet or singlet excitons is potentially applicable to "photon harvesting". Unfortunately many polymers possess pendent chromophores that absorb only in the UV. Experiments at 77K on poly(2-vinylnaphthalene) (P2VN) films doped with an electron acceptor (1,2,-4,5 tetracyanobenzene) and pyrene have demonstrated that the triplet exciton of P2VN can be sensitized via visible light absorption in the charge transfer band. In principle this approach could be used for a large range of polymer bound aromatic chromophores.

The phenomenon of excitonic energy transport in polymer films has been studied actively for the past decade (1). This photophysical process is relevant to photodegradation and photoconductivity in polymers, but in this contribution we wish to emphasize the potential application of polymer films as photon "harvesters" with subsequent transfer of energy to a reaction center, analogous to the so-called "antenna effect" in chloroplasts.

There are at least two ways in which one can imagine using polymer films in this fashion. A polymer film could be used to support a monomeric chromophore which undergoes facile energy transport via the Förster mechanism, or at higher chromophore concentrations, by an exchange mechanism (2,3). Alternatively the appropriate chromophore could be incorporated into the polymer chain which would most likely serve to reduce solubility problems that could be encountered in the first case. Obviously to have potential for solar energy collection the chromophore must absorb visible light, which is not the case for many of the more familiar chromophore-bearing polymers (e.g. polystyrene, polyvinylnaphthalene, poly(N-vinylcarbazole)). For many aromatic chromophores there is a very large energy gap between the S_1 and T_1 excited states. Consequently a triplet exciton could be produced via visible excitation if a suitable sensitizer could be found. These two approaches are represented in Figure 1.

0097–6156/83/0220–0457$06.00/0

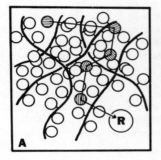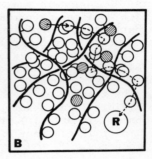

Figure 1.   Representation of the "antenna effect" for (a) energy
transfer between chromophores dissolved in film and (b) sensitization of exciton state.   The shaded circles represent the chromophore while the open circles are non-absorbing pendent groups
(which potentially can be sensitized).   R represents a "reaction
center".

The present paper is an example of the latter approach, applied to poly(2-vinylnaphthalene) (P2VN). However, rather than add a sensitizer molecule that absorbs in the visible but with a triplet state of higher energy than T_1 of P2VN, we have used 1,2,-4,5 tetracyanobenzene (TCNB) which forms a charge transfer complex with naphthalene (absorption from 340-475 nm). From earlier work in molecular crystals and low temperature glasses (4) it seemed likely that $^1CT_N^* > ^3CT_N^* > T_1$ (P2VN), where $^{2s+1}CT_N^*$ denotes a naphthalene-TCNB (N-TCNB) excited state. Thus the action spectrum for the production of the T_1 exciton in P2VN could be shifted from the UV ($\lambda \leq 330$ nm) into the visible. Analogous results would be expected for other chromophores (e.g. anthracene, pyrene) but stable polymers with these pendent chromophores are not readily available.

Unfortunately experimental proof that the T_1 exciton is produced via charge transfer excitation is not direct in the case of P2VN-TCNB. In undoped P2VN triplet-triplet annihilation produces a naphthalenic excimer delayed fluorescence, i.e.

$$T_m + T_d \rightarrow {}^1D_d^*$$ (1)

where T_m is a mobile triplet (triplet exciton), T_d is a trapped triplet and $^1D_d^*$ is a singlet excimer formed at the trapped triplet site. The process depicted in (1) can occur only if triplet excitons are produced. In the case of P2VN-TCNB the naphthalenic fluorescence is replaced by that of the $^1CT_N^*$ state. There are two mechanisms by which the P2VN fluorescence can sensitize the $^1CT^*$ state: (1) the singlet exciton can migrate into the vicinity of a naphthalene-TCNB pair, with concomitant formation of $^1CT^*$, (2) dipole-dipole (Förster) transfer from $^1D_d^*$ to $^1CT^*$ is "allowed" (i.e. there is suitable overlap of $^1D_d^*$ fluorescence and $^1CT^*$ absorption) although the rate constant is expected to be low because the $^1CT^*$ absorption is weak. The charge transfer fluorescence extensively overlaps the P2VN phosphorescence. Thus pyrene (Py) molecules were dissolved in P2VN-TCNB films to act as a triplet exciton trap. Phosphorescence from $^3Py^*$ was observed upon visible excitation of P2VN-TCNB-Py films, which we interpret as arising from the following process:

$$h\nu_{vis} \rightarrow {}^1CT_N^* \rightsquigarrow T_m \begin{cases} \nearrow T_d \rightarrow \text{P2VN phos.} \\ \searrow {}^3Py^* \rightarrow \text{Py phos.} \end{cases}$$ (2)

One complication with the use of Py as a probe for triplet excitons is that Py is a much better electron donor than naphthalene, such that formation of Py-TCNB complexes are favored over naphthalene-TCNB complexes (5). The former complexes were minimized since the mole fraction of Py was never higher than 0.016,

but as will be discussed below the formation of Py-TCNB complexes
was detected.

All experiments to be discussed herein were conducted on
films at 77K. This limitation was imposed for two reasons: (1)
the photophysics of polymer films (including P2VN) at 77K has been
relatively well characterized, and (2) thermally activated processes (including E-type delayed fluorescence) are minimized.

Experimental

The sample of P2VN was synthesized by free radical polymerization in benzene at 65°C using AIBN as initiator. The solutions
were outgassed by freeze-pump-thaw techniques and sealed before
polymerization. The 2-vinylnaphthalene monomer (from Aldrich
Chemical) was vacuum sublimed just before use. According to GPC
elution (Waters microstyragel columns, calibration based on polystyrene samples), the molecular weight of prepared P2VN was
approximately 80,000 (polydispersity ≈ 1.5). Previous experience
has shown that the viscosity molecular weight of P2VN is approximately twice that obtained from GPC elution curves. TCNB (Aldrich
Chemical) was purified by vacuum sublimination. Pyrene was re-
fluxed with maleic anhydride in toluene to eliminate anthracene
which is a common impurity of pyrene, recrystallized, and then
vacuum sublimed. In this fashion, the yellow color (due to impurity tetracene) was essentially eliminated. As will be seen
the fluorescence and phosphorescence spectra that correspond to
pure pyrene are observed.

Polymer films were cast inside a quartz tube for experimental
convenience. Benzene solutions of P2VN and various dopants were
evaporated in a rotating quartz tube at reduced pressure, outgassed under ~10^{-5} torr vacuum for ~12 hours and sealed off.

Films of pure P2VN and pyrene doped P2VN were colorless and
TCNB doped P2VN films were pale yellow. Upon addition of excess
pyrene to TNCB doped P2VN, the film color became a deeper orange,
presumably indicating the formation of Py-TCNB pairs (6). Homogeneity of films was inspected visually using a microscope (homogeneity for P2VN-TCNB and P2VN-Py was always satisfactory, unlike
P2VN-anthracene, where microcrystals of anthracene were easily observed under the microscope). Delayed emission spectra at 77K
were obtained on a homebuilt phosphorimeter (excitation-observa-
tion time ~2.4 ms) which utilizes a 200W high pressure Hg lamp as
an excitation source (7). A Corning 7-54 UV band-pass filter was
used to excite P2VN and blue band-pass interference (λ_{pass} = 430
nm) was used to excite P2VN-TCNB charge transfer complex.

Decay curves were obtained using a Fabri-Tek signal averager
connected to our phosphorimeter. Prompt emission spectra were
obtained using a SPEX Fluorolog with our own low-temperature
sample mounting modification. The SPEX Fluorolog was modified by
the addition of phase shifted choppers (1200 rpm with an observation period of approximately 6.3 ms) in the excitation and emis-

sion monochromators. By shifting the relative phase of these two
choppers the prompt fluorescence and delayed emission spectra
could be run sequentially without removing the sample from its
mounted position. The excitation intensity that results from the
150W Xe lamp and double excitation monochromator is too weak to
produce a significant biphotonic process, such as T-T annihilation.
All experiments designed to observe delayed fluorescence were
carried out on the homebuilt phosphorimeter mentioned above. It
should be noted that the photomultiplier on the phosphorimeter has
an S-20 response, which greatly enhances the ^{3}Py* phosphorescence
at 595 nm compared to the photomultiplier on the SPEX. We will
note these differences in spectra to be presented in the following section.
   All spectra were digitized and plotted using a HP-85 micro-
computer, which allows convenient scaling and comparison of different spectra.

Results

   As was mentioned earlier, we have used a pyrene dopant as a
probe for triplet excitons. Since we will require an analysis of
ternary P2VN-TCNB-Py mixtures (see below) it is necessary to consider the binary mixtures P2VN-Py and P2VN-TCNB first.

   P2VN-Py. Since pyrene is to be a probe in P2VN films it is
necessary to understand how the system P2VN-Py behaves. Films
containing pyrene up to 4 wt% showed no sign of pyrene aggrega-
tion by either microscopic inspection or by the characteristic
pyrene excimer fluorescence found in the crystalline state (this
excimer fluorescence is easily observed at ~10 wt%). The pyrene
fluorescence and phosphorescence are structured and easily distinguishable from the corresponding emission features of P2VN
films. Excitation of the film at ~290 nm yields fluorescence that
arises primarily from Py (but with a discernible P2VN component)
for the complete range of pyrene concentrations studied (wt% from
0.3 to 2.0, corresponding to a mole fraction of pyrene from 0.0025
to 0.017). Singlet energy transfer from naphthalene to pyrene is
very efficient, which when coupled with the high fluorescence
quantum yield of pyrene, explains why pyrene dominates the P2VN
excimer fluorescence. The shape of the prompt fluorescence spectrum in Figure 2 is typical of all values of X_{Py} studied, all
excitation wavelengths from 350 nm (direct excitation of pyrene)
to 280 nm (primary excitation of P2VN). There is no significant
change in this spectrum between 77K and room temperature.
   The delayed emission spectrum reveals several interesting
features (Figure 2) (note that these delayed emission spectra were
taken with a S-20 response photomultiplier to enhance the red
portion of the ^{3}Py* phosphorescence):
(1) At the lower pyrene concentrations no delayed fluorescence
     at all is observed. Evidently the trapping of the naphtha-

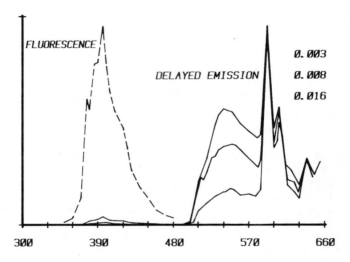

Figure 2. The prompt fluorescence (dashed line) and delayed emission (solid line) of P2VN-Py films at 77K. The mole fraction of pyrene for the corresponding delayed emission spectrum is indicated on the figure. Note that the delayed emission spectrum was obtained with a more red-sensitive photomultiplier than the prompt emission (see Experimental section) (1).

lene triplet exciton by the pyrene is so efficient that the
usual annihilation process (8), represented by equation (1),
is quenched. However, at higher pyrene concentrations a pyrene component of delayed fluorescence is observed. Thus it
appears that the following heterogeneous annihilation is
occurring,

$$T_m + {}^3Py* \rightarrow {}^1Py* \rightarrow Py \text{ fluorescence} \tag{3}$$

In this case $^1Py*$ fluorescence is more easily observed than
$^1D_d^*$ because of the difference in fluorescence quantum yield.
(2)   The phosphorescence is composed of two obvious components:
(a) P2VN phosphorescence from T_d (which is excimer-like)
peaking at approximately 540 nm and (b) the structured $^3Py*$
phosphorescence with an origin at ~595 nm. As the pyrene
concentration increases the relative intensity of the pyrene
component increases approximately linearly, as one would expect for either simple trapping or direct excitation.
        The presence of $^3Py*$ phosphorescence does not prove that
triplet exciton sensitization occurs since under the conditions of
our experiment direct excitation of $^1Py*$ can occur. However, the
observation of heterogeneous annihilation per reaction (3) does
demonstrate that triplet excitons (T_m) are not totally removed by
Py and that the $^3Py*$ molecules are accessible to T_m.
        The $^3Py*$ phosphorescence decay is fit quite adequately by a
single exponential ($\tau_{3Py*} \approx 500$ ms, typical of $^3Py*$). The P2VN
phosphorescence decay is multiexponential, as usual (8), but
there is little effect of Py on the decay rate. This is consistent with earlier work that suggests that the bulk of phosphorescence originates from T_d (8,9).

        P2VN-TCNB. P2VN-TCNB films are a light yellow (the CT ab-
sorption band extends from approximately 340 to 500 nm; below 320
nm the P2VN absorption prevents an assessment of the CT absorption). Similar to the case for P2VN-Py, excitation in the P2VN
absorption region yields a significant CT fluorescence, implying
efficient sensitization:

$$S_m \rightsquigarrow {}^1CT_N^* \rightarrow CT \text{ fluorescence} \tag{4}$$

where S_m is a singlet exciton. S_m can also be trapped at an excimer forming site producing $^1D*$ fluorescence. The CT fluorescence is red shifted away from the polymer film fluorescence such
that the mixed $^1D*$ and $^1CT_N^*$ fluorescence is easily observed. It
appears that the sensitization is not as efficient as for the case
of pyrene. Excitation at wavelengths longer than 340 nm yields
exclusively CT fluorescence (see Figure 3).
        The phosphorescence spectrum is similar to neat P2VN, but
slightly red shifted and broadened. The phosphorescence spectrum
can be excited in either the P2VN absorption band or in the CT

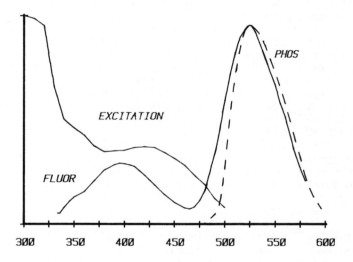

Figure 3.   Spectrum of typical P2VN–TCNB film at 77K (X_{TCNB} = 0.008).
The excitation spectrum is for the CT fluorescence at ~530 nm.   The
prompt fluorescence (solid line) and phosphorescence (dashed line)
were measured with the same photomultiplier (see text) ([1]).

absorption band.   In the latter case there is a very slight red-
shift in the phosphorescence.   In addition to CT complexes, it is
possible that new traps are created by the presence of TCNB (de-
noted T_d') that are slightly different than the normal P2VN traps,
T_d.   Undoubtedly the energy gaps are very small between $^3CT_N^*$, T_d
and T_d'.   Consequently the phosphorescence may originate from a
mixture of these species.   The phosphorescence decay is multiexponential with a slightly longer lifetime than for P2VN alone
and with a slight dependence on the wavelength of observation,
which further implies a multicomponent phosphorescence.   However,
we conclude that the phosphorescent triplet state is like that of
undoped P2VN films.   This is consistent with the earlier work on
naphthalene-TCNB which implies that the CT triplet is localized
on the naphthalene chromophores (4).

One complication in P2VN-TCNB films is that the CT fluorescence is very close to the phosphorescence emission (10).   This
is illustrated in Figure 3 in which the prompt fluorescence (solid
line) and phosphorescence (dashed line) are compared.   (The broad
feature peaking at 400 nm is the usual P2VN excimer fluorescence.)
These spectra were both taken on the same instrument (modified
SPEX Fluorolog) with in-phase and out-of-phase choppers respectively (see Experimental section).   In this manner no correction
for photomultiplier response is required and a direct comparison
of the spectra is meaningful.   The fact that the $^1CT_N^*$ emitting
state and the phosphorescent state(s) (possibly T_d' or $^3CT_N^*$) are
so close in energy makes it difficult to assess the significance
of T_m sensitization.

As in the case for P2VN-Py there is no evidence for homogeneous T-T annihilation, since neither an obvious delayed fluorescence component nor superlinear or sublinear excitation intensity dependence of any emission feature was observed (11).   However, the absence of these two effects only shows that annihila-
tive processes are kinetically unimportant; triplet excitons
could still be excited via the CT absorption.   It is to this point
that the experiments of the next section are addressed.

P2VN-TCNB-Py.   Films containing various mole fractions of
TCNB and pyrene were prepared.   The results to be discussed explicitly are for films in which $X_{TCNB} = 0.008$ and X_{Py} ranged from
0.003 to 0.016.   The prompt fluorescence spectrum of these films
is composed of a pyrene feature and a CT fluorescence (see Figure
4).   As the pyrene content is increased the CT fluorescence is
progressively red-shifted.   This almost certainly reflects an
additional component of $^1CT_{Py}^*$ fluorescence (see below).   As was
pointed out earlier, the formation of Py-TCNB CT complexes despite the relatively small mole fraction of pyrene is not surprising given the superior electron donating property of pyrene and
the quenching efficiency of $^1Py^*$ and TCNB (5).

The excitation spectra for fluorescence at 395 and 520 nm in
Figure 4 reflect in the presence of pyrene and a CT absorption

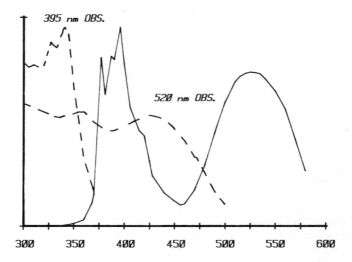

Figure 4. 77K fluorescence spectrum of P2VN-TCNB-Py (X_{Py} = 0.016) excited at 290 nm (solid line). The excitation spectra (dashed line) at the two observation wavelengths clearly illustrate the pyrene and N-TCNB absorption features, respectively (1).

very similar to P2VN-TCNB. At the highest pyrene concentration the 520 nm excitation spectrum does red-shift slightly, once again implicating Py-TCNB pairs. The comparison of the prompt fluorescence and the delayed emission using the modified SPEX Fluorolog (Figure 5) demonstrates the proximity of the CT fluorescence and the polymer phosphorescence (cf. Figure 3). The pyrene phosphorescence peaks are barely visible on the long wavelength side of the polymer phosphorescence. The weakness of these peaks is an artifact of the wavelength response of the photomultiplier used in the SPEX. This point is clearly illustrated by the series of phosphorescence spectra in Figure 6(A) excited at 430 nm (the CT absorption), and observed on the homebuilt phosphorimeter with a S-20 PMT. The decline of the polymer phosphorescence and growth of ^{3}Py* phosphorescence with increasing X_{Py} is evident. Because of the relative strength of the pyrene phosphorescence, the retention of its characteristic structure and slightly lengthened phosphorescence lifetime (12) we propose that the scheme given in equation (2) is operative. An alternative explanation for the observed ^{3}Py* phosphorescence is that 430 nm excitation is also effective for Py-TCNB pairs, and excitation of these pairs yields a "localized" ^{3}Py* state. The evidence against this is the following:

(1) The presence of pyrene only slightly modifies the excitation spectrum for the CT fluorescence in the 360-460 nm region. It would be rather fortuitous if the Py-TCNB absorption spectrum matched that of naphthalene-TCNB in this region (13).

(2) Excitation of P2VN-TCNB-Py films at wavelengths longer than 480 nm leads to a much weaker structureless phosphorescence feature that we assign to the ^{3}CT$^*_{Py}$ state (see Figure 6(B), dashed line portion) (14).

Summary and Discussion

In this paper we have explored the photophysics of P2VN films containing TCNB which forms charge transfer complexes with the pendent naphthalene groups. Using a pyrene dopant as a triplet trap we have tested the hypothesis that excitation in the CT absorption band can produce triplet excitons. The following systems were studied separately: P2VN-Py, P2VN-TCNB and P2VN-TCNB-Py. A schematic representation of the energy levels is presented in Figure 7 which unifies the observations. All states joined to the ground state by a solid line are spectroscopically active in absorption and/or emission. States connected by the dashed line are interconverted by some radiationless process (e.g. intersystem crossing, Förster energy transfer, sensitization, etc.). It must be true that all energy states are better represented as a band of states arising from different local environments. This is explicitly shown only for ^{3}CT$^*_N$. For most of our observations we do not need to invoke the presence of Py-TCNB pairs shown on the far right side of the energy level diagram. Thus we see that ^{1}N*

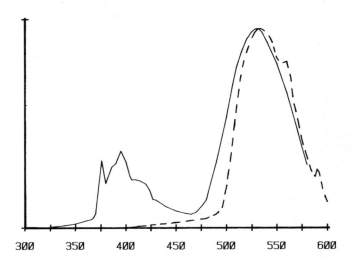

Figure 5.   77K fluorescence (solid line) and phosphorescence (dashed line) spectrum of P2VN-TCNB-Py (X_{Py} = 0.008) with an excitation wavelength of 290 nm (measured with same photomultiplier, see text) (1).

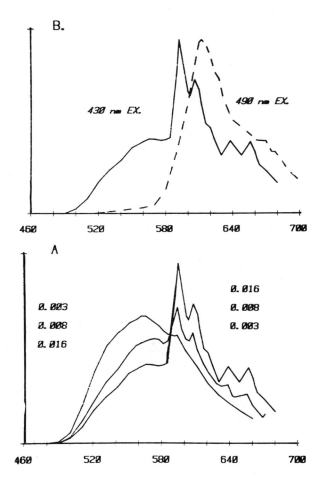

Figure 6. A. Phosphorescence of P2VN-TCNB-Py excited at 430 nm ordered by X_{Py} as shown (X_{TCNB} = 0.008 for all cases). B. Comparison of phosphorescence excited at 430 nm and 490 nm for P2VN-TCNB-Py with X_{Py} = 0.016 (all spectra with S-20 PMT) (1).

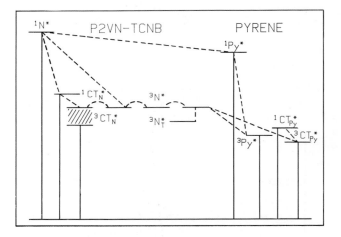

Figure 7. Representation of the energy level diagram and photophysical processes for P2VN films containing pyrene or TCNB.
Spectroscopically observed processes are indicated by solid lines
and radiationless processes by dashed lines (1).

(= S_m) can sensitize $^1Py*$ or $^1CT_N^*$, the former particularly efficiently. We (and others) have postulated that the P2VN phosphorescence originates from trapped naphthalene triplets ($^3N_T^*$) [8]. It also seems likely that some phosphorescence originates from trapped $^3CT_N^*$ states (or equivalently, new traps created by the presence of TCNB). We have also postulated that $^3CT_N^*$ can sensitize triplet excitons (= T_m) which in turn can sensitize the $^3Py*$ state. This phenomenon provides a model for producing a polymer film photoresponse in the visible that can efficiently sensitize a "reaction center" (pyrene in this case). However, our results do not prove unambiguously that the $^3N*$ state is produced by $^1CT_N^*$. For example, it is possible that the latter sensitizes weakly bound Py-TCNB pairs (via the Förster mechanism, for example) for which the triplet state is localized $^3Py*$ rather than $^3CT_{Py}^*$. However, for the reasons presented in Results we favor a mechanism like equation (2) for $^3Py*$ sensitization.

Acknowledgments

This work was supported by the Robert A. Welch Foundation (F-356) and the National Science Foundation (DMR-8013709) whose support is gratefully acknowledged.

Literature Cited

1. Based on a paper published in Macromol. 1982, 15, 430.
2. Renschler, C.L.; Faulkner, L.R. J. Am. Chem. Soc. (in press).
3. Johnson, G.E. Macromol. 1980, 13, 145.
4. See the general review, including CT crystals by Nagakura, S. "Excited States, Vol. 2", E.C. Lim, ed., Academic Press: New York, 1975; pp 322-95.
5. (a) The ionization potentials are: naphthalene, 8.12 eV; pyrene, 7.58 eV (Matsen, F.A. J. Chem. Phys. 1956, 24, 602); (b) The quenching constant of Py fluorescence TCNB is quite large (4600 M^{-1}) as measured by Grellmann, K.H.; Watkins, A.R.; Weller, A. J. Phys. Chem. 1972, 76, 3132.
6. While the color change is easily observed visually, the absorption spectrum shows a rather modest red shift.
7. This phosphorimeter has been described in an earlier publication (Pasch, N.F.; Webber, S.E. Chem. Phys. 1976, 16, 361).
8. Kim, N.; Webber, S.E. Macromol. 1980, 13, 1233.
9. For similar work on poly(1-vinylnaphthalene) films see Burkhart, R.D.; Aviles, R.G.; Magorini, K. Macromol. 1981, 14, 91.
10. This is a typical result for CT complexes. See reference 4, or McGlynn, S.P.; Azumi, T.; Kinoshita, M. "Molecular Spectroscopy of the Triplet State", Prentice-Hall: Englewood Cliffs, New Jersey, 1969; chapter 2, section 2(i), p. 88.
11. If any part of the emission spectrum was produced via TT annihilation to form an emitting singlet then the emission in-

tensity would be proportional to $(I_{ex})^n$ with $n > 1$; if TT annihilation depleted the phosphorescent triplet state then $n < 1$ would be observed. For a beautiful example of this type of behavior in polyvinylcarbazole films see Rippen, G.; Kaufmann, G.; Klöpffer, W. Chem. Phys. 1980, 52, 165 and references therein.

12. If $^3Py^*$ is populated by triplet excitons with a lifetime on the order of that of isolated $^3Py^*$ the effect will be an apparent lengthening of τ_{3Py^*}.

13. The absorption spectrum of vapor deposited CT complexes of Py-TCNB (no polymer matrix) is strongly red-shifted relative to naphthalene-TCNB (unpublished results of N. Kim).

14. According to H. Möhwald and E. Sackmann (Z. Naturforsch. 1974, 29a, 1216) the triplet state of Py-TCNB pairs in a 1:1 naphthalene-TCNB CT crystal is lower by 1400-2200 cm^{-1} than the naphthalene-TCNB triplet exciton. This energy gap is similar to what we observed for the P2VN phosphorescence ($\lambda_{max} \sim 540$ nm) and the structureless emission with $\lambda_{max} \sim 610$ nm.

RECEIVED November 22, 1982

Photoelectrochemical Catalysis with Polymer Electrodes

HOWARD D. METTEE

Youngstown State University, Chemistry Department, Youngstown, OH 44555

Polymer electrodes which absorb light and catalytically convert this energy into stored chemical potential, possibly with electrochemical assistance,
using endoergic reactions such as water splitting,
are of increasing interest as practical means of
solar energy storage are sought. Developments in
this field may be interpreted in terms of photocatalytic and electrochemical reactions catalysed
by polymers. Emphasis is placed upon the roles of
polymer films in protecting n-type semiconductors
from anodic dissolution and in helping to understand electron transfer mechanisms. Requirements
of free-standing polymeric photoelectrochemical (PEC)
catalysts are outlined.

In 1972 Fujishima and Honda (1) focussed attention on the
possibility of using visible light to photolytically sensitize
the splitting of water into hydrogen and oxygen. This light
driven, energy storing reaction has obvious attractions in that
it produces a clean burning, re-cyclable fuel and it ultimately
depends upon an infinite energy source. However, to couple absorbed solar energy to this thermodynamically uphill reaction
requires photoelectrochemical (PEC) catalysts which will simu-
taneously oxidize and reduce water.

$$PEC + h\nu_{solar} \longrightarrow PEC^*$$

$$2\ H_2O \xrightarrow{\quad PEC^* \quad} \longrightarrow O_2 + 4H^+ + 4e^-$$

$$4\ H_2O + 4e^- \xrightarrow{\quad PEC^* \quad} \longrightarrow 2H_2 + 4\ OH^-$$

These catalysts must not only cause the birth of reactive intermediates, which in some cases have been identified as H·, e⁻, and
OH· etc., but also prevent their back reaction until the O_2 and
H_2 have formed. Obviously they must maintain their own chemical

0097–6156/83/0220–0473$06.50/0

integrity in the process. Thus the functional requirements of
these PEC catalysts are more demanding than those of their photocatalytic counterparts, which serve only to accelerate energetically downhill reactions.

In the intervening decade since Fujishima and Honda's paper
a great deal of chemical effort has gone into the design and construction of these special PEC catalysts. The role of polymers
has been an important one at all levels of investigation whether
the system be colloidal micelles, vesicles, microemulsions or
"microelectrodes", or bulk semiconductors. Without polymeric support for the catalysts in colloidal systems for instance, important diagnostic reactions could not be detected. More significant for the polymer chemist, however, is the increasingly central role polymers are playing in the actual light absorption,
charge separation and particle flow dynamics that characterize
the intermediate chemistry prior to H_2 and O_2 formation.

A completely polymeric unit which accomplishes solar induced
water splitting has not yet been devised, nor indeed has any system which can successfully compete with the efficiency and longevity of photovoltaically-driven water electrolysis ($\emptyset \approx 10\%$).
Thus current efforts are addressed at selecting appropriate light
absorbing agents, charge creating and separating media, and catalytic environments which meet the necessary thermodynamic and
kinetic requirements. Of course the more practical considerations of reasonable quantum yield, durability and cost cannot be
ignored either. Broadly speaking, polymers have contributed both
to the stability of PEC catalysts and to the understanding and
control of charge migration and redox chemistry in these systems.
In the future, polymers offer an extra degree of synthetic freedom which may be exploited to enhance the quantum yields, durability and economic practicality of PEC systems still further.

A number of closely related reviews have been published recently which highlight a number of approaches to solar induced
water splitting. Bard (2) has summarized the semiconductor design criteria for example. More detailed reviews by Nozik (3)
and Wrighton (4) consider the interplay of the thermodynamic and
kinetic behavior of these semiconductor units, and Wrighton has
helped develop the concept of surface modified, semiconductor
electrodes. Calvin (5) and Porter (6) have emphasized the close
similarity of PEC water splitting to natural photosynthesis,
thereby leading to the study of biomimetic systems at the colloidal level. Whitten (7) has considered photoinduced electron
transfer in homogeneous solutions, very frequently involving
chlorophyll-like sensitizers.

Publication of the proceedings of three recent symposia dealing with PEC processes in the past two years indicates the high
level of chemical interest in these systems. Two in the A.C.S.
Symposium series (8,9) consider the importance of interfaces in
PEC systems, while a Faraday Discussion Volume (10) contains
principally semiconductor contributions. It appears that the

semiconductor approach is a dominant one at the moment, and it
should not be at all surprising to find out that polymers have
contributed mainly to research in this area. This is not to suggest that work like that of Regen and co-workers (11a), who have
photopolymerized vesicle walls and thereby extended the typical
vesicle lifetime from 48 hours to more than two weeks, is incidental. Indeed no one can foretell at this moment what the final
forms of successful PEC cells will be, and colloidal surfactant
systems may well be among them. However, polymers have been most
extensively applied to bulk photoelectrodes and it is this subject that principally occupies this review.

Operational Background

        Figure 1 depicts the physical arrangement of a typical PEC
cell with a polymer coated photoanode (e^- pass from the electrolyte through the polymer layer to the working electrode), or photocathode. The polymer layer may be either covalently attached
or physically adsorbed to the substrate, and it frequently contains a redox couple to be conductive. The conductivity of the
polymer layer may be due to a metal-like wide energy band (eg.
polypyrrole (11b)), or may occur through a narrow energy window
resulting from specific electroactive sites within the film (eg.
ferrocene). The substrate may be a noble metal or graphite, in
which case the sensitizer is embedded in the film, or a semiconductor which may then assume the role of sensitizer. Metal electrodes are often used in control experiments to distinguish film
behavior from that of semiconductors.
        An illustration of a PEC cell of this type, which operates
in reverse, may be found in the work of Rubinstein and Bard (12).
The well known duPont polymer Nafion was dip coated on a graphite
electrode for this experiment, and then immersed in a solution
of $Ru(bpy)_3Cl_2$ (bpy = 2,2'-bipyridine). Cyclic voltammograms of
this treated electrode showed broad oxidative and reductive waves
close to 1.2V vs NHE of the potential sweep, characteristic of
the presence of the $Ru(bpy)_3^{2+}$ complex sequestered in the polymer
film. However, when oxalate ion was added to the supporting
electrolyte, which ion is only oxidized at higher potentials by
plain Nafion, three effects were noted; namely, an enhanced oxidative current (more $Ru(bpy)_3^{2+} \rightarrow Ru(bpy)_3^{3+}$), a damped return
wave (less $Ru(bpy)_3^{3+} \rightarrow Ru(bpy)_3^{2+}$), and an intense orange luminescence due to $Ru(bpy)_3^{2+(*)}$. Thus it appears that the power-
ful $Ru(bpy)_3^{3+}$ oxidant catalytically oxidizes oxalate with sufficient excess energy to produce an electronically excited state.

        Electrochemical Polymers. PEC experiments with polymer in-
terfaces arise from electrochemical work by Anson (13a, b, c)
Kuwana (14), Miller (15) and others. This field has been reviewed by Murray (16) and Snell and Keenan (17). A study by Anson
and Oyama (13b) illustrates several general principles of films.

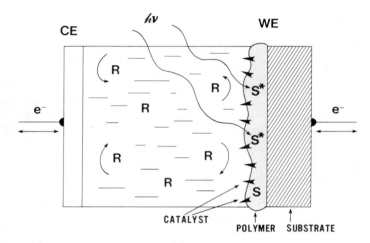

Figure 1.   A photoelectrochemical cell with a polymer/electrolyte
            interface containing a light absorbing sensitizer (S)
            embedded in the polymer. Light absorption may enable
            a redox reaction of (R) dissolved in the electrolyte.
            When a semiconductor is the substrate, it is also
            often the sensitizer. (WE and CE denote working and
            counter electrodes).

Their work included coating pyrolytic graphite electrodes with
poly(4-vinylpyridine) and immersing them in aqueous solutions of
Ru^{III}(edta)OH_2. A slow and steady growth of the redox waves of
the $Ru^{III/II}$ couple was observed as the complex entered what was
otherwise an electrochemically quiet film. Control experiments
showed that when in the film, the Ru^{III} ion chemically coordina-
ted with pyridine by ligand substitution. This work showed that
thick durable polymer films could conveniently be made, and that
their electrochemical activity could result from the redox couples sequestered from the ambient solution.

The principle of "mediated" electron transfer, whereby electrons are passed from the reduced form of a relatively negative
redox couple to the oxidized form of a relatively positive couple,
has been demonstrated to occur between two polymer layers of
slightly different Ru^{II}(bpy)$_3$ complex polymers by Murray and co-
workers (18). This kind of stepwise, unidirectional electron
transfer may be very significant in future polymer coated PEC
cells which seek to separate charge, and of additional interest,
Ru^{II}(bpy)$_3$ complexes are frequently used as cyclic PEC catalysts
in water splitting experiments. Some details of this experiment
are thus informative.

To demonstrate unidirectional charge flow via electron mediation, Murray's group electrochemically polymerized complexes
[Ru(bpy)$_2$(vpy)$_2$]$^{2+}$, A, and [Ru(bpy)$_2$(vpy)Cl]$^+$, B, on Pt electrodes in CH_3CN. (vpy is 4-vinylpyridine.) The order of deposi-
tion of the films is crucial of course since $E^{\circ \prime}_{surf}$ [$Ru^{3+/2+}$(A)] =
+1.23 and $E^{\circ \prime}_{surf}$ [$Ru^{3+/2+}$(B)] = +.76 V vs SSCE and the inner film
mediator (poly(A)) would not be expected to move electrons uphill.
The results are summarized in Figure 2, where it is clear that
both redox waves associated with the outer film couple (poly (B))
are missing in the dual layer system (Fig 2(b)). The (A + B)
copolymerized single film electrode (Fig 2(c)) shows the electronic presence of both couples at the Pt/polymer interface.
Thus the inner polymer layer has seemingly screened communication
between the hole (h^+) at the Pt surface and the redox couple in
the outer film.

Electronic mediation, the passage of e^- from one $Ru^{3+/2+}$ couple to another, is shown by considering the details of Figure 2(b).
One important feature is the oxidative spike, or prewave, in the
first oxidative scan, and its absence from subsequent scans if
the return potential is not swept negative past about -1 V. A
second is the pronounced cathodic spike on the large reduction
wave if this negative scan is carried out <u>subsequent</u> to a positive
one.

In the first anodic scan, the Ru ions in both polymer layers
begin as Ru^{2+}. When the electrode potential first passes
$\sim$ +1.15 V, the outer layer Ru^{2+}(B) $\rightarrow$ Ru^{3+}(B) is signalled as a
sharp spike, and the inner layer Ru^{2+}(A) $\rightarrow$ Ru^{3+}(A) acts as a
mediator. When $\sim$ +1.23 V is passed, the remaining
Ru^{2+}(A) $\rightarrow$ Ru^{3+}(A) occurs. However, on the return scan, only

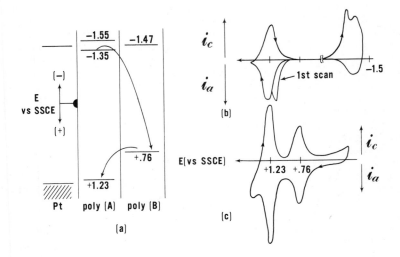

Figure 2.   a) A double polymer layer on a Pt substrate with
                energy levels used to interpret voltammograms (b)
                and (c).   Scan (c) is that of a single layer of co-
                polymerized complex A and B.   Adapted from Murray
                et al. (18).

the inner layer reduction $Ru^{3+}(A) \rightarrow Ru^{2+}(A)$ occurs since the $Ru^{3+}(B)$ is energetically and spatially inaccessible (cf Fig. 21(a)). Unless electrons whose potential is more negative than +.76 V can be supplied to the outer film, the $Ru^{3+}(B)$ sites will remain trapped there. Subsequent anodic scans do not reveal the presence of $Ru^{2+}(B)$ in the outer layer, as few if any are present.

The outer layer holes at +.76 V may be discharged, and the oxidative prewave returned, however, if either the monomer of $Ru^{2+}(B)$ is added to the electrolyte $(E_m^{\circ \prime}[Ru^{3+/2+}(B)] = +.76 \text{ V})$ or if the potential is swept negative enough to reduce $bpy^{\circ} \rightarrow bpy^{-}$ in the inner layer ligands. (See redox values in Figure 2(a)). In this latter case a pronounced cathodic spike signals the reduction of $Ru^{3+}(B) \rightarrow Ru^{2+}(B)$ in the outer layer, at a potential nearly 2 V more negative than necessary to effect the reduction!

One obvious conclusion to be drawn from these experiments is that these polymer films are essentially insulating, except in the rather narrow energy range associated with their included redox couple. Another is that charge rectification may be sustained in a laminate polymer system, thereby mimicking the role played by the potential gradient within the depletion layer of conventional p-n semiconductor junctions. Further discussion of the oxidative and reductive prewaves may be found in the work of Meyer (19). Examples of the propagation of a redox reaction through polymers films may be found in the work of Miller (15a).

Photocatalytic Polymers. The electrochemical experiments cited above were chosen from a vast body of recent polymer coated electrode work. Likewise the field of polymer supported photo-redox catalysts is also broad and has a more extended history. Possibly a common linkage between electrochemical and photochemical catalyses, assisted by polymers, can be traced to oxidation of ascorbic acid (AH_2). In 1966 Davidov (20a) found that light exposed, polyacrylonitrile (PAN) containing solutions of AH_2 consumed oxygen in a measurably different manner than similar solutions without AH_2. (Simple photoabsorption of O_2 also occurs). In 1977 Kuwana and coworkers (14) found that RF discharged vapors of benzidine (BZ) led to the formation of a film on graphite electrodes (PBZ) which electrochemically catalysed the same oxidation. (Naked graphite electrodes functioned very poorly).

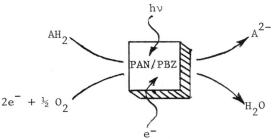

It may be that this coincidence is nothing more than fortuitous, but it is worth noting that whether a chemical oxidizing or reducing potential is produced at the molecular level by the action of light or an electrical field, the result is essentially the same.

An important difference remains, however, between electro- and photocatalytic processes. In the electrocatalytic case, the creation of the charge gradient is the initial act, the one leading to charge separation and thereby chemical potential. On the other hand, photolytically excited molecules may undergo numerous dissipating processes such as fluorescence and internal conversion, not to mention irreversible photochemistry. Thus it is essential to reduce these dissipative processes to a minimum, and to the extent that polymers can do this and promote rapid charge separation, polymers will continue to find a useful function.

It is useful to note here that nature has already accomplished the goal of light induced charge separation in the reaction centers of Rps. sphaeroides, a photosynthetic bacteria. These large macromolecular assemblies contain organized units of cytochrome C (cyt C), bacteriochlorophyll dimers ($(BChl)_2$), bacteriopheophytin (BPh), a metal-ion free chlorophyll, an iron complexed quinone (FeQ) and a second quinone molecule (Q_2). The light absorbing and charge separating sequence, creating in the end both oxidizing and reducing power on opposite sides of a membrane, is currently vizualized as follows (20b), where the primary donor and acceptor appear in [ ], and the time scale is given on the right.

Species	Time (picoseconds)
cyt C,[$(BChl)_2$,BPh],QFe,Q_2	0
↓ hν	
cyt C,[$(BChl)_2^*$,BPh],QFe,Q_2	0
↓	
cyt C,[$(BChl)_2^{\ddagger}$,BPh⁻],QFe,Q_2	<10
↓	
cyt C,[$(BChl)_2^{\ddagger}$,BPh],Q⁻Fe,Q_2	100-200
↓	
cyt C⁺,[$(BChl)_2$,BPh],Q⁻Fe,Q_2	10^6
↓	
cyt C⁺,[$(BChl)_2$,BPh],QFe,$Q_2^{\cdot -}$	10^8

It is probable that the very rapid separation of charge is facilitated by the close juxtaposition of these units within the reaction center, as well as their relative redox levels. In comparison to these natural systems, synthetic polymer PEC catalysts are an ambitious undertaking, wherein the distinction between photo- and electrocatalytic events may not always be clear.

Unfortunately it appears that photocatalysis using polymeric supports is more difficult to control than electrocatalysis.   The examples of Wamser (21) and Hautala and Little (22) are suggestive.   The latter workers covalently attached dimethylaminobenzophenone (DIMBOP) to both polystyrene and silica.   DIMBOP by itself light sensitizes the energy-storing norbornadiene to quadricyclene conversion, which may be reversed catalytically.   Unfortunately the high quantum efficiency of nearly 100% conversion in homogeneous solution is reduced to 78% with DIMBOP on polystyrene and 24% on silica.   A reduced triplet yield is held responsible for these losses.

Wamser and Wagner's study (21) points to a critical role of the solvent in determining polymer conformation, which in turn effects the dissipative loss of molecular excited states.   Polystyrene beads derivatized with cyclopentyl- and cyclohexylacetophenones, attached through the aromatic ring, were irradiated and the yield of the Type II photoelimination product (cyclopentene or cyclohexene) was found to decline slightly in pentane (compared to the homogeneous solution) and precipitously (to zero) in ethanol.   Although the solvent change by itself does lower the yield appreciably in the homogeneous cases as well, binding the chromophore to the polymer tends to greatly exaggerate the difference between pentane, a swelling solvent, and ethanol, a binding solvent.   Thus self-association of the chromophores tends to reduce their photodecomposition.   Daum and Murray (23) have discussed the importance of polymer swelling in polyvinylferrocene films in terms of their electrochemical behavior.

Solvent effects may in certain cases contribute to the charge conduction mechanism with polymer films, however, as the work of Kaufman et al (24) has shown.   On the basis of spectral data, these workers have concluded that mixed valence states of tetrathiafulvalene (TTF^{2+}, TTF^{+}, etc.) are not essential to charge-conduction in poly TTF films, but that electron hopping modulated by the solvent-induced pendent group collisions is.   An additional phenomenon related to the electrolyte noted by this group is that ion flow into the polymer phase appeared to limit the kinetics of oxidation reactions in these films.

Finally the photocatalytic activity known for many years in molecular films of phthalocyanines (25) may find a new form in the conducting molecular stacks of partially oxidized polyphthalocyanines of Marks (26).   These face-to-face phthalocyanine rings are connected by oxo-bridges between their central atoms, and show a semiconductor like dependence of their conductivity on temperature.   The thin film work of Tachikawa and Faulkner (27a) using molecular phthalocyanines to photoassist mediated electrode reactions suggests that polyphthalocyanines might likewise be similarly employed.   Fluorometalopolyphthalocyanines have recently been reported to be conductive after 3 months exposure to air and thus at least show remarkable stability (27b).

Polymer Assisted Photoelectrochemistry

An Overview. The uses of polymers in solar energy conversion and storage include specialized materials designed as light
collectors, coatings of optical equipment, encapsulating agents,
pottants and solar-pond liners. A great deal of additional research effort has been dedicated to understanding the photooxidation, photohydrolysis and photodegradation of polymers. Finally
the long known semiconducting properties of polymers such as PAN
are being adapted to the surge of interest in solar energy.
    Relatively scant attention has been given to the synthesis
of new polymeric materials which could photoelectrochemically
split water. An obvious reason for this is the consideration
that if everyday insulating polymers are subject to photodegradation in routine environmental exposures, what chance would a semiconducting or conducting polymer, with more reactive centers such
$\geq$c$=$c$\leq$, have to survive under yet more harsh chemical conditions.
Such odds have not deterred polymer chemists in the past, however, and now that the attention of more chemists has been sti-
mulated, rapid developments in this area may be anticipated. In
fact even the labile polyacetylene has been found to be significantly stabilized when physically mixed with polyethylene (27c) or
when Cl$^-$ is available in the contacting electrolyte (27d).
    Semiconductors contacted with an electrolyte are able to ab-
sorb light and rapidly separate the resulting (e$^-$/h$^+$) pairs to
produce useable chemical potential. They perform this function
seemingly without many of the complications faced by collections
of molecules assembled for the same purpose. Were it not for the
facts that:

    i)   (e$^-$/h$^+$) pairs recombine degrading the light into heat;

    ii) the (h$^+$) often oxidize visible-light absorbing semi- .
         conductors (28); and

    iii) much of the light energy is wasted in overcoming over-
         voltage requirements;
semiconductors would be perfect units for solar induced water
splitting. To overcome these problems, surfaced attached polymers are designed to kinetically assist the semiconductor in applying its chemical potential to molecules in the electrolyte
solution.
    Much of the present interest in polymer derivatized semi-
conductor electrodes stems from the work of Wrighton and co-workers at M.I.T. They demonstrated (29,30) that n-Si (a photoanode)
could be successfully protected from self-oxidation by attaching
a surface film of ferrocene, which later came to be regarded as
oligomeric in nature. More recently this group (31,32,33), and
Bard and Abruña (34) derivatized p-Si (a photocathode) with a
polymeric form of MV^{2+} to catalytically enhance this electrode's
reducing capability. Figure 3 summarizes an arrangement of redox levels within semiconductor units used to split water, with

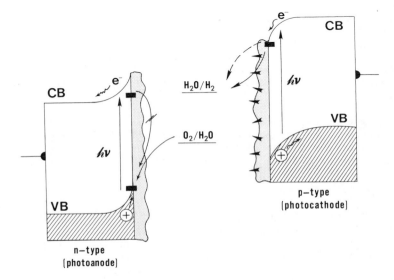

Figure 3.   Schematic illustration of the established roles of
polymers in anodic protection and cathodic acceleration in semiconductor photoelectrodes. The polymer
films reduce self-oxidation of anodes and the overvoltage required for reduction of solution species at
the cathode.

the surface attached polymers mediating the kinetics. (For details on the operation of such systems the reader is referred to the reviews cited earlier.)

However, not all polymer activity in photoelectrochemistry has centered on altering surface kinetics. For example, an approach of Ghosh and Spiro (35,36) uses a $Ru(bpy)_3^{2+}$ oligomeric film on the surface of $n-SnO_2$ to photosensitize the injection of electrons into the conduction band. This study is based on the well accepted fact that transition metal oxides are stable to photoanodic dissolution, which makes them desirable PEC catalysts, but they do not absorb visible light significantly due to their large band gap. Thus the experimental strategy is to overlay a polymer film which contains chromophores to shift the absorption into the visible range.

Finally there is an experiment by Shirakawa et al (37) which combines the polymer and semiconductor functions in a single polymer phase of p-type, transpolyacetylene. Thus polymers appear to have gained an ever expanding role in PEC energy conversion and storage.

Redox Couples. Because of their frequent occurrence in the PEC literature, the redox couples given in Table I may prove useful. Many of the values of $E°^{\prime}$ refer to monomeric units in a particular solution, but these are not usually drastically altered by incorporating these units into polymers in the same solvent.

Photoanodes. Two major improvements have been brought about to stabilize the photoanodic decomposition of n-Si. One is the derivitization of the surface by attaching oligomeric ferrocene films, as shown by Wrighton et al (29,30). The other strategy is to photo/electrodeposit thin films of polypyrrole, which has been developed by Fan et al (41), Noufi, Frank and Nozik (42) and Skotheim (43). A number of interesting details regarding the mechanism of protection and design of possible future PEC interfaces have thus become available. Bolts et al (29) were the first to provide direct proof of mediated electron transfer at any derivatized electrode. Using n-Si derivatized by ferrocene and immersed in aqueous solutions of $NaClO_4$, an oxidation wave corresponding to the $Fe(C_2H_5)_2^{+/0}$ couple in the film was seen at the illuminated electrode. Since this wave is seen at +0.45 V vs SCE at a Pt electrode, and it appears at +0.27 V vs SCE on illuminated n-Si, a photoassistance of close to 200 mV is noted. Figure 4(b) shows a great enhancement of the anodic wave when dilute $Fe(CN)_6^{4-}$ is added to the electrolyte ($E°^{\prime}$ = +.2V vs SCE) indicating mediation by the more positive ferrocene couple. On the dark return scans, holes are no longer created at the semiconductor surface and electrons flow back into the now oxidized ferrocene film with about 200 mV greater ease than if the light were on. Of importance is the great loss of intensity in the cathodic

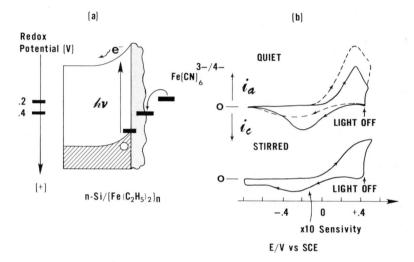

Figure 4.   The mediation effect of polyferrocene films on n-Si
            shown schematically and with the experimental evidence
            of Bolts et al.(29).   (a) The relative placement of
            the redox levels.   (b) Quiet solutions
            0.1 M $NaClO_4/H_2O$ of the coated electrode (———) and
            with dil. $K_4Fe(CN)_6$ (---).   The stirred solution is
            dil. $K_4Fe(CN)_6$ and the anodic wave is recorded at
            ½ the sensitivity of the quiet solution, while the
            cathodic wave is 10X the anodic sensitivity.

Table I. Redox Couples Frequently Used in PEC Work [a]

Redox Couple	Electrode	Solvent (pH)	E°/V vs SCE	Reference
$MV^{2+/+}$	Pt, Au	H_2O (7)	$-.69 \pm .01$	33
	Hg, n-Si(dark)	H_2O (1-7)	$-.69 \pm .01$	33
	Pt, Au, n-Si	CH_3CN	$-.45 \pm .01$	33
	p-Si(illum)	CH_3CN	$-.9$	32
$(MV^{2+/+})_n$ [b]	Pt, Au, n-Si	H_2O (1-8.6)	$-.55 \pm .01$	33
$MV^{+/0}$		CH_3CN	$-.45 \pm .01$	33
$(MV^{+/0})_n$ [b]	p-Si(illum)	CH_3CN	$\sim-.9$	32
$Fe(C_2H_5)^{+/0}$	Pt, Au	H_2O	$+.45$	30
		CH_3CN	$+.40$	19
		EtOH	$+.32$	38
	n-Si(illum)	H_2O	$+.27$	30
		EtOH	$+.30$	30
		CH_2Cl_2	$+.4$	39
$(Fe(C_2H_5)_2^{+/0})_n$ [b]	Au	CH_3CN	$+.47^c$	38
	Pt		$+.53 \pm .04^c$	38
			$+.35$	23
	n-Si(dark)		$+.6^c$	38
	n-Si		$+.1^c$	38
		CH_2Cl_2	$+.46$	39

Redox Couple	Electrode	Solvent (pH)	E°/V vs SCE	Reference
$Ru(bpy)_3^{3+/2+}$	Pt	CH_3CN	+1.24	18
	$n-SnO_2$	H_2O	+1.0	35
$(Ru(bpy)_3^{3+/2+})_n^b$	Pt	CH_3CN	+1.22	18
			+1.23	19
			+1.23	36
$Ru(bpy)^{2+/+}$	SnO_2	H_2O	−1.5	35
$Ru(bpy)_3^{3+/2+(*)d}$	SnO_2	H_2O	−1.1	35
$Ru(bpy)_3^{2+/(*)/+d}$	SnO_2	H_2O	+.6	35
H_2O/H_2	Pt	pH = 0	−.24	40
		pH = 7	−.66	
		pH = 14	−1.08	
O_2/H_2O	Pt	pH = 0	+.99	
		pH = 7	+.57	
		pH = 14	+.15	

[a] Abbreviations: Ferrocene, $Fe(C_2H_5)_2$; ruthenium(II)tris (2,2'-bipyridine), $Ru(bpy)^{2+}$; Methylviologen, MV^{2+};   saturated calomel electrode, SCE.

[b] polymeric unit contains functional groups not in monomer.

[c] depends somewhat on film history.

[d] $Ru(bpy)_3^{2+(*)}$ will reduce the oxidized form of any couple whose redox potential is more (+) than −1.1 V vs SCE (eg H_2O, pH = 1−14).

[d] $Ru(bpy)_3^{2+(*)}$ will oxidize the reduced form of any couple whose redox potential is more (−) than +.6 V vs SCE (eg H_2O, pH≈7).

return scan due to $Fe(CN)_6^{4-}$ reducing the $Fe(C_2H_5)_2^+$ at the poly-
mer/electrolyte interface. If the solution is stirred, the anodic
wave grows still farther, and the dark cathodic wave likewise
diminishes. These observations are consistent with a more rapid
exchange of electrolyte species with the polymer film.

In addition to enabling the study of the electron mediation
mechanism, these films increase the operational lifetime of the
naked n-Si electrodes from minutes to hours due to the inhibition
of insulating SiO_x growth at the semiconductor/polymer interface
(30). It is worth noting, however, that when the potentials of
the illuminated, derivatized electrodes are swept more positive
than +.6 V vs SCE (to oxidize water), more rapid film deterioration results. Making the solution basic which reduces the voltage necessary to oxidize water, unfortunately also hydrolyses the
polyferrocene (29). As of this time the polyferrocene derivatized
electrodes of n-Si have not yet sustained PEC water oxidation.

An example of the protective value of polypyrrole is to be
found in the work of Skotheim et al (43) who stabilized n-Si with
thin films of this material. These conductive films were photoelectrochemically generated and they extended the operational
lifetime of illuminated n-Si from about four hours to 6 days in
the presence of I_3^-/I^- couple in aqueous solution. The cell
showed no decay after running continuously at anodic current densities of 9 ma/cm^2 for this long period. Noufi et al (42) found
similar enhanced stability of n-Si coated with polypyrrole in the
presence of the $Fe^{3+/2+}$ couple, and for n-GaAs in CH3CN solutions (45).

Two further improvements in cell performance have occurred
using polypyrrole films. One noted by Fan et al (41) precoats
an n-Si surface with a thin film of Au(~15Å) followed by a thicker
film of polypyrrole (3300Å). Such a combined electrode substantially improves stability in aqueous electrolytes containing the
$Fe^{3+/2+}$ couple. The other improvement is due to Cooper et al (44)
who sputter deposited a 72Å "film" of Pt on the surface of a polypyrrole film covering a Ta substrate. (Ta is known to form surface oxides like n-Si). They were then able to sustain O_2 evolution by water electrolysis in a manner indistinguishable from the
use of naked Pt electrodes. Their results show that small deposits of Pt at the polymer/electrolyte lower the overvoltage necessary for water oxidation compared to that needed using non-
platinized polypyrrole.

The experiment of White, Abruña and Bard (46a) sheds some
light on how polymers may enhance the PEC response of semiconductor electrodes. n-WSe$_2$ and n-MoSe$_2$ showed quite variable oxidation waves when immersed in aqueous solutions of $Fe(CN)_6^{3-/4-}$
and I_3^-/I^- couples. In general, those electrodes with numerous
surface imperfections (edges, steps) showed the poorest photo-
voltages and photocurrents. Surface imperfections historically
have been considered to be centers for the recombination of
(e^-/h^+) pairs, and they may act as catalysts for dark "back

reactions" (eg. $n-WSe_2(-) + Fe(CN)_6^{3-} \rightarrow n-WSe_2 + Fe(CN)_6^{4-}$).
When these electrodes were partially covered with electrochemically polymerized o-phenylenediamine (OPD), polymerization occurred only at the surface imperfections. The most marked improvements in PEC response came from those crystals which were the
poorest performers to begin with. Either direct (e^-/h^+) pair recombination within the semiconductor, or e^- leakage through a surface state to reduce an oxidized form of the solution redox couple
(generated by discharging an optically induced hole in the valence
band) were considered as loss mechanisms. Since the dark current
of returning cathodic scans were reduced, indicating a blocked
"back reaction," upon polymerization, the authors suggested that
OPD blocked e^- leakage through a surface state.

Although O_2 generation at illuminated, polymer derivatized
semiconductor electrodes has proved elusive, it has not been im-
possible. Frank and Honda (46b) used polypyrrole stabilized
n-CdS in a recent experiment to sustain visible light induced water cleavage yielding molecular oxygen. They further noted that
when the polypyrrole film was coated with a thin film containing
the known oxygen generating catalyst $RuO_2(s)$, a more than 3 fold
improvement in the oxygen yield occurred.

Photocathodes. It is a fact that O_2 generation is much more
difficult than hydrogen generation in solar photoelectrochemistry.
This is true in the colloidal artificial membrane systems just as
much in their bulk semiconductor counterparts. The difficulty of
catalytically managing four oxidizing equivalents $(4e^-)$ per O_2
molecule from water, compared to only two reducing equivalents
for H_2 production (beginning with $2H^+$), may account for this.
Thus the remarkable achievement of Kiwi et al (47), who simultaneously oxidized and reduced water with a single bifunctional catalyst (TiO_2 particles treated with Pt and RuO_2), stands as a mile-
stone which will be difficult for polymeric systems to emulate.
Few systems indeed have accomplished this using just visible light
(48 and references therein).

For semiconductors, cathodic dissolution is not a problem.
In fact cathodic protection is a well known electrochemical procedure to prevent slow oxidation. Thus p-type semiconductors do
not suffer the magnitude of corrosion problems that n-type photoanodes do.

Generation of H_2 at p-type surfaces seems mainly to be a problem of reducing overvoltages, which of course means providing
additional kinetic pathways to make use of the available chemical
potential. The methylviologen couple, $MV^{2+/+}$, has been as ubi-
quitously used in reductions as $Ru(bpy)_3^{2+}$ has in oxidations. It
is capable of reducing water in neutral and acidic solutions with
a maximum efficienty pH≈4.

Bookbinder and Wrighton (31) reported the light assisted reduction of a surface-confined MV^{2+} derivative on p-Si, showing
voltammographic waves for both $MV^{2+/+}$ and $MV^{+/0}$ reductions in

CH_3CN. Both waves were shifted positively +.5 V over the surface-confined MV^{2+} on Pt, a fairly dramatic light effect. (The band gap in Si is 1.1eV). The addition of $Fe(C_2H_5)_2^+$ to the electrolyte produced an enhancement of the cathodic wave and a diminution of the dark return (anodic) wave, the latter being nearly extinguished when the solution was stirred. These results parallel closely those on n-Si and are consistent with mediated reduction of $Fe(C_2H_5)_2^+$ by MV^+.

When $PtCl_6^{2-}$ ion is incorporated into the surface film of poly-MV^{2+} and reduced to Pt(0), a uniform distribution of Pt(0) is found throughout the film using Auger spectroscopy of the slowly sputtered surface (32). Interestingly, this depth profile analysis revealed a thin (20Å) coat of SiO_x between the semiconductor surface and polymer layer. The most significant finding, however, was the successful reduction of water mediated by these illuminated electrodes. Bard and Abruña (34) conducted nearly parallel experiments on p-Si, but their poly-MV^{2+} films were deposited by evaporation or electroprecipitation of the performed polymer. They also platinized their films and showed photoassisted electrochemical reduction of water to produce H_2.

The work of Dominey et al (33) expands upon the work of Bookbinder and Wrighton to include detailed pH dependence and light vs. dark effects to help characterize these surfaces. They noted that the photovoltage developed at different pH's helped clarify the importance of Fermi level pinning (49) in these systems, which is very significant in understanding the role surface states play in PEC processes.

Polymer Photoelectrochemistry

Perhaps because of the difficulty of finding at least semiconductive polymers which are stable under the conditions of water photoelectrolysis, there are few examples of wholly polymeric PEC processes. A metal-like polymer, such as heavily doped polyacetylene (50,51), would have some difficulty separating charge without a space-charge layer, and these systems are quite air sensitive. Polypyrrole seems a marked exception in terms of stability.

The brief but trenchant note of Shirakawa and coworkers (37) demonstrates that p-type semiconducting films of trans-polyacetylene can photoelectrochemically reduce MV^{2+} in aqueous solution, though with low quantum efficiency of electron production ($\emptyset \sim 10^{-3}$). The photovoltage at open circuit, V_{oc}, is 0.3 V which is quite comparable with other PEC systems. The photo response curves do not match the visible absorption spectrum of these films, which is taken to indicate that photogenerated h^+ near the illuminated surface are predominately trapped. At wavelenghts where the extinction coefficient is lower, the h^+ have a greater chance of transiting the space charge layer and being registered as carriers in the external circuit. The location of traps,

referred to as "solitons", within the band gap of polyacetylenes
is theoretically predicted by Chien (52), and these traps account
for a number of other properties, such as cross-linking, observed
for these materials.

    This experiment may presage the growth of PEC systems where
the light absorption, charge separation and resultant electro-
chemistry occur entirely within the polymeric substrate, or at
the polymer/electrolyte interface.

Conclusions

    PEC catalysis using polymeric assistance is considered to be
a natural outgrowth of polymer assisted electrochemical and photochemical catalysis. Early experiments, mostly combining electroactive polymer films with small band gap semiconductors have
demonstrated the feasibility of using such systems in the PEC
decomposition of water. Both gaseous H_2 and O_2 have been separately generated from polymer coated photoelectrodes: H_2 from
poly-MV^{2+} films on p-Si and O_2 from polypyrrole films on n-CdS.
The major success of polymer coated photoanodes is to
protect them against photooxidation, which rapidly
degrades the performance of many small band gap n-type
semiconductors.

    Combining the functions of light absorption, charge creation
and separation, followed by catalytic transduction of chemical
potential into new products, all in a polymeric device is a future
goal for chemists interested in solar induced water splitting.
New experiments have made strides in this direction. It is worth
remembering that nature has accomplished this remarkable feat for
hundreds of thousands of years in the photosystems of natural
chloroplasts. For man, however, the journey has just begun.

Literature Cited

1.   Fujishima, A., Honda, K., *Nature*, 1972, *238*, 37.
2.   Bard, A.J., *J. Phys. Chem.*, 1982, *86*, 172.
3.   Nozik, A.J., *Ann. Rev. Phys. Chem.*, 1978, *29*, 189.
4.   Wrighton, M.S., *Acct. Chem. Res.*, 1979, *12*, 303.
5.   Calvin, M., *Ener. Res.*, 1979, *3*, 73.
6.   Porter, G., *Proc. Roy. Soc. Lond. (A)*, 1978, *362*, 281.
7.   Whitten, D.G., *Acct. Chem. Res.*, 1980, *13*, 83.
8.   Wrighton, M.S., ed., "Interfacial Photoprocesses: Energy
     Conversion and Synthesis," *Adv. Chem. Ser.*, 1980, *184*.
9.   Nozik, A.J., ed., "Photoeffects at Semiconductor-Electrolyte
     Interfaces," *ACS Symp. Ser.*, 1981, *146*.
10. Young, D.A., ed., "Photoelectrochemistry," *Faraday Disc.
     Chem. Soc.*, 1980, *70*.
11. (a) Regen, S.L., Singh, A., Oehme, G., Singh, M., *J. Am.
         Chem. Soc.*, 1982, *104*, 791;

(b) Street, G. B., Abstr. 28th Macromol. Symp. (IUPAC),
    1982, 418.
12. Rubinstein, I., Bard, A.J., J. Am. Chem. Soc., 1980, 102,
    6641.
13. (a) Oyama, N., Anson, F.C., J. Electrochem. Soc., 1980, 127,
       247;
    (b) Oyama, N., Anson, F.C., J. Am. Chem. Soc., 1979, 101,
       3450;
    (c) Oyama, N., Anson, F.C., J. Am. Chem. Soc., 1979, 101,
       739.
14. Evans, J.F., Kuwana, T., Henne, M.T., Royer, G.P., J. Elec-
    troanal. Chem. 1977, 80, 409.
15. (a) Fukui, M., Kitani, A., Degrand, C., Miller, L.L.,
       J. Am. Chem. Soc., 1982, 104, 28;
    (b) Van DeMark, M.R., Miller, L.L., J. Am. Chem. Soc., 1978,
       100, 3323.
16. Murray, R.W., Acct. Chem. Res., 1980, 13, 135.
17. Snell, K.D., Keenan, A.G., Chem. Soc. Revs., 1979, 8, 259.
18. Abruña, H.D., Denisevich, P., Umaña, M., Meyer, T.J.,
    Murray, R.W., J. Am. Chem. Soc., 1981, 103, 1.
19. Ellis, C.D., Murphy, W.R. Jr., Meyer, T.J., J. Am. Chem.
    Soc., 1981, 103, 7840.
20. (a) Davydov, B.E., Ph.D. Dissert., USSR Acad. Sci., 1966;
       Ref. (40) in Davydov, B.E., Krentsel, B.A., Adv. Polym.
       Sci., 1977, 25, 1.
    (b) Dutton, P.L., Prince, R.C., Tiede, D.M., Photochem.
       Photobiol., 1978, 28, 939.
21. Wamser, C.C., Wagner, W.R., J. Am. Chem. Soc., 1981, 103,
    7232.
22. Hautala, R.R., Little, J.L., Ref. (8), 1.
23. Daum, P., Murray, R.W., J. Electroanal. Chem., 1979, 103,
    289.
24. Kaufman, F.B., Schroeder, A.H., Engler, E.M., Kramer, S.R.,
    Chambers, J.Q., J. Am. Chem. Soc., 1980, 102, 483.
25. Krasnovskii, A.A., Brin, G.P., Dokl. Akad. Nauk. SSSR, 53,
    447; Ref. (268) in Daydov, B.E., Krentsel, B.A., Ref. (20a)
    here.
26. (a) Schoch, K.F., Jr., Kundalkar, B.R. Marks, T.J., J. Am.
       Chem. Soc., 1979, 191, 7071;
    (b) Dirk, C.W., Schoch, K.F. Jr., Marks, T.J., U.S. Dept.
       Comm. NITS, AD-A099 520, 1981.
27. (a) Tachikawa, H., Faulkner, L.R., J. Am. Chem. Soc., 1978,
       100, 4379.
    (b) Wynne, K.J., Brant, P., Nohr, R.S., Weber, D., and
       Haupt, S., Abstr. 28th Macromol. Symp. (IUPAC), 1982,
       405.
    (c) Wnek, G.E., ibid., p. 409.
    (d) Guiseppi-Elie, A., and Wnek, G.E., ibid., p. 442.
28. Bard, A.J., Wrighton, M.S., J. Electrochem. Soc., 1977, 124,
    1706.

29. Bolts, J.M., Bocarsly, A.B., Palazzotto, M.C., Walton, E.G., Lewis, N.S., Wrighton, M.S., J. Am. Chem. Soc., 1979, 101, 1378.
30. Bocarsly, A.B., Walton, E.G., Wrighton, M.S., J. Am. Chem. Soc., 1980, 102, 3390.
31. Bookbinder, D.C., Wrighton, M.S., J. Am. Chem. Soc., 1980, 102, 5123.
32. Bookbinder, D.C., Brace, J.A., Dominey, R.N., Lewis, N.S., Wrighton, M.S., Proc. Nat. Acad. Sci. USA, 1980, 77, 6280.
33. Dominey, R.N., Lewis, N.S., Bruce, J.A., Bookbinder, D.C., Wrighton, M.S., J. Am. Chem. Soc., 1982, 104, 467.
34. Abruña, H.D., Bard, A.J., J. Am. Chem. Soc., 1981, 103, 6898.
35. Ghosh, P.K., Spiro, T.G., J. Am. Chem. Soc., 1980, 102, 5543.
36. Ghosh, P.K., Spiro, T.G., J. Electrochem. Soc., 1981, 128, 1281.
37. Yamase, T., Harada, H., Ikawa, T., Ikeda, S., Shirakawa, H., Bull. Chem. Soc. Jpn., 1981, 54, 2817.
38. Wrighton, M.S., Bocarsly, A.B., Bolts, J.M., Bradley, M.G., Fischer, A.B., Lewis, N.S., Palazzotto, M.C., Walton, E.G., Ref (8), 295.
39. Merz, A., Bard, A.J., J. Am. Chem. Soc., 1978, 100, 3222.
40. Handbook. Chem. Phys., 59th ed., CRC, 1978, D-193.
41. Fan, F.F., Wheeler, B.L., Bard, A.J., Noufi, R.N., J. Electrochem. Soc., 1981, 128, 2042.
42. Noufi, R., Frank, A.J., Nozik, A.J., J. Am. Chem. Soc., 1981, 103, 1849.
43. Skotheim, T., Lundstrom, I., Prejza, J., J. Electrochem. Soc., 1981, 128, 1625.
44. Cooper, G., Noufi, R., Frank, A.J., Nozik, A.J., Nature, 1982, 295, 578.
45. Noufi, R., Tench, D., Warren, L.F., J. Electrochem. Soc., 1980, 127, 2311.
46. (a) White, H.S., Abruña, H.D., Bard, A.J., J. Electrochem. Soc., 1982, 129, 265.
     (b) Frank, A.J., and Honda, K., J. Phys. Chem., 1982, 86, 1933.
47. Kiwi, J., Borgarello, E., Pelizzetti, E., Visca, M., Grätzel, M., Angew Chem. Int. Ed. Engl., 1980, 19, 646.
48. Mettee, H., Otvos, J.W., Calvin, M., Sol. Ener. Mater., 1981, 4, 443.
49. Bard, A.J., Bocarsly, A.B., Fan, F.F., Walton, E.G., Wrighton, M.S., J. Am. Chem. Soc., 1980, 102, 3671.
50. Chaing, C.K., Druy, M.A., Gau, S.C., Heeger, A.J., Louis, E.J., MacDiarmid, A.G., Park, Y.W., Shirakawa, H., J. Am. Chem. Soc., 1978, 100, 1014.
51. Nigrey, P.J., MacInnes, D., Nairns, D.P., MacDiarmid, A.G., Heeger, A.J., J. Electrochem. Soc., 1981, 128, 1651.
52. Chien, J.C.W., J. Polym. Sci., Polym. Letts., 1981, 19, 249.

RECEIVED November 22, 1982

INDEX

INDEX

B

Jacket design by Kathleen Schaner
Indexing and production by Florence Edwards and Paula Bérard

Elements typeset by Service Composition Co., Baltimore, MD
Printed and bound by Maple Press Co., York, PA